E–Learning for Geographers:
Online Materials, Resources, and Repositories

Philip Rees
University of Leeds, UK

Louise Mackay
University of Leeds, UK

David Martin
University of Southampton, UK

Helen Durham
University of Leeds, UK

INFORMATION SCIENCE REFERENCE

Hershey · New York

Director of Editorial Content: Kristin Klinger
Development Editor: Julia Mosemann
Director of Production: Jennifer Neidig
Managing Editor: Jamie Snavely
Assistant Managing Editor: Carole Coulson
Copy Editor: Alana Bubnis
Typesetter: Carole Coulson
Cover Design: Lisa Tosheff
Printed at: Yurchak Printing Inc.

Published in the United States of America by
       Information Science Reference (an imprint of IGI Global)
       701 E. Chocolate Avenue, Suite 200
       Hershey PA 17033
       Tel: 717-533-8845
       Fax: 717-533-8661
       E-mail: cust@igi-global.com
       Web site: http://www.igi-global.com

and in the United Kingdom by
       Information Science Reference (an imprint of IGI Global)
       3 Henrietta Street
       Covent Garden
       London WC2E 8LU
       Tel: 44 20 7240 0856
       Fax: 44 20 7379 0609
       Web site: http://www.eurospanbookstore.com

Library of Congress Cataloging-in-Publication Data

E-learning for geographers : online materials, resources, and repositories / Philip Rees ... [et al.].

   p. cm.

  Includes bibliographical references and index.

  Summary: "This book draws lessons from a unique collaboration of an international team of geographers, educators and computer scientists in developing learning materials and guidance to all who intend to work in the field of electronic/online learning"--Provided by publisher.

  ISBN 978-1-59904-980-9 (hardcover) -- ISBN 978-1-60566-981-6 (ebook)

  1. Geography--Computer-assisted instruction. 2. Geography--Study and teaching--Audio-visual aids. 3. Internet in education. 4. Distance education. I. Rees, P. H. (Philip H.), 1944-

  G76.E44 2009

  910.78'54678--dc22

               2008024204

British Cataloguing in Publication Data
A Cataloguing in Publication record for this book is available from the British Library.

To our families, for putting up with our absences during the DialogPLUS project.

List of Reviewers

Katherine Arrell
University of Leeds, UK

Andrew Booth
University of Leeds, UK

Lee Brown
University of Leeds, UK

Graham Clarke
University of Leeds, UK

Samantha Cockings
University of Southampton, UK

Grainne Conole
Open University, UK

Frances Deepwell
University of Coventry, UK

Adam Dennett
University of Leeds, UK

David DiBiase
The Pennsylvania State University, USA

Oliver Duke-Williams
University of Leeds, UK

Andrew Evans
University of Leeds, UK

Melissa Highton
University of Leeds, UK

James Hogg
University of Leeds, UK

Samuel Leung
University of Southampton, UK

David Martin
University of Southampton & ESRC Census Programme, UK

Karon McBride
University of Edinburgh, UK

Andrew Nelson
Joint Research Centre of the European Commission, Ispra, Italy

Paul Norman
University of Leeds, UK

John O'Donoghue
University of Central Lancashire, UK

James Petch
University of Manchester, UK

Linda See
University of Leeds, UK

Yim Ling Siu
University of Leeds, UK

Mark Stiles
The Learning Development Centre, Staffordshire University, UK

Richard Treves
University of Southampton, UK

Table of Contents

Section I
Collaboration in E-Learning Development

Chapter I

  Philip Rees, University of Leeds, UK
  Louise Mackay, University of Leeds, UK
  David Martin, University of Southampton, UK
  Gráinne Conole, The Open University, UK
  Hugh Davis, University of Southampton, UK

Chapter II

  Samuel Leung, University of Southampton, UK
  David Martin, University of Southampton, UK
  Richard Treves, University of Southampton, UK
  Oliver Duke-Williams, University of Leeds, UK

Chapter III

  Helen Durham, University of Leeds, UK
  Katherine Arrell, University of Leeds, UK
  David DiBiase, The Pennsylvania State University, USA

Section II
Geography Exemplars

Section III
Software Support for Learning Materials

Section IV
Additional Selected Readings

Detailed Table of Contents

Section I
Collaboration in E-Learning Development

Chapter I
> *Philip Rees, University of Leeds, UK*
> *Louise Mackay, University of Leeds, UK*
> *David Martin, University of Southampton, UK*
> *Gráinne Conole, The Open University, UK*
> *Hugh Davis, University of Southampton, UK*

Chapter I outlines the learning philosophies and learning strategies that inform the development of e-learning materials, focusing on a particular discipline context. The chapter authors come from a range of disciplines: geography, education, and computer science. Out of this inter-disciplinary collaboration has come new understanding of the range of approaches to learning (by the geographers) and new understanding of the enthusiasm of subject specialists (by the non-geographers).

Chapter II
> *Samuel Leung, University of Southampton, UK*
> *David Martin, University of Southampton, UK*
> *Richard Treves, University of Southampton, UK*
> *Oliver Duke-Williams, University of Leeds, UK*

In contrast to other Web-based resources, e-learning materials are not always exchangeable and shareable. Although transferring electronic documents between networked computers has become almost effortless, the materials may often require careful design and a great deal of adaptation before they can be reused in a meaningful manner. This process involves consideration of pedagogic issues such as course curricula, learning outcomes, and intended audience, as well as technological factors including local institutional virtual learning environments (VLE) and any relevant learning technology standards. Chapter II illustrates how these issues have been addressed resulting in the successful exchange of e-learning resources at three levels: (1) at content level, where learning nuggets are created and packaged in a standards-compliant format to guarantee interoperability; (2) at the user level, whereby learners or tutors, rather than the resources, are transferred between VLEs; (3) at a higher system level, where the emerging Web Single Sign-On technology of federated access management is being used to enable truly cross-institutional authentication allowing learners to roam freely in different learning environments.

Section II
Geography Exemplars

Chapter III

Helen Durham, University of Leeds, UK
Katherine Arrell, University of Leeds, UK
David DiBiase, The Pennsylvania State University, USA

Collaborative learning activity design (CLAD) is a multi-institution approach to the creation of e-learning material from the design phase through the development stage and onto the embedding of learning activities into existing modules at higher education institutions on both sides of the Atlantic. This was the approach taken by a group of academic and e-learning material developers at the Pennsylvania State University and the University of Leeds to develop a series of learning activities to support the use and understanding of the global positioning system (GPS). Aided by concept mapping, a Guidance Toolkit and Web conferencing facilities, the group worked seamlessly at producing a series of e-learning resources, including the basics of turning on a GPS unit and obtaining a spatial location, GPS data properties and GPS components, differential correction, and sources of GPS error and error correction. Chapter III reflects on the success of this project.

Chapter IV

David Martin, University of Southampton, UK
Philip Rees, University of Leeds, UK
Helen Durham, University of Leeds, UK
Stephen A. Matthews, The Pennsylvania State University, USA

Chapter IV presents the development of a series of shared learning materials prepared to facilitate teaching in human geography. Our teaching exemplars include those designed to develop students' understanding of the data collection process, for example through the use of an online census questionnaire; analysis methods, through the provision of visualization tools to show demographic trends through time; and substantive examples, by comparison between urban social geographies in the USA and UK. Particular challenges are presented by the different nature (format, content, detail) and licensing arrangements for the census data available for student use in the UK and USA.

In Chapter V we outline the issues involved in developing, delivering, and evaluating a Level 2 undergraduate module in fluvial geomorphology. The central concept of the module was the use of online digital library resources, comprising both data and numerical models, to foster an appreciation of physical processes influencing the evolution of drainage basins. The aim of the module was to develop the learners' knowledge and understanding of drainage basin geomorphology, while simultaneously developing their abilities to (i) access spatial data resources and (ii) provide a focus for developing skills in scientific data analysis and modeling. The module adopts a global perspective, drawing on examples from around the world. We discuss the process of course and assessment design, our teaching experiences, involving a particular combination of "face-to-face" lectures and online sessions, complemented by independent online learning, and supported by the associated virtual learning environment. Finally, we discuss the issues highlighted by a comprehensive module evaluation.

There is an inherent antithesis between environmental management as professional practice and as concept or philosophy. In Chapter VI the undergraduate focus of a module on Upland Catchment Management and on environmental management is compared with e-learning for postgraduate delivery (a module on GIS for Environmental Management). The differing styles of delivery highlight the flexibility of e-learning as a vehicle for acquiring skills and knowledge, and underpin the claim that the result is an enhanced engagement with the practice of informed management.

Chapter VII

In our experience of earth observation (EO) online learning we highlight the usefulness of the World Wide Web in terms of its software, functionality, and user accessibility for developing and delivering a range of activities and delivery modes to both undergraduate and advanced learners. Through the mechanism of developing teaching materials and adapting them for the online classroom, EO learning can become highly interactive and well-illustrated by linking to online image processing software and relevant image data, make use of the Web's graphical interface to reinvigorate DOS-based remote sensing programs to be more student-friendly, and with the advent of collaborative Web software, such as Wiki, provide a networked community for EO learners. Chapter VII showcases a variety of delivery modes for our EO materials - online lectures delivered within a blended learning module for the undergraduate to individual online activities (remote sensing practical exercises and an electronic learning diary) for the advanced EO learner.

Chapter VIII

Helen Durham, University of Leeds, UK
Samuel Leung, University of Southampton, UK
David DiBiase, The Pennsylvania State University, USA

Academic integrity (AI) is of relevance across all academic disciplines, both from the perspective of the educator and the student. From the former perspective there is the need to increase the awareness of AI amongst the student population whilst monitoring and enforcing the rules and regulation regarding plagiarism within their institution. On the other hand, students need a full appreciation of the importance of AI and a clear recognition of the penalties for flouting the regulations in order to steer a successful passage through higher education and on into their professional career. By repurposing learning materials originally developed by the Pennsylvania State University (USA), the Universities of Southampton and Leeds (UK) have developed academic integrity guidelines to support students in their studies and provide an assessment of their understanding of AI concepts.Chapter VIII describes the development of these learning activities and examines the technical and content issues of repurposing materials for three different institutions.

Section III
Software Support for Learning Materials

Chapter IX

Karen Fill, KataliSys Ltd, UK
Gráinne Conole, The Open University, UK
Chris Bailey, University of Bristol, UK

The DialogPLUS Toolkit is a web-based application that guides the design of learning activities. Developed to support the project's geographers, it incorporates well-researched pedagogic taxonomies that are presented as drop-down lists with associated 'help' pages. Toolkit users are encouraged to consider and specify factors including learning and teaching approach, environment, aims and outcomes, assessment methods, learner and tutor roles and requisite skills as they design any number of tasks within a learning activity and select the tools and resources needed to undertake them. The output from the toolkit is a design template that can then be used to guide the instantiation and implementation of online learning activities. The designs are saved within the toolkit, forming a database of designs, which other toolkit users can view. Chapter IX presents the rationale for the toolkit and the detailed taxonomies.

Chapter X

David DiBiase, The Pennsylvania State University, USA
Mark Gahegan, The University of Auckland, New Zealand

Chapter X investigates the problem of connecting advanced domain knowledge (from geography educators in this instance) with the strong pedagogic descriptions provided by colleagues from the University of Southampton, as described in Chapter IX, and then adding to this the learning materials that together comprise a learning object. Specifically, the chapter describes our efforts to enhance our open-source concept mapping tool (ConceptVista) with a variety of tools and methods that support the visualization, integration, packaging, and publishing of learning objects. We give examples of learning objects created from existing course materials, but enhanced with formal descriptions of both domain content and pedagogy. We explain how the resulting learning objects might be deployed within next-generation digital libraries that provide rich search languages to help educators locate useful learning objects from vast collections of learning materials.

Chapter XI

Terence R. Smith, University of California at Santa Barbara, USA
Marcia Lei Zeng, Kent State University, USA

Chapter XI describes a digital learning environment (DLE) organized around sets of concepts that represent a specific domain of knowledge. A prototype DLE developed by the Alexandria Project currently supports teaching at the University of California at Santa Barbara. Its distinguishing strength is an underlying abstract model of key aspects of any concept and its relationship to other concepts. Our strongly-structured model (SSM) of concepts is based on the viewpoint that scientific concepts and their interrelationships provide the most powerful level of granularity with which to support effective access and use of knowledge in DLE's. The DLE integrates various semantic tools facilitating the creation, merging, and use of heterogeneous learning materials from distributed sources, as well as their access in terms of our SSM of concepts by both instructors and students. Evidence indicates that undergraduate instructional activities are enhanced with the use of such integrated semantic tools.

Chapter XII

Richard Treves, University of Southampton, UK
David Martin, University of Southampton, UK

Teaching geography at university level involves students in study of complex diagrams and maps. These can be made easier to understand if split into parts. Chapter XII reports the work of a team writing a series of courses in geographic information systems (GIS) and their solution to the problem, which involved authoring simple multimedia animations using Microsoft PowerPoint™ software. The animations were authored by those writing the courses with little input from the Web specialist supporting the team. The techniques that the team used to produce the animations are explained, as are the nine points of best practice that were developed and how the animations were used with other non-animated content. Three sub-categories of these animations are described and explained and the issues of maintenance and reuse of the animated content is considered.

Chapter XIII is concerned with the evaluation of learning materials and activities developed as part of the DialogPLUS project. A range of evaluation activities was undertaken, focusing on the experiences of students, teaching staff, and the entire project team. Student evaluations included both quantitative and qualitative approaches, particularly using a questionnaire design drawing on a specific methodology and generic quality criteria, facilitating comparative analysis of results. Discussion of the student evaluations is focused on specific taught modules from both human and physical geography. Results of these evaluations were discussed with teaching staff and contributed to improvements in the various online resources. Both internal and external evaluators were involved in interviewing key project staff and their different perspectives are presented. The chapter concludes by reflecting on the effectiveness and impact of different DialogPLUS activities, highlighting the principal impacts of the project as perceived by the students and staff involved.

In this book we have illustrated the materials, software, and experience of developing and delivering geography e-learning courses and learning activities. In Chapter XIV we summarize how the teaching of a variety of geography topics has benefited from the following set of activities: creating media-rich online materials that take full advantage of linking to digital libraries; developing and adapting online, collaborative, and design software; and internationalizing materials through geography teachers in different countries working together. We take a moment to reflect on the experience of material development and the prospects for facilitating exchange of resources and student access. We provide advice to the aspiring geography e-tutor and describe how to access the wealth of materials that have been introduced in the preceding chapters. We then explain how the materials created will continue to be relevant beyond this book. We envisage that teachers, including ourselves, will download and then adapt the materials, borrowing content, techniques for presentation, or learning style. There will be an ongoing process of teaching and review that incorporates tutor and student feedback. The material, its delivery, and its style will not remain static but we hope new developments will be shared via learning repositories. It is important to sustain good online resources. This can be achieved by readers updating the geography e-learning materials and depositing improved versions in the new UK academic learning material depository Jorum.

Section IV
Additional Selected Readings

Chapter XV discusses the design, technical development, delivery, and evaluation of two online learning activities in environmental geography. A "blended" approach was adopted in order to best integrate the new materials within the existing unit. The primary aim of these online activities was to provide students with opportunities to develop and demonstrate valuable practical skills, while increasing their understanding of environmental management. A purpose-built system was created in order to overcome initial technological challenges. The online activities have already been delivered successfully to a large number of students over two academic years. Evaluation and staff reflection highlight the benefits and limitations of the new activities, and the chapter concludes with recommendations for others wishing to adopt a similar approach.

Chapter XVI describes G-Portal, a DL of geospatial and georeferenced resources. G-Portal is designed to support collaborative learning among its users. This is achieved through personalized project spaces, in which individuals or groups gather and organize collections of resources drawn from the DL's holdings that are relevant to specific learning tasks. In addition, G-Portal provides facilities for classification and visualization of resources, spatial searching, annotations and resource sharing across projects.

Chapter XVII

Shivanand Balram, Simon Fraser University, Canada
Suzana Dragićević, Simon Fraser University, Canada

Chapter XVII describes the origins, boundaries, and structures of collaborative geographic information systems (CGIS). A working definition is proposed, together with a discussion about the subtle collaborative vs. cooperative distinction, and culminating in a philosophical description of the research area. The literatures on planning and policy analysis, decision support systems, and geographic information systems (GIS) and science (GIScience) are used to construct a historical footprint. The conceptual linkages between GIScience, public participation GIS (PPGIS), participatory GIS (PGIS), and CGIS are also outlined. The conclusion is that collaborative GIS is centrally positioned on a participation spectrum that ranges from the individual to the general public, and that an important goal is to use argumentation, deliberation, and maps to clearly structure and reconcile differences between representative interest groups. Hence, collaborative GIS must give consideration to integrating experts with the general public in synchronous and asynchronous space-time interactions. Collaborative GIS provides a theoretical and application foundation to conceptualize a distributive turn to planning, problem solving, and decision making.

Foreword

A combination of drivers have encouraged the introduction of e-learning into a wide range of educational contexts—from global and economic imperatives, through national policy directives to institutional strategies and, sometimes, the commitment and enthusiasm of individual teachers and learners. What can sometimes get lost is a consideration of the underlying pedagogic purpose in implementing technologies to support and enhance learning. It is a significant challenge for teachers in higher education (HE) to take the time to re-think their curricula, articulate the underlying pedagogic approaches, and consider how the range of available technologies can support this. It is far easier to upload existing tried and tested materials into the institutional virtual learning environment or course management system.

This book tells a different story and is the result of some excellent work led by clear pedagogic intent in the geography discipline. It presents a range of approaches taken by this multi-national, multi-disciplinary team to enhance the student learning experience. Working collaboratively across disciplines and cultures presents many challenges, particularly in developing common understandings, and is often a huge learning experience for those involved. Creating, developing and embedding learning materials within a curriculum is a complex process involving a range of individuals, and the DialogPLUS team developed imaginative approaches to facilitate this across the Atlantic, where the usual technical, legal and cultural constraints to sharing content were magnified.

Detailed practical case studies in human geography, environmental management, geomorphology, and earth observation are provided. These describe the various approaches to making the resources available in different learning contexts. Whilst clearly of significant interest to those teaching in geography and allied subject areas, the case studies also have genuine relevance to anyone interested in enhancing the learner experience of finding, using, and critically engaging with real data to carry out authentic tasks.

The Digital Libraries in the Classroom Programme was a joint venture between the Joint Information Systems Committee (JISC) in the United Kingdom and the National Science Foundation (NSF) in the United States of America. The four projects aimed to transform learning and teaching by bringing emerging technologies and readily available digital content into mainstream educational use. What is evident from the work of all the projects, and what is illustrated well in this book, are the innovative and wide ranging outcomes that span digital repositories, learning resources, pedagogy, and staff development, implementing institutional change and facilitating globally networked learning opportunities.

It is always tempting to focus on the tangible outputs from development projects and for this team these were significant. As well as the "learning nuggets" themselves, they developed the DialogPLUS toolkit, which provides a practical tool to take teachers through the learning design process and reflects the team's thoughtful and collaborative approach to curriculum design. What excited me about this work was the value of the learning process for the members of the team and the evidence of significant shifts in perception for both the academic team and the educational technologists. If as a reader you are prepared to be as reflective and thoughtful about your own practice you could benefit from much more than the hugely practical content.

Lou McGill
University of Bristol
UK

Lou McGill was formerly a programme manager on the JISC eLearning team, responsible for two large programmes of work which have focussed on transforming the learning experience by implementing innovative e-learning solutions to a range of institutional or subject discipline challenges – the JISC/NSF Digital Libraries in the Classroom Programme and the Scottish Funding Council e-Learning Transformation Programme. Prior to joining JISC, McGill worked at the University of Strathclyde for the DIDET Project (digital libraries for global distributed innovative design and teamwork), one of the JISC/NSF Digital Libraries in the Classroom initiatives. Whilst at Strathclyde she was a member of the VLE Implementation Group and was involved in developing a consultation process for the development of the University Teaching and Learning through Technology Strategy. She was a member of the JISC eLearning and pedagogy Experts Group and was part of a team that carried out the "Research study on the effectiveness of resources, tools and support services used by practitioners in designing and delivering e-learning activities." Prior to this McGill had 18 years experience as a professional information specialist in the adult education world, working in both institutional and academic libraries. She has always worked closely with educators to ensure that libraries and librarians are considered essential components of the learning environment. As a teacher of information literacy she has employed and evaluated a variety of e-learning technologies to enhance and support student learning.

Preface

ABSTRACT

This chapter introduces the book, setting it in the context of developments in online or electronic learning. We define electronic learning as that which occurs when a computer is used as the means of transmitting the learning materials and organizing the activities. Electronic learning has proved invaluable in distance learning: the numbers of online degree programs is growing steadily. But electronic learning is also a developing feature of the undergraduate curriculum. Most universities now provide their students with a wealth of electronic information via their libraries and via their administrative Web sites. Most universities provide a vehicle for the delivery of course materials. Known as virtual learning environments in the UK, and as course management systems in the U.S., these software platforms are now big business. Students expect course syllabi, reading lists, course exercises, data, old exam papers, and timetables to be online. The next step in this e-learning evolution is to put the lectures and tutorials and student activities online. For most geography teachers (of a certain age) this is a daunting challenge involving radical re-skilling and much effort. What are the best ways to do this? What tools are available to help in the process? What is the best mix of online learning and face-to-face learning? This book relates the experience of and describes the resources developed by a team of academics funded by the UK Joint Information Systems Committee, a body charged by the Higher Education Funding Councils to encourage new electronic learning and knowledge and skills in using digital materials. We hope that this account (including both successes and challenges) will inspire the already electronically literate geography teacher to experiment and develop their own materials or persuade the less electronically confident to take the first steps on this road to, we believe, enhanced learning for our students. The chapter includes a summary of what the parts and chapters cover and a guide on how to use the book and associated materials in an intelligent and time efficient way.

INTRODUCTION

Electronic learning (e-learning), which occurs when a computer is used as the means of accessing learning resources and activities, provides today's university student with an unprecedented array of learning materials. Most universities in the developed world now provide their students with a wealth of electronic information via libraries, administrative Web sites, and virtual learning environments. Students expect course syllabi, reading lists, exercises, data, old exam papers, and timetables to be online. A logical next step in this learning evolution is to put the lectures, tutorials, and student activities online. As working online can be viewed as a new context for learning and not just a learning tool, for most geography teachers, especially those accustomed to traditional delivery methods, this is a daunting challenge involving radical re-skilling and much effort. What are the best ways to do this? What is the best mix of online and face-to-face learning? What tools are available to help in the process? Through the experiences of an international team of geographers, educationalists, and computer scientists we aim to answer these questions.

This book makes a unique contribution to methods of collaboration between partners in developing common materials, linking courses, and sharing students, pointing to a joined up and networked future of learning. We describe the learning resources developed in the geography topic areas of human geography, environmental management, geomorphology, and earth observation, suitable for use by students of the physical or man-made environment. Through a number of carefully explained case studies, connected to digital libraries of geographic resources and preserved for future re-use in the Jorum digital repository (www.jorum.ac.uk) developed by the Joint Information Systems Committee (JISC), we provide guidance to geography teachers and educational professionals intending to work in the vibrant and growing field of electronic/online learning. For consistency we use the term e-learning throughout this book, but we do not seek to assert any special differences between this term and technology-enhanced learning (TeL), also currently in widespread use.

The contributors to the book were all involved in the JISC/NSF funded Digital Libraries in the Classroom Programme as members of the Digital Libraries in Support of Innovative Approaches to Learning and Teaching in Geography project (DialogPLUS, http://www.dialogPLUS.org). The authors are experienced teachers, researchers, and e-learning developers at Pennsylvania State University, the University of Leeds, the University of California at Santa Barbara, and the University of Southampton.

BACKGROUND

Geography and E-Learning

Many geographers have enthusiastically embraced the Internet as a tool for supporting learning. The discipline of geography, by its very nature, lends itself more than most to multimedia teaching, as it is a spatial science, which greatly benefits from a visual narrative. E-learning has been warmly embraced by geographers since the mid-1990s with the general advent of Web-based learning technologies allowing for flexible delivery of courses. Studies on e-learning and geography higher education discuss the value of Web-based resources, the role of virtual case studies, and the development of study skills in Web-based environments (Gardner, 2003; Stainfield et al., 2000; Goett & Foote, 2000). The role of multimedia is often championed and widely supported as enhancing learning and teaching in geography (Castleford et al., 1998; Lemke & Ritter 2000; Jain & Getis, 2003). However, few geographers have developed and delivered materials within a context of successful international collaboration (for material development, delivery, and access) with strong pedagogic and design support and rigorous evaluation as we have with our material development.

The development of e-learning for geographers falls within a wider series of e-learning initiatives in HE teaching and learning. For example, within the UK the Higher Education Funding Council for England (HEFCE)'s ten year e-learning strategy (from 2005) refers to the need to embed e-learning into teaching and learning. Implemented by HEFCE's two lead partners, the Higher Education Academy (HEA) and JISC identify their e-learning strategy as contributing to the aims of increasing student numbers, retaining a more diverse student group, facilitating high quality learning for students, and promoting lifelong learning (Hamburg, 2007).

The growth of information and communication technology in higher education has paved the way for experiments in teaching and learning, including the practice of linking faculty and students worldwide using Web technologies. Allied to this is the significant development in the internationalization of the university curriculum, a trend that many scholars view as likely to become a defining force in higher education in the twenty-first century (Goodman, 1999). It is argued that in the light of shrinking levels of public funding for higher education, many academic institutions will find themselves tapping the revenue-generating potential of online courses and degree programs and increasingly doing so to an international market (Solem et al., 2003). We have seen, since the 1990s, geographers engaged in collaborative and international education (Hurley et al., 1999; Solem et al., 2003). This is seen as providing numerous advantages for the student including access to knowledge and resources and development of cross-cultural relationships - all providing the skills they need to compete in the global economy (Shepherd et al., 2000). The internationalization of geography teaching is also seen at the learner level where students use communication technologies such as electronic discussion boards to support collaboration at a distance. For example Warf et al. (1999) taught a course concurrently in the USA, UK, and Ireland in which students shared their views on issues facing developing regions.

There is not a strong base of evidence to support the theoretical advantages of computer-supported collaborative learning, particularly in the context of international education (Bonk & Cunningham, 1998). It is argued that due to the lack of resources, training, and research, it is difficult for many geographers to understand how to plan and carry out international teaching collaborations using multimedia (Shepherd et al., 2000). However, in this book we showcase the successful development of collaborative (and international) material creation ranging from the simplest of interactions between colleagues to the use of more sophisticated, pedagogically-supported design and concept tools.

The DialogPLUS Project

This book has developed out of the DialogPLUS project and serves to showcase the outcomes of our endeavors. The project aimed to develop e-learning materials for university courses and to do so for four geography exemplar topics: human geography, geomorphology/environmental management, geographical information systems (GIS), and earth observation. These materials were to be linked with digital libraries of resources (particularly those funded by JISC) and preserved in a suitable repository (JISC's Jorum repository). Allied to this was the additional experimentation of collaborations between universities within the UK and in the USA. Specifically, the collaborations involved the development of materials, the exchange of topic materials within modules, of generic materials, of whole modules and of students. These collaborations form the focus of Section I of the book.

At the heart of the project lies the creation of learning objects (L'Allier, 1997; Wiley, 2002; Gibbons et al., 2000). Although already familiar to the educationalists and computer scientists involved in the DialogPLUS project, the academic geographers found the concept to be rather too restrictive and preferred instead to use the rather more flexible concept of a *nugget*, a re-usable and exchangeable e-learning object. The concept of the nugget used here is scaleable - part of a lecture, a practical or a whole module. Our nuggets come in a variety of type and size: full courses developed from e-learning materials; individual learning activities, for example, census nuggets; and generic learning materials, for example, an academic integrity nugget. This terminology allowed the geographers to collaborate over elements of learning that were readily identifiable in practical teaching situations without initially attending to issues of granularity, metadata, and formal definition. We also experimented with various styles of delivery of the nuggets, from Web-enabled individual activities delivered within an existing face-to-face course to a range of blended learning scenarios within an entire module or course where a mixture of face-to-face and online delivered activities consolidate and support each other. Our nuggets are mostly stored and accessed from institutional virtual learning environments (VLEs) for particular teaching needs. However, in order to benefit the wider HE/FE community we have maintained a project

Web site and placed nuggets on the Jorum repository. The lessons learned from the development and varied delivery of our nuggets provide invaluable insights to e-learning provision within a curriculum - this forms the contribution of Section II of the book.

As well as outputs in terms of learning materials and the knowledge gained from their creation and delivery, the project has aimed to provide tools for materials design. We developed a *Learning Activity Toolkit* to guide teachers through a process of articulating their information needs in order to produce a "lesson plan" for a particular learning activity. Teachers can then modify learning activities from the toolkit's suggested pedagogical approaches and tools and resources. We also developed a conceptual mapping approach to aid the process of collaborative learning activity design. Both these tools can aid the tutor in designing their own e-learning nuggets. The introduction of these and other learning design tools, and the process of evaluating our teaching endeavors using our new nuggets, forms Section III of the book.

HOW TO USE THE BOOK

What Type of Reader are You?

This book is intended for a readership spanning the community of higher and further education academics who are engaged with or moving to online learning. We identify several types of users for this book and here make suggestions as to how they may find the book to be most helpful.

Section I: Chapters I, II, and III

As a director of learning and teaching programs you may find it most helpful to begin with our introduction to e-learning within the academic field of geography. This addresses the ways in which we have successfully collaborated to produce new materials, repurpose old materials, and provide access to students and programs both nationally and internationally. We would direct you initially to Section I of the book.

Section II: Chapters IV to VIII

As a faculty user you may be most interested to begin with our case studies in Section II, which cover the development and delivery of a range of online learning materials spanning human and physical geography topics. In particular, you may fall into one of the following categories:

1. A *lecturer* interested in developing e-learning materials in your teaching; ranging from the enthusiastic champion of e-learning, to a passing interest in new teaching methods, to the technophobe not wishing to change in style from their much valued traditional face-to-face lecturing (but slowly realizing that change is afoot)!
2. A *drop-in tutor* who requires topic-specific online activities for a particular teaching exercise.
3. A *research student* thinking of becoming a lecturer and faced with the new e-learning environment.
4. *Master's students* in education and geography who are referred to materials by their tutors.

If one of these descriptions sounds like you, we would suggest that you start with the chapters in Section II.

Section III: Chapters IX to XIV

As a prospective user of your own learning materials we would also direct you to Section III of the book. As a novice online tutor you may greatly benefit from acquiring new teaching skills and examples by accessing our material design tools introduced within these chapters.

Many of our chapters are useful to a variety of users and not just those with a background in geography. Several of our chapters are of generic use across all fields; for example collaboration and exchange of e-learning resources are covered in Chapter II; instilling academic integrity in your students is addressed in Chapter VIII; and conceptual models and tools for learning design are discussed in Chapters X and XI.

Section IV: Additional Selected Reading

Our publisher has arranged for us to include in the book three additional chapters from previously published collections. Chapter XV will be of interest to teachers of geography, environmental science and related subjects who wish to get their students engaged with current online materials pertinent for courses on the environment. Chapter XVI introduces a useful portal library developed by colleagues in South East Asia, who show how those resources can be used for geographic learning. Chapter XVII presents a view of "collaboration" in employing GIS that supplements our account in Chapter III. Collaboration in Chapter XVII is concerned with encouraging participation by "customers" in opinion formation, while the earlier Chapter III stresses collaboration in design of materials by a team of authors.

Accessing the Online Learning Materials

Web Site

All of our e-learning materials are available through the DialogPLUS project Web site: http://www.dialogPLUS.org. We are allowing free access to this site and the learning objects it holds and invite users to take the materials and repurpose them in their teaching.

Online Repository: Jorum

The project e-learning materials are also housed on the new JISC online repository Jorum (http://www.jorum.ac.uk/). The repository is freely available to registered members of the UK HE/FE academic community. Materials can be downloaded and repurposed for teaching use, the site allows for user feedback in order for authors to review and update their materials. Each exemplar chapter in Section II lists the materials deposited on Jorum.

BOOK SUMMARY

The book describes the experience of developing electronic learning materials in geography topic areas. In particular we address three main parts: (i) collaboration in e-learning; (ii) developing and delivering learning resources ranging from the use of census data to understand population distributions to the principles of earth observation; and (iii) use and creation of material design aids.

Section I. Collaboration in E-Learning Development

In Chapter I, Rees et al. outline the learning philosophies, styles, and strategies that inform the development of e-learning materials. The "learning object" is defined and critiqued, emphasizing the importance that geographers place on their students' need to access a wide variety of digital resources that inform them about the world. From this point the authors define a learning material unit termed the "nugget" (discussed in a previous section) of materials for student use with one or more activities designed to develop understanding combined with student evaluation of the knowledge gained (tests, exercises, reflections). Nuggets connect to external digital resources held in libraries, repositories or Web sites. The chapter illustrates the different styles possible for putting together nuggets from a number of geography exemplars. Out of the inter-disciplinary collaboration of the book authors has come new understanding of the range of approaches to learning (by the geographers) and new understanding of the enthusiasm of subject specialists (by the non-geographers). The authors also report on the understanding gained through working with colleagues in another country. The chapter also provides an example of how e-learning was developed in one master's program. Lessons are drawn from this experience.

In Chapter II, Leung et al. explore the exchange of e-learning materials, modules, and students. The authors illustrate how e-learning materials require careful design and repurposing before they can be adapted and reused in a meaningful manner by considering pedagogic issues such as course curricula, learning outcomes and intended audience as well as technological aspects including local VLEs and any agreed learning technology standards. The authors illustrate how these issues have been addressed in the successful exchange of e-learning resources at three levels: (1) at the content level, where nuggets are packaged for interoperability; (2) at the institutional level, where learners are transferred from their home institution's VLE to the collaborating institution's VLE; and (3) at a higher system level, using an emerging Web single sign-on technology known as federated access management to allow cross-institutional authentication for learners to roam freely in different learning environments.

Durham, Arrell, and DiBiase in Chapter III reflect on the experience of a multi-institution approach to the creation of e-learning material from the design phase through the development stage and onto the embedding of learning activities into existing modules at HE institutions on both sides of the Atlantic. In Chapter III we are introduced to Collaborative Learning Activity Design (CLAD), the approach taken by a group of academic and e-learning material developers at Pennsylvania State University and the University of Leeds to develop a series of learning activities to support the use and understanding of the global positioning system (GPS). The authors reflect on the impact of identifying mutual learning aims and objectives at design phase and their collaborative experience aided through the use of concept mapping and Web-conferencing facilities.

Section II. Geography Exemplars

In part two of the book we showcase our online geography topics: how they were developed, delivered, what they contain, and how they can be accessed.

In Chapter IV, Martin et al. present the development of a series of shared learning materials developed to facilitate teaching in census and population topics in human geography. The principal focus of this work is demographic, with the created materials intended to support teaching of social geography and demographic analysis. As much of the data describing population characteristics are themselves published online, the teaching exemplars are particularly well-suited to blended learning; for example, understanding census content through the use of an online census questionnaire, analysis methods through the provision of visualization tools to show demographic trends through time and substantive examples by comparison between urban social geographies in the USA and UK. The authors highlight the particular challenges presented by the different nature (format, content, detail) of the census data available for student use in the UK and USA.

In Chapter V, Darby et al. outline the issues involved in developing, delivering, and evaluating a module designed to support undergraduate learning in fluvial geomorphology. The central concept of the module, which was designed to be delivered in a blended mode, involving a combination of traditional lectures and online learning activities, was the use of a suite of online digital library (DL) resources, comprising both data and numerical models. The module is designed to develop learners' knowledge of the physical processes influencing the evolution of drainage basins, while simultaneously developing their abilities to (i) access spatial data resources within the digital library and (ii) provide a focus for developing skills in scientific data analysis and modeling. The authors discuss the process of course and assessment design, explaining the pedagogy underlying the decision to deliver a unit in a blended learning mode, and the issues highlighted by course evaluation.

Wright et al. (ChapterVI) argue that e-learning offers a mechanism through which to bridge the potential gap between environmental management as professional practice and as concept or philosophy, by allowing enhanced skill and understanding through *virtual* practical experience. The work describes the delivery of information and challenge (question or problem) to students through the communications and decision-support technologies that they will increasingly find taken for granted in professional practice. In this chapter the undergraduate focus of DialogPLUS (a module on Upland Catchment Management) is compared with e-learning projects for postgraduate delivery (a Worldwide Universities Network [WUN] module on GIS for Environmental Management) and a suite of e-learning modules on river management designed for professional training in the Environment Agency (for England).

In Chapter VII, Mackay, Leung, and Milton illustrate the versatility and variety of earth observation as a geography topic in its translation to e-learning-from content-orientation covering a full lecture course in the physical principles of earth observation to smaller topic-specific e-activities such as atmospheric modeling. The authors discuss the development and delivery of a blended learning module where lectures are delivered online and show how learning benefits from the delivery of media-rich materials that can link to earth observation image databases. Equally, the Web is of benefit as a highly adaptable medium where legacy and DOS-based remote sensing programs and courseware can be renovated and reprogrammed to provide a more user-friendly interface. Examples are discussed in this chapter to show exactly how the Web has improved teaching of earth observation.

In Chapter VIII, Durham et al. showcase the development of a generic learning material designed to introduce the development of academic integrity (AI) in students, highlighting its relevance across all academic disciplines. Students need an increased awareness of AI and a clear recognition of the penalties of plagiarism throughout higher education and on into their professional career. At the same time, educators need means to monitor and enforce the rules and regulations regarding plagiarism within their institution. This chapter describes the development and embedding of repurposed AI learning materials from Pennsylvania State University (USA) into the geography programs of the Universities of Southampton and Leeds (UK). The use of the online plagiarism detection service, *Turnitin,* to police plagiarism cases is described; and the effect on skills gained by the geography students from their AI experience is evaluated.

Section III. Software Support for Learning Material Design

In the third and final part of the book we introduce examples of tools and concepts in learning material design: thus providing the novice e-tutor with a wealth of resources for learning material creation and provide insight to the benefits gained from evaluation of materials for improving material revision.

Fill et al. (Chapter IX) discuss the development of a Web-based application that guides the design of learning activities - *The DialogPLUS Toolkit*. Developed to support the project's geographers, toolkit users are encouraged to consider and specify factors including learning and teaching approach, environment, aims and outcomes, assessment methods, learner and tutor roles, and requisite skills as they design any number of tasks within a learning activity and select the tools and resources needed to undertake them. The output from the toolkit is a design template that can then be used to guide the instantiation and implementation of online learning activities and is saved for repurposing. The chapter authors present the rationale for the toolkit and its taxonomies; they describe the software design, development and implementation, including the approach to contextual "help" and then provide examples of learning activity designs created using the toolkit with feedback from users.

Chapter X (Gahegan et al.) investigates the problem of connecting advanced domain knowledge (from geography educators in this instance) with the strong pedagogic descriptions provided by colleagues from the University of Southampton, as described in Chapter IX, and then adding to this the learning materials that together comprise a learning object. The chapter describes the authors' efforts to enhance their open-source concept mapping tool (*ConceptVista*) with a variety of tools and methods that support the visualization, integration, packaging, and publishing of learning objects. Learning objects enhanced with formal descriptions of domain content and pedagogy are exemplified and used to illustrate improved communication of educational aims and processes. It is proposed that such learning objects might be deployed within next-generation digital libraries that provide rich search languages to help educators locate useful learning objects from vast collections of learning materials.

Smith and Zeng (Chapter XI) describe the design and implementation of digital learning environments (DLEs) organized around sets of concepts selected by an instructor to represent a specific domain of knowledge. The chapter provides an overview of a DLE that has been developed and implemented by the Alexandria Digital Earth Prototype (ADEPT) and is currently in operational use for teaching geography courses at the University of California at Santa Barbara. The DLE involves the design and implementation of various semantic tools facilitating the creation, integration, and use of heterogeneous learning materials from many distributed sources as well as their organization and access in terms of the authors' Strongly-Structured Model (SSM) of concepts. Evidence indicates that undergraduate instructional activities that are based on the application of these ideas in digital library environments are greatly facilitated with the use of such integrated semantic tools.

Learning style theory suggests students will have a preferred technique (or techniques) of learning such as factual, theoretical, or spatial. Since tertiary level geography students have chosen an area of study intimately related to maps, the majority can be assumed to prefer to learn spatially, that is, using maps, diagrams, and schematics. The design of data presentation in diagram form is traditionally presented on paper with an accompanying text commentary. Advances in software have enabled a new multimedia approach to be taken where the diagram is "built" in successive layers with a narrative text commentary. This enables concepts with a high intrinsic cognitive load to be understood whilst lessening the extraneous cognitive load. Recent software development has enabled any author with good skills in PowerPoint™ to be able to create Web friendly content directly. Best practice in learning design is discussed in Treves' Chapter XII in relation to a real case study of a series of distance learning, master's level GIS courses that use this multimedia technique.

Earlier chapters describe and discuss some of the online materials and activities developed to enhance learning for geography students. In Chapter XIII, Fill and Mackay explore the student-focused evaluation of these innovations using quantitative and qualitative methods including questionnaires, observation, interviews, and analysis of online discussion board activity. Subsequently, the analysis of student reflections leads to improvements in the resources. Additionally key project staff members were interviewed towards the end of the project about their experiences by both internal and external evaluators. The authors reflect on the lessons learned from the approaches and results of both the student and staff evaluations.

Section IV: Additional Selected Reading

In the spirit of re-use and re-purposing of e-learning materials, which is one of the themes running through this book, we have included cognate, linked materials from earlier books concerned with e-learning, with the permission of the publisher and the consent of the authors. These additional chapters add different perspectives from experts in e-learning, which would otherwise be missing from our account.

CONCLUSION

In our final chapter (Chapter XIV), the editors take a moment to reflect upon the experience and lessons to be learnt from the development and delivery of geography e-learning courses and learning activities. The benefits to teaching are summarized: creation of media-rich online materials that take full advantage of linking to digital libraries; development and adaptation of online, collaborative, and design software; and internationalization of materials through geography teachers in different countries working together. The chapter highlights the prospects for facilitating exchange of resources and student access and provides advice to the aspiring geography e-tutor. We champion the relevancy of the created materials beyond this book and propose that material, its delivery, and its style will not remain static but new developments will be shared via learning repositories (such as the new UK academic learning material depository Jorum), where readers can update geography e-learning materials and deposit improved versions.

The book will be of considerable interest to the expanding community of HE/FE academics who are experimenting with online learning in their courses and encouraging their students to make more intelligent use of the Web to access geographic resources. The book provides students with a guide to valuable learning materials prepared by the book authors, which are available to all in the HE/FE community. In the world of online learning, which is ever expanding and sometimes overwhelming to the user, the book provides a structured introduction to a set of valuable products available for re-use by geographers and educationalists.

We hope that this account, including both our successes and challenges, will inspire already e-literate geography teachers to experiment and develop their own materials and persuade the less electronically confident to take the first steps on this road to, we believe, enhanced learning for all students.

REFERENCES

Bonk, C., & Cunningham, D. (1998). Searching for learner-centered, constructivist, and sociocultural components of collaborative learning tools. In C. Bonk & K. King (Eds.), *Electronic collaborators: Learner-centered technologies for literacy, apprenticeship, and discourse* (pp. 25-50). London, Lawrence Erlbaum.

Castleford, J., Robinson, G., Charman, D., Elmes, A., Towse, R. J., Garside, P., Browne, T., Funnell, D., Burkill, S., Gratton, J., Whalley, W. B., Rea, B. R., & Crampton, J. (1998). Arena symposium. Evaluating IT-based resources for supporting learning and teaching in geography: Some case studies. *Journal of Geography in Higher Education, 22*(3), 375-423.

Gardner, A. (2003). Discovering networked information in the Internet age: The JISC resource guide to geography and the environment. *Journal of Geography in Higher Education, 27*(1), 103-108.

Gibbons, A. S., Nelson, J., & Richards, R. (2000). The nature and origin of instructional objects. In D.A. Wiley (Ed.), *The instructional use of learning objects.* Bloomington, IN: Association for Educational Communications and Technology.

Goett, J. A., & Foote, K. E. (2000). Cultivating student research and study skills in Web-based learning environments. *Journal of Geography in Higher Education, 24*(1), 92-99.

Goodman, A. E. (1999). From the American century to the education century: Driving forces behind the internationalization of higher education. *Journal of Intensive English Studies, 13*, 108-114.

Hamburg, L. (2007). *E-learning in HE: Where are we now and what does the future hold?* Retrieved April 27, 2008, from:http://www.heacademy.ac.uk/assets/York/documents/resources/publications/exchange/web0183_exchange_1_hamburg.pdf

Hurley, J. M., Proctor, J. D., & Ford, R. E. (1999). Collaborative inquiry at a distance: Using the Internet in geography education. *Journal of Geography, 98*(3), 128-140.

Jain, C., & Getis, A. (2003). The effectiveness of Internet-based instruction: An experiment in physical geography. *Journal of Geography in Higher Education, 27*(2), 153-167.

L'Allier, J. J. (1997). *Frame of reference: NETg's map to the products, their structure and core beliefs.* NetG. Retrieved April 27, 2008, from: http://www.netg.com/research/whitepapers/frameref.asp

Lemke, K. A., & Ritter, M. E. (2000). Virtual geographies and the use of the Internet for learning and teaching geography in higher education. *Journal of Geography in Higher Education, 24*(1), 87-91.

Shepherd, I., Monk, J., & Fortuijn, J. (2000). Internationalisation of geography in higher education: Towards a conceptual framework. *Journal of Geography in Higher Education, 24*(2), 285-298.

Solem, M. N., Bell, S., Fournier, E., Gillespie, C. L., Lewitsky, M., & Lockton, H. (2003). Using the Internet to support international collaborations for global geography education. *Journal of Geography in Higher Education, 27*(3), 239-253

Stainfield, J., Fisher, P., Robinson, G., & Bednarz, R. S. (2000). International virtual fieldtrips: A new direction? *Journal of Geography in Higher Education, 24*, 255-262.

Warf, B., Vincent, P., & Purcell, D. (1999). International collaborative learning on the World Wide Web. *Journal of Geography, 98*(3), 141-148.

Wiley, D. A. (2002). Connecting learning objects to instructional design theory: A definition, a metaphor, and a taxonomy. In D. A. Wiley (Ed.), *The instructional use of learning objects*. Bloomington, IN: Agency for Instructional Technology.

Acknowledgment

This book is the result of a collaborative project called *Digital libraries in support of innovative approaches to learning and teaching in geography* funded by the Joint Information Systems Committee (JISC) over the period February 1, 2003 to January 31, 2008, as part of the Digital Libraries in the Classroom Programme. At the University of Leeds and the University of Southampton we are very grateful for JISC's support over that period and for their commitment to the development of e-learning and the use of the rich digital library resources, which they have built up. Equivalent support for researchers at Penn State University and the University of California at Santa Barbara was provided by the National Science Foundation.

The project was called **DialogPLUS** to recognize that it was a partnership between Pennsylvania State University, the University of Leeds, the University of California at Santa Barbara and the University of Southampton. The project was supported fully by our respective universities. All of the book's authors participated in the project, at one stage or another. The editors are very appreciative of the efforts and support, in particular, of Hugh Davies who led the British team and to David DiBiase who provided intellectual leadership for the American team. Our thanks are also due to our project manager at JISC, Susan Eales, who helped steer the project through at times choppy waters and organized relations with the other projects in the Digital Libraries programme.

There were other staff who contributed to the project who are not authors. At Penn State, David DiBiase was assisted by Steven Weaver, Khusro Kidwai, and Mark Wherley. At Leeds these include: Andy Nelson, who did a super job preparing our academic integrity nugget; Andrew Booth, author of the Virtual Learning Environment, Bodington, which we used throughout the project, who was always encouraging and knew the answers to technical questions well beyond our competences; Jon Maber, principal software engineer for the Bodington VLE who provided valuable technical advice; Richard Hardy, who worked on the physical geography exemplar, Upland Catchment Management, and delivered an online course to Leeds geography undergraduates; Stuart Lane, who guided the development of the physical geography exemplar and who invented the DialogPLUS acronym on a flight to Penn State for our first project meeting; and Anthony Lowe of the Learning Development Unit who gave valuable advice on the academic integrity nugget. At the University of California at Santa Barbara Mike Freeston and Linda Rose set up and maintained the project Web site and swiki for the majority of the project. At Southampton all DialogPLUS project staff members are represented among the book's authors.

We are very appreciative of the skilled attention to converting the figures into a clear and common format that James Heggie of the Graphics Unit (Geography, University of Leeds) provided. We are also grateful for the work of the numerous referees who reviewed the chapters of the book, often to short deadlines, and gave valuable advice on improving the texts.

The book reports on the use of extensive digital library resources which are cited in the relevant chapters with the appropriate acknowledgements of copyright. The effort that goes into building up these digital resources is rarely recognized but without it e-learning would be rather empty of content. In the UK both JISC and ESRC (the Economic and Social Research Council) have supported the development of digital libraries such as the ESRC Census Programme. In the USA the NSF has funded major activity in the digital library field.

We have also received much valuable help from the information technology services at our universities, for example, in setting up federated access management for our online programs. We should not take for granted the skills and expertise that are employed everyday to keep us academics always connected.

So, this book is about collaboration in e-learning made possible through investment by two national funding bodies in both a major project and the supporting infrastructure. We hope our largely positive experience encourages others to take up the challenge of e-learning, connected to digital libraries of information and knowledge.

REFERENCES

JISC Programme and Project details are provided at: http://www.jisc.ac.uk/whatwedo/programmes/programme_dlitc/project_dialogplus.aspx.

The DialogPLUS Project website can be found at http://www.dialogplus.org/.

The ESRC Census Programme resources can be accessed at: http://www.census.ac.uk.

Section I
Collaboration in E–Learning Development

Chapter I
Developing E-Learning in Geography

Philip Rees
University of Leeds, UK

Louise Mackay
University of Leeds, UK

David Martin
University of Southampton, UK

Gráinne Conole
The Open University, UK

Hugh Davis
University of Southampton, UK

ABSTRACT

Technologies offer a range of tantalizing potentials for education—in terms of providing access to media-rich context and for students to visualize and interact with learning materials, as well as a variety of mechanisms for students to communicate and collaborate with their peers and tutors. This book describes the findings of an interdisciplinary research project, which provides a contextualized case study of a concerted attempt to integrate e-learning in one discipline, geography, across an international context. This chapter outlines the learning philosophies and learning strategies that inform the development of e-learning materials, focusing on a particular discipline context. The chapter authors come from a range of disciplines: geography, education, and computer science. Out of this inter-disciplinary collaboration has come new understanding of the range of approaches to learning (by the geographers) and new understanding of the enthusiasm of subject specialists (by the non-geographers). We will also report on understanding developed through working with colleagues in another country. In particular we have gained valuable insights into the challenges associated with carrying out interdisciplinary research in this area, as well as working in an international context. At the heart of the work reported here is the notion of creation and use of learning materials for geography. We set down some definitions of learning materials to begin with. We critique the widely used "learning object" concept as being computationally convenient, but restrictive, and argue for a more specialized term that better describes the discipline context. Some definitions demand that a learning object stands alone without reference to external resources. Geography teachers usually want their learners to engage with Web-based materials. Geographers want their students to tap into a wide variety of digital resources out there in cyberspace that inform them about the world. They wish to guide the students through the resources and their uses, empowering them to make their own explorations in the future. To import materials and hermetically seal them within learning objects potentially sterilizes them and presents an oversimplified view of the world. This argument leads to the definition of a learning material unit ("nugget" was the shorthand we debated and developed in the JISC-funded DialogPLUS project, part of the Digital Libraries in the Classroom program) as materials for student use with one or more activities designed to develop understanding, combined with student evaluation of the knowledge gained (tests, exercises, reflections). Nuggets connect to external digital resources held in libraries, repositories, or Web sites. This chapter also illustrates how e-learning has developed over time within a master's program, initially in one university but now involving collaboration between three. We conclude by drawing lessons for developing e-learning in geography.

INTRODUCTION

The aim of this chapter is to review the learning philosophies and strategies that geographers need to be aware of when preparing e-learning materials. Our review is informed by collaborations between geographers, educationalists, and computer scientists in the course of the DialogPLUS transnational project, supported by the Joint Information Systems Committee (JISC) of the UK Higher Education Funding Councils and the National Science Foundation (NSF) in the United States, involving four institutions, Pennsylvania State University, the University of Leeds, University of California at Santa Barbara, and the University of Southampton. More details of this project are described in the Preface to the book.

In the chapter we present our perspectives on the issues, controversies, and problems as they relate to e-learning in geography. We compare and contrast the different approaches to e-learning as exemplified in the contributions and main themes of the book. We discuss solutions and make recommendations in dealing with the issues, controversies, and problems presented.

We begin with a definition of e-learning, which is expanded to explore the variety of forms and settings that it can take. Then we review the value of e-learning to the teacher and student. All this draws on the collective experience of DialogPLUS project participants. In the second part of the chapter we outline learning philosophies and strategies of which geographers developing e-learning should have some knowledge, even if they decide to retain a large part of the approach that has served them well in the classroom. In the third part of the chapter we discuss and critique the nature of the learning object, which has been a focus of thinking in e-learning development. We then report on our use of the more general concept of the learning nugget, which has been developed and tested out in the international collaborative project, DialogPLUS, in which most of the book's contributors were involved. In the fourth part of the chapter we describe how e-learning has developed in one master's program. We conclude with some lessons from our experience in developing e-learning, which might be of use to geographical colleagues embarking on new e-learning experiments.

What is E-Learning?

E-learning as a term has a variety of connotations and shifting meanings (Conole & Oliver, 2007, p. 4) but can be construed as literally meaning "electronic learning," which is learning using electronic devices, often computers, or electronic media to convey or develop knowledge in learners (E-Learning Centre, 2008). The term is often used to imply that some kind of advanced learning technology (computer-based) has been incorporated. However, the spectrum of technological sophistication is very wide indeed. At one extreme we might place the advanced simulators used increasingly in industries that employ high levels of technology. Pilots will have learnt initially to fly aircraft of a given type using simulators, which are a combination of sophisticated computer programs that simulate the effect of pilot actions on the aircraft's flight, together with mechanisms that move and shake the simulator to mimic aircraft behavior. At the other extreme, a traditional university lecturer resistant to new technologies might consent to have their reading list uploaded to the library Web site for student use and linked to the electronic catalog entries or downloadable electronic resources.

The list of electronic learning technologies available is long and growing. It includes learning management systems (LMSs), also known as course management systems (CMSs) or virtual learning environments (VLEs), in which course materials are offered via the Web in an ordered environment, accessible to learners who are registered to use the materials. These systems can offer discussion boards, e-mail facilities, quizzes, and interactive learning technology. In this book, the term VLE is used generically when referring to this type of system. The VLE is normally accessed by the student from a personal computer via the Internet, traditionally from a purposed computer lab facility but increasingly by the student from their own laptop using wireless technology. In addition, so-called Web 2.0 technologies, which emphasize the social dimensions of technologies, networking possibilities, and user control, are having an increasing impact.

The definition of e-learning can be extended to mean interactive learning in which the learning content is available online and provides feedback to the student's learning activities. Online communication with others may or may not be included. Traditionally, the focus of online learning has usually been more on the learning content than on communication between learners and tutors. However, this traditional view is being replaced by one that sees e-learning as capitalizing on the potential of technology for education in which communication between learner and tutor is a vital component.

The term e-learning is often used more broadly as a synonym for online education. Kaplan-Leiserson has, for example, developed an online e-learning glossary, which provides this definition:

E-learning (electronic learning): Term covering a wide set of applications and processes, such as Web-based learning, computer-based learning, virtual classrooms, and digital collaboration. It includes the delivery of content via Internet, intranet/extranet (LAN/WAN), audio- and videotape, satellite broadcast, interactive TV, CD-ROM, and more (Learning Circuits, 2000).

In the glossary of elearningeuropa.info (2008), the definition of e-learning includes learning by facilitating access to resources and services as well as remote exchanges and collaboration. Students are encouraged to seek and use available

Internet resources. The term e-learning means many things to many authors and is in a constant state of flux and change, along with related terms such as "learning technologies," "educational technologies," and "technology-enhanced learning;" each term has a socio-cultural context and an inevitable political dimension.

Learning is just one element of education and is often embedded in the wider umbrella of online education. The term online education covers a much broader range of services than the term e-learning. The specialist companies that claim to deliver e-learning focus on course content, while online education institutions cover the whole range of educational services. There are many synonyms for online education. Among those commonly used are: virtual education, Internet-based education, Web-based education, and education via computer-mediated communication. Online education is characterized by:

- The separation of teachers and learners, which distinguishes it from face-to-face education
- The influence of an educational organization, which distinguishes it from self-study and private tutoring
- The use of a computer network to present or distribute some educational content
- The provision of two-way communication via a computer network so that students may benefit from communication with each other, teachers, and staff

The second, third, and fourth attributes are now becoming characteristic of many face-to-face, campus-based courses.

E-Learning Standards

Because the means of delivery of e-learning involves software and the Internet, it has been necessary to develop and use standards, both of a general nature or specific to e-learning to ensure interoperability – that is, that tools and content can be exchanged between systems. There exist a number of organizations that define e-learning standards of a technical nature. These include the Advanced Distributed Learning Network (ADLNet), the Aviation Industry CBT Committee (AICC), the Instructional Management Systems Project (IMS), Microsoft's Learning Resource Interchange (LRN) and the IEEE Learning Technology Standards Committee (LTSC). Widely used are standards such as the Sharable Course Object Reference Model (SCORM) or the extensible markup language (XML). These standards are being used to help in the exchange and sharing of e-learning materials between VLEs and other content management systems. Many of these standards are further explored in subsequent chapters of this book. However, pedagogically the development of standards is still in its infancy – success to date has primarily been in the development of technical standards. Attempts to codify educational knowledge and representation through the Learning Object Metadata (LOM) and the Learning Design specification have been less successful. Some would argue that standards can never completely capture and represent the educational aspects of learning content and design.

Does the geographer developing e-learning materials need to understand the technical details of such standards? We would hope not because of the computer science knowledge and skills they necessitate. However, an institution delivering e-learning does need people who can advise subject teachers on standards and help them deliver e-learning in the most accessible way.

Virtual Learning Environments

One way in which e-learning is delivered is through VLEs. They are becoming increasingly powerful and integrated with the institutional systems for registering and progressing students. The geography teacher and the geography student need to learn about the appropriate functionality of their institution's VLE and to re-train when the institution changes its VLE.

Paulsen (2003a) provides an excellent overview of the competing systems in the marketplace. Some 50 commercial systems were included in a survey of VLE users across Europe. The most used commercial systems were North American in origin (e.g., BlackBoard, FirstClass, Lotus Learning Space, Desire2Learn, and WebCT), but where English was not the medium of instruction four European systems (TopClass, ClassFronter, LU-VIT, and Tutor2000) were strong competitors. The open source product Moodle has become increasingly popular in recent years, as demonstrated by the decision of the Open University in the UK to adopt Moodle as its VLE solution to cater for its 200,000 students (Sclater, 2008). Self-developed systems were reported in Paulsen's survey as better adapted to local needs, written in the language of instruction and cheaper than the commercial systems. However, self-developed systems face sustainability challenges: they are vulnerable to the situations and ageing of their inventors and to the lack of continuing investment. For example, at the University of Leeds a decision was taken in 2007 to move from a successful home-grown system called Bodington to a commercial system because of the sustainability issue.

Higher and further education institutions now see investment in the development or licensing of a VLE as essential to their educational mission and need to provide for considerable support for its operation (in terms of hardware, software, help), including the training of staff and students in its use. The necessary induction of a geography teacher at a university will increasingly include training in the institutional VLE. Within each VLE there will be, *de rigueur*, e-learning systems to introduce users of different kinds to the functionality. There will be increasing integration of the VLE with the institution's student management system, starting with the automatic connection of a student's registration with the profile of modules/courses being taken. Institutions may also take the next step of insisting that all modules/courses have an existence within the VLE, whether the teacher is an e-learning practitioner or not. This in turn will raise student expectations about delivery of learning materials online, use of discussion rooms and posting of answers to FAQs. There is increasingly no hiding place for the ICT *refusnik*.

However, the emergence of Web 2.0 technologies and numerous free tools for enabling different forms of communication (tools for creating wikis and blogs for example) and for managing and visualizing content (such as Compendium and Freemind for visualization and representation of knowledge and Google docs for content storage and sharing) runs counter to the notion of centralized, institutionally controlled VLEs. Many are now beginning to question the need for central control as an approach and are arguing for a more user-centered and controlled approach based around loosely coupled Web 2.0 tools.

Mobile Learning

Our e-learning systems will have to evolve in other ways in face of technical change, in particular to the increasing availability of wireless communication. Dye et al. (2003) argued that wireless Internet with a high data transfer rate must be widespread for mobile-learning (m-learning) to be practical. M-learning is here defined as learning that can take place anytime, anywhere with the help of a mobile computing device. The device must be capable of presenting learning content and providing two-way wireless communication.

This requirement for wireless broadband is now being rapidly realized. Whereas in the past, campus-based e-learning took place in a wired laboratory, it is likely to move rapidly out of the laboratory to anywhere convenient for the learners. We will have entered the world of mobile learning.

Dye et al. (2003) forecast that increased flexibility would place additional demands on both teacher and student. It requires the student to have a high level of discipline in order to achieve his or her academic goals, while the teacher might have no clear-cut division between working hours and leisure time. In 2008 this scenario has arrived and both teachers and students are adapting to the challenges. Managing expectations about speed and nature of student-teacher interactions will be crucial. Recent research into students' actual use and experience of using technologies to support their learning suggests that they are indeed embracing technologies and harnessing them to support all aspects of their learning. They adapt and personalize the technologies to suit their individual needs and consider technologies to be a core tool for learning (Conole et al., 2008a).

There were expectations that the device used for mobile learning would be the mobile phone or personal digital assistant. This has been true in some environments but is unlikely to be the universal model because of costs and lack of data transfer capacity to date. What is happening on higher education campuses is the adoption of ever-cheaper laptops by most students combined with provision of high bandwidth to registered students and staff. The traditional lab practical class may be replaced by Internet available work programs combined with surgery sessions and tutorials with tutors for those with problems needing face-to-face advice.

For geographers, there is enormous potential added benefit from the development of widespread m-learning with respect to mobile computing devices in fieldwork environments. This concerns the move of e-learning from the campus or student residence into the field setting, which is actually the focus of geographical education and where the mobile device can variously help with landscape visualization and interpretation or with data logging and sharing. Rieger and Gay (1997) and Fletcher et al. (2003) deal with this specific application. In Chapter III one of

the nuggets of learning materials on the global positioning system (GPS) has been converted into a form that can be loaded into a personal digital assistant. The Personal Inquiry project is exploring the use of mobile technologies across formal and informal settings to promote inquiry-based learning for science and geography. The project has developed four models for inquiry-learning, along with a clarification of the different forms of representation that are needed to articulate different educational scenarios in this area (Conole et al., 2008b).

E-Learning Methods, Techniques, and Devices

E-learning needs a certain level of technological investment in e-learning systems to succeed. These are computers, Internet connections, VLEs, specialist software, digital or computerized e-learning materials, teachers who are able to utilize these resources, and students who are able to use such systems. We illustrate how this infrastructure underpins e-learning by comparing e-learning in the late 1970s with e-learning in the late 2000s.

Let us think back to computer systems and their use in e-learning in the late 1970s. Typically, in a university geography department there would be a specialist computer lab with terminals connected to a mainframe computer. Students would use the computers to access software and databases in the lab. They would mainly run software written by their teacher or colleagues to carry out exercises that implemented techniques taught in lecture classes. The student of social geography might, for example, run a program that computed indices of dissimilarity for comparing the spatial distributions of two population groups across wards in a city. The student prepared the data file using a simple text editor following instructions in a manual for the program. The teacher and his assistants would rush around the class trying to keep up with the forest of hands in the air as students encountered problems. The course evaluations would come in with appreciation for the efforts of the teacher but with remarks such as "I never want to see a computer again". There were further disadvantages. The student could not take the program that computed the indices home with them for further use, nor could the technique be used after graduation. The university might switch mainframes and operating systems in mid-course and the software and exercises would need to be re-written for the new system.

Let us now fast forward to the late 2000s. The same exercise was still being used to demonstrate techniques of census, spatial, and social analysis (Unit 6 of the GEOG5101 Leeds Module described in Chapter IV). However, the students were now using personal computers using a version of the Microsoft Windows operating system, implementing their own programs of computation using ubiquitous spreadsheet software and downloading the necessary census information from remote Web sites using intuitive menus for data extraction. The theory was explained in e-learning documents in universal portable document format downloaded from the university virtual learning environment. By doing the exercise they were learning some generic skills in translating formula into spreadsheet commands and the skills of diagnosing and fixing the inevitable errors that occur in such work. The output from the spreadsheet could be easily copied into reports for interpretation and submission for assessment. The lecture class covered the principles behind the technique rather than the reams of ephemeral technical detail of the mainframe era. The students are able to reuse and adapt the spreadsheet techniques learnt in the exercise to other problems.

Table 1. Teaching methods, devices and techniques used in e-learning

Teaching Methods	Teaching Techniques	Teaching Devices
One-online	Online Learning Documents Online Databases Online Publications Online Software Applications Online Interest Groups Interviews	Information Retrieval Systems, mainly Web Browsers linking to Learning Management Systems or Web Resources
One-to-one	Learning contracts Apprenticeships Internships Correspondence Studies Teacher-student questions/answers	E-mail Systems
One-to-many	Lectures Symposiums Skits	Learning Management Systems
Many-to-many	Debates Simulations or Games Role Plays Case Studies Discussions Transcript-based Assignments Brainstormings Delphi Techniques Nominal Group Techniques Forums Projects Student Presentations	Computer Conferencing Systems

Adapted from Paulsen (2003b)

The techniques and examples are available to them after graduation.

The e-learning was already leaving the fixed lab location as students migrated to their own wireless-enabled laptops, now near-universal on campuses. The exponential growth of computer labs across campus had come to a halt and instead flexible e-learning environments in libraries, coffee shops, or university foyers were replacing them as preferred locations for study. And of course, the exercise was now part of a distance learning module being studied by students from the UK, Jamaica, and Canada. Underpinning these changes are the ubiquity of the portable computers with only a small number of operating systems, the invention and development of the Web and search engines, the development of the high speed fibre optic Internet, and the spread of wireless hotspots. Perhaps the most important of these developments is the ubiquity of the Web and the availability of browsers from a variety of devices. The world of mobile learning has arrived (*The Economist*, 2008).

Table 1 taken from Paulsen (2003b) shows the range of methods, techniques, and devices available to the e-learning developer in the late 2000s. Here there is no mention of computer hardware, operating systems, or universal software. These are now taken for granted like the availability of water, food, and shelter in rich countries if not poor.

What is the Value of E-Learning for the Teacher?

It is clear that both the institution and the teacher must make considerable investments of resources and time to deliver e-learning successfully. Evidence to date from DialogPLUS partner institutions indicates that adopting this new technology is not cheap in either money or time. To motivate the teacher to make the transition, there must be some innovations and some incentives. Here we make some observations about developments based on our own campus experiences.

One strategy would be to create new professional careers of e-learning teachers and learning technologists. While the latter are essential to both re-train and support mainline teachers in making the transition to e-learning, there has been little enthusiasm for creating new e-lecturers except in wholly dedicated institutions or centres. Staff appointed as e-lecturers tend to aspire to normal careers rather than a ghetto status, with the result that they switch over as opportunities arise. So, if e-learning is a strategy that should be adopted by mainstream teachers, what is in it for them?

There is usually a need for significant investment in the development of high quality materials that will survive in an e-learning environment. A conventional Powerpoint™ lecture presentation is no longer fit for purpose: it needs to be fully fleshed out with explanations, narrative text, and activities. There are several payoffs to the lecturer for doing this.

- They will be forced to improve the quality of the learning materials and experience.
- They will also "materialize" their teaching. The e-learning materials will preserve the intellectual product outside of the classroom in the institution's VLE or repository, or in community repositories. This will raise the status of teaching in the long run in competition with research where the product has long been preserved in print and now e-print.
- They will be able to reuse the materials in several alternative contexts without excessive revision effort. For example, materials developed for face-to-face modules can be re-purposed for distance learning. This happened to the *Census Analysis and GIS* module at the University of Leeds (see Chapter IV for details). Originally converted in 2001 to be part of a University of Leeds distance program in GIS, it has become a module in a World Wide Universities program offered

by the University of Leeds, the University of Southampton, and The Pennsylvania State University. Individual units are frequently made available to students requiring knowledge of one topic (e.g., units dealing with the construction of social segregation indices, offered by both the University of Leeds and Pennsylvania State University).

• Modules that might otherwise be unviable as face-to-face courses because of low student numbers can be offered to small classes several times a year to suit student schedules.

However, only certain kinds of curricular materials/course are suitable for this kind of treatment, that is, material that has some reasonable shelf life. The concepts and data involved do not change radically from year to year.

What is the Value of E-Learning for the Student?

The value of e-learning to the student will depend very much on both expectations and circumstance.

Distance e-learning enables continuing professional development through taking courses while still earning and pursuing a career. It can be regarded as part of life-long learning, to which many societies are committed. The key feature of distance learners is that they have signed up to the distance learning experience.

Use of distance learning in the undergraduate curriculum is likely to be problematic, as experience in delivering an Earth Observation half module at the University of Leeds proved (see Chapter VII for the substance of this and Chapter XIII for an account of its evaluation). Despite being alerted to the distance nature of the first part of the module, undergraduate students were simply not prepared for or confident in the distance e-learning situation. The experiment was not repeated after its initial offering.

Experience in blended learning has been much more positive. Here the e-learning is combined with lectures and tutorials or with surgery, or problem solving, sessions. With the increasing IT literacy and personal equipment of geography students, we may expect successful use of blended learning. It is now necessary to make all e-learning materials Web-accessible, and to plan lessons and activities with the mobile, laptop engaged student in mind.

However, this view of the role of e-learning may be conservative as a prediction of the future. Prensky (2005) has, for example, argued that teachers are now way behind their students in digital "nous" and that they need to catch up fast.

LEARNING PHILOSOPHIES AND STRATEGIES

It is useful to take the plunge here and review for geographers some of the key thinking in the educational field as it applies to e-learning. There is an enormous literature on learning philosophies and strategies as used in distance learning and e-learning. We draw on the reviews provided by the Delphi project at Nettskolen Distance Education (NKI) in Norway (Paulsen, 2003a). Dyke et al. (2007) provide a recent review of learning theories and their relationship to e-learning. They distinguish three main distinctive learning theory perspectives: associative (such as didactic and behavioral approaches), cognitive, and socio-constructivist. They distill the essence of the characteristics of learning across these different theoretical perspectives, to propose that learning is about (1) thinking and reflection, (2) experience and activity, and (3) conversation and interaction. The different theoretical perspectives foreground these key characteristics.

Paulsen (2003a) provides a review of online teaching and learning philosophies. These are based on traditional learning theories supplemented by theoretical perspectives on distance education. We pick out here the main points of interest to teachers and students of geography using e-learning. Many different theories describing the processes at the heart of distance education have been presented during recent decades. Keegan (1988) identifies three sets of theories, which still have relevance: theories of autonomy and independence, theories of industrialization, and theories of interaction and communication.

Theories of Autonomy and Independence

Several authors stress that distance learners are characterized by a high degree of autonomy and independence in their learning goals and strategies (see Moore & Anderson, 2003, 2007). They need to be able to navigate through their educational programs without much, if any, support from peers. Peer support may be provided through discussion forums or other forms of asynchronous communication but this is not a pre-requisite for success. Distance learners also need some maturity to negotiate with their teachers at a distance, particularly about the extensions to assignment deadlines, which often occur because of the pressures of work and family responsibilities. Distance learning programs vary considerably in the degree to which they offer a very structured syllabus through which the student progresses or whether there are more frequent opportunities for dialogue between student and instructor.

Theories of Industrialization: A Failure of Distance E-Learning Based on the Industrial Model

In industrial organizations, including universities, there is a continuous drive in a competitive environment to improve products and processes. Improved products will command a higher price in the market (higher student fees) while improved processes may lead to lower costs (and hence to

higher profits). The ambition of educational institutions to reduce the high costs of education by developing distance e-learning for large cohorts of eager students has not been realized. The ambition was probably unrealistic in the first place.

One such attempt was the UK eUniversities (UKeU) initiative, an ambitious project to deliver a wide range of UK courses electronically to a worldwide market. The project was cancelled in 2004 after three years of investment of £62 million of public money (Computing, 2004). The House of Commons Education and Skills Committee (2005) report is damning. The project was technology-led rather than education-led. An independent evaluation showed that the project was flawed on a number of fronts – lack of a robust needs analysis, a disjuncture between those from an academic background and those from a business perspective, and no pedagogical framework to drive the developments (Conole et al., 2006). There was no consultation with participating universities and no analysis of the potential market or the needs of users. The bonuses paid to senior staff were "wholly unacceptable and morally indefensible" (House of Commons Education and Skills Committee, 2005, p.4). There was a failure to investigate and learn from the pioneers in the e-learning field in institutions such as Pennsylvania State University (World Campus) and NKI in Norway.

Peters (1993, translated and reproduced in Keegan 1994) suggests that a new model for distance learning is needed that incorporates more interaction between student and teacher and between student groups and a much greater control by students of their own learning. These features have been built into modern VLEs, though their use will still depend on instructor preference and the articulation of student demand. Younger instructors, more used to "always on and available" communication, find this kind of online intensive interaction with their students fairly natural. Older instructors used to thinking about and considering carefully their responses to student questions are more resistant. Over time, ageing and replacement of instructor generations will mean that such online interactions become second nature.

Theories of Interaction and Communication

The Guided Teaching-Learning Conversation

Holmberg (1960, 2001) has stressed the importance of the dialogue between instructor and student, which he termed in his later paper the "teaching-learning conversation." Central to online/distance learners is the relationship with their instructors at the institution delivering their course. If the learning material is designed to speak to the individual student, then the student will feel much more motivated. Text that engages with the student is likely to be more effective. For example, *"Congratulations! You have succeeded in dissolving the output area boundaries from the census to provide the ward boundaries for mapping the ethnic groups indicators you developed earlier. The next step is for you to ..."* would generally be preferred to *"On completion of the OA boundary dissolve exercise, proceed to the thematic mapping exercise."* Most geography teachers will have developed these kinds of skills in a classroom, seminar, or tutorial context (particularly in the latter). The trick is to incorporate the equivalent skills into formal e-learning materials.

It is not only the instructor who needs to get involved in the conversation with the student. The director of the e-learning degree program and the e-learning program administrator need also to be involved as well. One device that is sometimes used to ensure involvement of both these players in the process is to delegate approval of course work extensions of deadlines to the program director rather than the instructor. At first the instructor may feel loss of autonomy but the program director will know the student and their learning career to date and so can make a judgment based on broader knowledge and empathy with the student.

One desire that is reported from student satisfaction survey after survey is that feedback on work submitted for assessment or review is returned in a short time period. Students also want quick turnaround to their queries. Here there will be institutional norms in place, which set standards for speed of feedback. This puts considerable pressure on instructors who will, as academics, have many other calls on their time. To cope with this pressure there are considerable advantages in having a team deliver a course/module. When one person is away on business, another instructor can respond. The development of "always on, mobile communications" also facilitates meeting the agreed program norms.

One caution must be sounded, however. There is in any learning activity some hard work of reading, note taking, practicing, and writing to be done and no amount of empathy from the instructor will help if the student is not motivated to progress their own learning. Within human geography there has been a drift away from learning and applying techniques based on mathematics, statistics, computing science, and the scientific method toward qualitative and cultural analysis. It is vital for human geography students to understand both approaches.

Educational Transactions and Control

Garrison (1989, 1993) has argued that successful e-learning provides the student with a good measure of control over their learning program. The advantage of e-learning is that it frees students from immediate time and place constraints (e.g.. the 9 a.m. lecture in the 300 seat lecture theatre with 100 of their peers). However, much e-learning is prescriptive (and has to be) in terms of the knowledge that must be acquired. Garrison argues that control must be shared between teacher and student: if either party assumes control then the process becomes stressful for the party not in control. This independence of action and planning is one of the factors identified more generally as contributing to good health.

FROM LEARNING OBJECT TO USE OF LEARNING NUGGETS

One of the debates in the field of e-learning is on the usefulness of the learning-object concept in the development of e-learning. The authors discussed this issue in many of our DialogPLUS consortium meetings because the project proposal had used more vague concept, which we called the (learning) nugget. We here consider what we have learned.

The Learning Object Concept

Learning objects (Wiley, 2000) are tightly defined bundles of learning materials (text, graphics, interactions, exercises, assessments). Learning objects have the following attributes: accessibility, interoperability, reusability, durability, and granularity.

Accessibility of a learning object is important because instructional components are often sourced from one remote location (a VLE) and delivered to many other locations, such as the students at their mobile learning devices.

Interoperability means that learning objects, developed at one place, with one set of tools or platform, can be used at another location, with a different set of tools or platform. When learning objects are transferred from one VLE to another, they must still work at the destination. If learning objects have linked standard metadata, this helps the process of transfer.

Reusability is a vital characteristic of well-designed learning objects. Users should be able to incorporate it into new applications without much added work. Storing, searching, and retrieving learning resources pose challenges in traditional electronic teaching and learning media (audio-

tapes, videotapes, CDs, DVDs), though textbooks continue to be intensively used. The Internet and learning object repositories enable the distribution and reuse of knowledge sources. In the UK, for example, the Higher Education Funding Councils have funded the development of a repository of e-learning materials called Jorum. In April 2008 this resource was converted from access only to registered institution members to open access via the Web. There is also considerable effort being devoted to the creation of institutional repositories of learning material.

Durability means that the learning materials can still be used when base technology changes, without redesign or re-programming. Both teachers and learners are aware of the speed of technical change. A learning object that can be easily updated will have a longer shelf life.

Granularity refers to how learning objects are defined and stored. A learning object can be a program, a course, a module, a lesson, or a lesson segment. The finer the granularity, the more re-usable the learning object. But a greater number of small objects needs more cataloguing effort and incur greater management costs.

Some authors and e-learning developers have argued that learning objects have to be broad enough to be meaningful to students and useful to teachers while being granular enough to be reused. This argument has caused conflict between e-learning developers and teachers in some situations. In the unsuccessful UK eUniversities project a set of rigid rules were laid down for the design of learning objects that were to be incorporated into modules: they had to be completely portable and to stand alone as learning units. This was highly uncomfortable for the geographers involved in the project. Many of the e-learning units in our modules contained exercises using resources located elsewhere on the Internet. Most e-learning units were part of developmental sequences: students needed to proceed through the units in order to build their knowledge cumulatively. To successfully negotiate later modules, the student needed to have completed earlier units. When the UKeU enterprise folded some instructors felt a sense of relief that they had regained control of their own e-learning course development. The campus learning management systems to which we returned were for the most part enabling rather than prescriptive.

The Learning Nugget Concept

This was conceived in our collaborative Dialog-PLUS project as a means of transferring e-learning materials between modules taught at different partner institutions. The idea was to review our module syllabi and identify where learning nuggets could be developed collaboratively and inserted at the appropriate point in each other's modules. A second objective was to learn how to develop the metadata needed for easy transference of e-learning materials from one institution's VLE to another's VLE. At first we constructed planning spreadsheets that defined the contents/syllabi of courses/modules across partners. We then identified common elements on which it would be useful to work. These common elements we called nuggets: objects that contained learning materials, student assignments, and evaluations of student achievement.

What was our experience in practice? The human geography collaborations are described in Chapter IV. Most were successful but sometimes things did not work out. So, for example, there was a common desire to show students the reality of census questionnaires by having them complete these online via Web interfaces. Such an online version was developed at Southampton by David Martin using .asp scripting, but there were software compatibility problems in transferring these online forms to Leeds. In fact, the Leeds module persisted in using the document form (in PDF format) in introducing the census and exploited the ESRC Census Programme portal's very useful collection of 1991 and 2001 Census

questionnaires. We hoped to develop parallel expositions of how to download census data from the Web site repositories in both the U.S. and UK and then embed them in our respective courses. Nuggets were developed that enabled students in the UK and U.S. to access census data in each other's country. But the U.S. course did not go ahead at that time, so the UK census data were never used in the U.S.

So, what did work? The nuggets that were developed and shared most successfully were of two kinds: highly technical materials related to the global positioning system and its use described in Chapter III, and highly generic materials such as those on academic integrity discussed in Chapter VIII. What also worked was the sharing of whole modules and therefore students across campuses as described in Chapter II. This collaboration in academic programs has been made possible by new developments (federated access management), which delegate the authentication and authorization of students wishing to access modules or nuggets back to their home institution.

What also worked well was the sharing of e-learning materials that linked to resources on open access across the Internet. So, for example, it was easy for UK students to access U.S. census data on the U.S. Census Bureau's open access American Factfinder Web site (U.S. Bureau of the Census, 2008). U.S. students could also access UK census data on the open access Web sites of the three UK national statistics agencies (National Statistics, 2008, GROS, 2008, NISRA, 2008). What proved problematic was getting U.S. students access to UK census data provided through the ESRC Census Programme (2008). These resources are provided by a network of units located in universities across the UK (CCSR, 2008; CDU, 2008; CeLSIUS, 2008; CIDER, 2008; EDINA, 2008; LSCS, 2008; UK Data Archive, 2008), but these resources are not openly accessible. This applies to many valuable resources provided by the UK higher education community due to the UK's approach to data licensing and intellectual property rights. However, the tide may be turning on the issue of open access:

It was announced [on 21 April, 2008] that Jorum, the UK national repository for learning and teaching materials funded by JISC, is to offer open educational resources. This will make it easier for lecturers and teaching staff to share and reuse each other's teaching resources. JorumOpen - as it will be called - will also provide a showcase for UK universities and colleges on the international stage (JISC, 2008).

Furthermore the Open Educational Resource (OER) movement has expanded dramatically in recent years since the initial announcement of MIT to make a significant amount of its educational content freely available. OER programs such as those at MIT, Carnegie Mellon and OpenLearn at the Open University in the UK have given us a better understanding of the issues and complexity associated with sharing and repurposing of educational content. Conole and Weller (in press) describe a new learning design methodology, which is attempting to provide a means of visualizing, scaffolding, and sharing designs to help in unlocking the potential of OERs.

How has the nugget concept been useful? The nugget guides the student through the jungle of the Web of useful educational resources. We learnt, however, that it was difficult to transfer nuggets, without careful and considerable re-working of the content and style to suit the teacher and the students at the receiving institution.

The nugget concept as played out in the DialogPLUS project was useful in demonstrating that it is essential that it be an open object in which students are sent off to look at, retrieve or work with external resources. Those resources are most usefully organized in digital libraries where the structure of resources is clear and guidance to content is given.

LEARNING PRACTICE IN E-CONTENT DEVELOPMENT AND DELIVERY

To date in the chapter we have discussed the issues that are central to the development of e-learning in general terms. Geographers may find this discussion hard to connect with, if they have not had exposure to educational theory. So, in this section of the chapter, we reflect on the way in which e-learning has developed in one geography department over the past fifteen years, focusing on the University of Leeds' master's in GIS program. The reader who is a geography teacher immersed in similar development of e-learning may find encouragement that others have faced and overcome the challenges they face.

A Face-to-Face Master's Program

In 1996/97 the School of Geography at Leeds launched a master's program in geographic information systems (GIS), a project pushed hard by the Head of School (John Stillwell assisted by program director Steve Carver). This was a face-to-face course, aimed at students wanting to develop skills in computer aided mapping and handling geographical information that would be useful in the workplace. GIS software was developing fast and becoming part of the tool kit used in private companies and public bodies involved in a wide range of activities from selling groceries to planning pipeline networks. The program stressed GIS applications rather than GIScience (the theory behind the GIS algorithms) and sought to build on staff experience in modeling for business and in using spatial analysis in geographic research. GIS was seen as a vital front window to display and to learn from the underlying modeling, spatial, and statistical analysis.

The degree lasts for an intensive eleven months, consists of four compulsory first semester modules and four optional second semester modules followed by a summer thesis based on three months' research. The degree and its variants recruit about 20-25 students per year, who pay their own fees and living expenses, though some overseas students have national scholarships. This has proved to be a stable market although there is intense competition around the world from similar offerings. E-learning was built into the degree program from the start in the sense that each module had intensive hands-on practical labs in which the students build applications and do analyses using the relevant methods and software, the principles of which are covered in lectures. Materials were delivered at first through shared disk facilities (difficult to find and search without guidance), but as the campus VLE developed resources were shifted there for much greater ease of access.

An Online Distance Master's Program

In 2000/1 the School launched an online master's in GIS by distance learning. The intention was to build on the experience and materials of the face-to-face program and tap into a much wider market of students whose work or family commitments prevented them taking a year out to do the face-to-face master's. It took some effort on the part of two successive Heads of School (Graham Clarke, Phil Rees) to persuade staff to take on this extra burden of teaching, although substantial external funding was obtained to employ additional e-learning teachers during a development period. The School benefitted enormously from the enthusiasm and hard work of successive directors of the online master's in GIS (Myles Gould, Linda See).

There was a lot to learn about program, module, and e-learning design in the first couple of years. We learned of course that a PowerPoint™ presentation was no substitute for a fully argued exposition of a subject, that every exercise needed to be thoroughly tested and not left to on-the-day fixes in class, and that activities and quizzes needed to be included to enliven the learning and

motivate the students. The program had to be designed with a series of exit points (certificate, diploma, and master's) and offered as a part-time degree taking 35 months rather than 11 months, in order that students could both work and study simultaneously. The original intention was that students would be recruited twice a year and that these bi-annual cohorts would proceed through time together. These neat schemes of progression proved to be less than realistic as students had to obtain extensions to complete module assignments and as students had to take breaks because of work or family commitments. We now have up to four cohorts per year and students switch between cohorts depending on pace through the program.

The development of the e-learning master's was aided by the parallel development of a campus VLE, written by university staff in another department (Andrew Booth, Jon Maber), who gave help and encouragement. This campus software is in process in 2008/9 of being replaced by a commercial VLE. The pioneers in e-learning have now handed over to a specialist team of system support people. The VLE is now a vital component of face-to-face learning as well as distance learning. The School's Director of Learning and Teaching chose the following *bon mots* to get this message across: *"I cannot stress enough that the VLE ain't one of your momma's teaching initiatives - this is going to be core to University business from now on. You need to engage"* (Evans, 2008).

An Inter-University Online Master's Program

The most recent development has been collaboration between the Universities of Leeds and Southampton, joined later by Pennsylvania State, in a shared online master's in GIS. Originally, this master's degree was supported by the UK eUniversities initiative discussed above. This initiative foundered, due variously to unrealistic expectations about growth in student numbers, a VLE without a proven track record, a restrictive learning object definition, and excessive management costs. At the same time as the UK e-Universities was rising and falling, a set of universities across the world was building a network for collaboration in research and teaching called WUN (Worldwide Universities Network, 2008). WUN is a consortium of American, Australian, British, Canadian, Chinese, Dutch, and Norwegian Universities. The online distance e-learning program was re-structured as a WUN master's in GIS (University of Leeds, University of Southampton, and Pennsylvania State, 2008). The e-learning materials are hosted on each institution's VLE (Bodington/Blackboard in Leeds, Blackboard in Southampton, Angel in Penn State) and students use their origin university registration to be authenticated and authorized for access to the e-learning materials. This new program has a majority of UK students but has reached out across the world to afford students access to a GIS course. Students who have studied or are studying the census analysis module, for example, live in Brazil, Canada, Jamaica, Nigeria, Sudan, United Kingdom, United States, and Venezuela.

We now draw out lessons from our discussion of e-learning theory and practice.

LESSONS FOR GEOGRAPHERS DEVELOPING E-LEARNING

We have come a long way in e-learning in geography in a short time. We would not have predicted even ten years ago the situation that we see today. Can we draw from this experience lessons for geographers developing e-learning? We make eight concluding points in answering this question. The lessons are generic across many disciplines rather than specific to geography, but nonetheless useful, we hope, for geographers.

1. The *investment* that is required to develop, launch, and maintain e-learning should not be under-estimated. It needs to be regarded as a long-term investment that probably takes at least five years to break even for the institution and for the individual.

2. The *quality* of the teaching materials must be raised substantially to compensate for lack of direct student-teacher contact.

3. The *returns* to investment are greatest where the learning content is relatively stable. So e-learning is well-suited to teaching techniques and methods that are tried and trusted rather than experimental.

4. The twin *drivers* of e-learning are technical innovations and student demand. Universities are now committed to developing the infrastructure to support e-learning and realize that the mobile/laptop age student must be fully supported. If you are a geography teacher or student, you will need to get involved.

5. There will, of course, be *refusniks*, teachers who dislike electronic media and think the book/journal is still the source of knowledge. But two developments will break down resistance to e-learning. The first is the inevitable ageing of refusniks and their replacement by the always-on, everywhere generation. The second is the impossibility of the paper model of knowledge repositories. Our libraries simply do not have the shelf space to hold the ever-expanding volume of publications. Resources must become electronic. Consider what has happened to journals in the past ten years. Academics protest loudly when a journal has not digitized its back catalogue and older articles can only be located in the recesses of the library basement. The same may in time happen to books, although the resistance is stronger there.

6. E-learning does afford the promise of *immortality* for good teachers because deposited e-learning materials have as good a chance of lasting as research materials.

7.   There are new possibilities for *collaboration* in the development of e-learning materials not just within institutions but also across institutions. Shared authorship of e-learning materials is challenging but rewarding in terms of exposure to new paradigms and ways of thinking about learning. International collaboration in teaching is now possible, although a long way behind international collaboration in research.

8.   The different *approaches* to collaboration have both merit and demerits. Direct exchange of e-learning nuggets or direct use of repository material is rarely possible. Adaptation is a necessity. Collaborative design is really exciting but again is rarely possible because of differences in teaching/learning objectives and contexts. Generic materials are easiest to share. Where the subject is specialist it is easier to share students.

E-learning in geography is here today on campus and on the Internet. Its use will expand because that is the way society is changing. It offers the opportunity to improve the quality of learning but will stretch all who use it.

REFERENCES

CeLSIUS. (2008). *Welcome to CeLSIUS.* London: Centre for Longitudinal Study Information and User Support, London School of Hygiene and Tropical Medicine. Retrieved April 27, 2008, from http://www.celsius.lshtm.ac.uk/

Census Dissemination Unit (CDU). (2008). *Accessing the census data.* Manchester: Census Dissemination Unit, University of Manchester. Retrieved April 27, 2008, from http://www.census.ac.uk/cdu/

Centre for Census and Survey Research (CCSR). (2008). *The samples of anonymised records.* Manchester: Cathie Marsh Centre for Census and Survey Research, University of Manchester. Retrieved April 27, 2008, from http://www.ccsr.ac.uk/sars/

Centre for Interaction Data Estimation and Research (CIDER). (2008). *Welcome to CIDER.* Leeds: Centre for Interaction Data Estimation and Research, University of Leeds. Retrieved April 27, 2008, from http://www.census.ac.uk/cids/

Computing. (2004). *The failure of UKeU: Computing's high-profile investigation into the government's disastrous £62m e-learning scheme.* Retrieved March 30, 2008, from http://www.computing.co.uk/computing/specials/2071853/failure-ukeu

Conole, G., Carusi, A., de Laat, M., Wilcox, P., & Darby, J. (2006). Managing differences in stakeholder relationships and organizational cultures in e-learning development: Lessons from the UK eUniversity experience. *Studies of Continuing Education, 28*(2), 135-150.

Conole, G., & Oliver, M. (2007). Introduction. In G. Conole & M. Oliver (Eds.), *Contemporary perspectives in e-learning research: themes, methods and impact on practice.* London: Routledge.

Conole, G., De Laat, M., Dillon, T., & Darby, J. (2008a). Disruptive technologies,' 'pedagogical innovation:' What's new? Findings from an in-depth study of students' use and perception of technology. *Computers and Education, 50*(2), 511-524.

Conole, G., Scanlon, E., Kerawalla, C., Mullholland, P., Anastopulou, S., & Blake, C. (2008b). *From design to narrative: The development of inquiry-based learning models.* Ed-Media Conference, July 2008, Vienna.

Conole, G., & Weller, M. (2008). Using learning design as a framework for supporting the design and reuse of OER. *Journal of Interactive Media in Education,* 2008(5), 1-12. Retrieved from http://jime.open.ac.uk/2008/05/jime-2008-05.pdf

Dye, A., Jones, B., & Kismihok, G. (2003). Mobile learning: The next generation of learning exploring online services in a mobile environment. Retrieved March 29, 2008, from http://www.dye.no/articles/mlearning/exploring_online_services_in_a_mobile_environmnet.pdf

Dyke, M., Conole, G., Ravenscroft, A., & de Freitas, S. (2007). Learning theories and their application to e-learning. In G. Conole & M. Oliver (Eds.), *Contemporary perspectives in e-learning research: themes, methods and impact on practice.* London: Routledge.

EDINA. (2008). *Welcome to UKBORDERS.* Edinburgh: Edinburgh Data Library, University of Edinburgh. Retrieved April 27, 2008, from http://www.edina.ac.uk/ukborders/

E-Learning Centre. (2008). *What is e-learning?* Sheffield: Learning Light Ltd. Retrieved March 25, 2008, from http://www.e-learningcentre.co.uk/eclipse/Resources/whatise.htm

Elearningeuropa.info. (2008). *Directory.* Retrieved March 29, 2008, from: http://www.elearningeuropa.info/directory/index.php?page=home

ESRC Census Programme. (2008). *The ESRC Census Programme.* Retrieved April 27, 2008, from http://census.ac.uk/censusprogramme/Default.aspx

Evans, A. (2008, April 24). *Happier, more productive...*(E-mail to the School of Geography, University of Leeds).

Garrison, D. R. (1989). *Understanding distance education.* London/New York: Routledge.

General Register Office for Scotland (GROS). (2008). *Welcome to SCROL: Scotland's Census Results Online.* Edinburgh: General Register Office for Scotland. Retrieved April 27, 2008, from: http://www.gro-scotland.gov.uk/

Fletcher, S., France, D., Moore, K., & Robinson, G. (2003). Technology before pedagogy? A GEES C&IT perspective. *Planet Special Edition 5 - Part B Pedagogic Research in Geography, Earth and Environmental Sciences,* pp. 52-55.

Garrison, D.R. (1993). Quality and access in distance education: Theoretical considerations. In D. Keegan (Ed.), *Theoretical principles of distance education.* London/New York: Routledge.

Holmberg, B. (1960). *On the methods of teaching by correspondence.* Lund: Gleerup.

Holmberg, B. (2001) A theory of distance education based on empathy. In M. Moore & W. G. Anderson (Eds.) *Handbook of distance education* (pp.79-86). Mahwah, NJ: Lawrence Erlbaum Associates.

House of Commons Education & Skills Committee. (2005). *UK e-University: Third report of session 2004-5.* London: The Stationery Office. Retrieved March 30, 2008, from http://www.publications.parliament.uk/pa/cm200405/cmselect/cmeduski/205/205.pdf

JISC. (2008). *Jorum to move to open access.* Bristol: Joint Information Systems Committee of the Higher Education Funding Councils of the UK. Retrieved April 25, 2008, from http://www.jisc.ac.uk/Home/news/stories/2008/04/jorumopen.aspx

Keegan, D. (1988). Theories of distance education. In D. Sewart, D. Keegan, & B. Holmberg (Ed.), *Distance education: International perspectives* (pp 63-67). London: Routledge.

Keegan, D. (1988). Problems in defining the field of distance education. *The American Journal of Distance Education, 2*(2), 4-11.

Keegan, D. (Ed.) (1994). *Otto Peters on distance education: The industrialization of teaching and learning.* London: Routledge.

Learning Circuits. (2000). *Glossary compiled by Eva Kaplan-Leiserson.* Alexandria, VA: American Society for Training and Developments.

Retrieved March 29, 2008, from http://www. learningcircuits.org/glossary

Longitudinal Studies Centre – Scotland (LSCS). (2008). *Linking lives through time.* St.Andrews: Longitudinal Studies Centre – Scotland, University of St. Andrews. Retrieved April 27, 2008, from http://www.lscs.ac.uk/

Moore, M., & Anderson, W.G. (Eds.) (2007). *Handbook of distance education.* London: Lawrence Erlbaum Associates.

National Statistics. (2008) *The census in England and Wales.* London, Newport, Southport, Titchfield: Office for National Statistics. Retrieved April 27, 2008, from http://www.statistics.gov. uk/census/

Northern Ireland Statistics and Research Agency (NISRA). (2008). *Welcome.* Belfast: Northern Ireland Statistics and Research Agency. Retrieved April 27, 2008, from http://www.nisra.gov.uk/

Paulsen, M. F. (2003b). E-learning - the state of the art. *NKI Distance Education, March 2003 Work Package One: The Delphi Project.* Retrieved December 12, 2007, from http://home.nettskolen. nki.no/~morten/E-learning/Teaching%20and%2 0learning%20philosophy.htm

Paulsen, M.F. (2003a). *WEB-EDU, Web education systems: A study of learning management systems for online education.* Powerpoint™ presentation. Retrieved March 29, 2008, from http://home.net-tskolen.com/~morten/pp/Web-edu.ppt

Prensky, M. (2005-2006). Listen to the natives. *Educational Leadership, 63*(4), 8-13. Retrieved May 12, 2008, from http://www.ascd.org/authors/ ed_lead/el200512_prensky.html

Sclater, N. (2008, June 10). Large scale open source e-learning systems at the Open University UK. EDUCAUSE Center for Applied Research. *Research Bulletin*, (12).

The Economist. (2008, April 12-18). Nomads at last: a special report on mobile telecoms. *The Economist.* Retrieved April 25, 2008, from http://www.economist.co.uk/specialreports/dis-playstory.cfm?story_id=10950394&CFID=3191 279&CFTOKEN=31474922

UK Data Archive. (2008). *Census.ac.uk: Moving you closer to the data.* Colchester: UK Data Archive, University of Essex. Retrieved April 27, 2008, from: http://www.census.ac.uk/

University of Leeds, University of Southampton, & Pennsylvania State University. (2008). *GIS online learning: MSc in geographic information systems.* Retrieved April 25, 2008, from http:// www.geog.leeds.ac.uk/odl/

U.S. Census Bureau. (2008) *American Factfinder.* Washington, DC: U.S. Census Bureau. Retrieved April 27, 2008, from http://factfinder.census. gov/home/saff/main.html?_lang=en

Wiley, D. A. (Ed.) (2000). *The instructional use of learning objects: Online version.* Retrieved March 27, 2008, from http://reusability.org/read/ chapters/wiley.doc

Wiley, D. A. (2002). Connecting learning objects to instructional design theory: A definition, a metaphor, and a taxonomy. In D.A. Wiley (Ed.), *The instructional use of learning objects.* Bloomington, IN: Association for Educational Communications & Technology. Retrieved December 10, 2007, from http://reusability.org/read/ chapters/wiley.doc

Worldwide Universities Network. (2008). *WUN: Worldwide Universities Network.* Retrieved April 25, 2008, from http://www.wun.ac.uk/

Chapter II
Exchanging E-Learning Materials, Modules, and Students

Samuel Leung
University of Southampton, UK

David Martin
University of Southampton, UK

Richard Treves
University of Southampton, UK

Oliver Duke-Williams
University of Leeds, UK

ABSTRACT

In contrast to other Web-based resources, e-learning materials are not always exchangeable and share-able. Although transferring electronic documents between networked computers has become almost effortless, the materials may often require careful design and a great deal of adaptation before they can be reused in a meaningful manner. This process involves consideration of pedagogic issues such as course curricula, learning outcomes, and intended audience, as well as technological factors including local institutional virtual learning environments (VLE) and any relevant learning technology standards. This chapter illustrates how these issues have been addressed resulting in the successful exchange of e-learning resources at three levels: (1) at content level, where learning nuggets are created and packaged in a standards-compliant format to guarantee interoperability; (2) at the user level, whereby learners or tutors, rather than the resources, are transferred between VLEs; (3) at a higher system level, where the emerging Web Single Sign-On technology of federated access management is being used to enable truly cross-institutional authentication allowing learners to roam freely in different learning environments.

INTRODUCTION

This chapter discusses the experience of Pennsylvania State University (Penn State) in the USA and the Universities of Leeds and Southampton in the UK in sharing teaching practices and learning resources in geography at undergraduate and graduate levels. This experience has clearly confirmed the need for internationally agreed standards at all stages of resource creation, assembly, and description. That said, a standards-compliant content object does not always remain meaningful let alone reusable after leaving its "birthplace" for another institution. In fact, format compliance is only one of the many contributing factors in any successful exchange process. The fate of any "migrant" content object is highly dependent on its adaptability to different learning contexts, data accessibility rights, and, above all, the unique characteristics of the intended audience in terms of needs, skills, and learning styles. This chapter will examine how e-learning content can best be exchanged at the content, users, and system levels. We consider the recent development of learning technology standards, a wider awareness of pedagogic requirements and the will to utilize information and communication technology (ICT) to enhance the education experience.

SHAREABLE GEOGRAPHY LEARNING AND TEACHING

The DialogPLUS project was originally set up in the expectation that the four collaborating institutions would develop and share innovative approaches in the learning and teaching of geography (Martin & Treves, 2007). The first phase of the project saw the geography teaching staff at each institution comparing common areas within their existing curricula and identifying elements that could potentially be shared and reused. This was undertaken in a face-to-face project meeting, facilitated by spreadsheet-based summaries of the characteristics of potential materials from each institution. To enable a meaningful curriculum mapping and avoid the distraction of the learning object debate, it was decided the project team would equate a discrete and self-contained learning activity with a "learning nugget," as discussed more fully in the Preface of this book and in Davis and Fill (2007). Readers who are interested in the origins of and controversy surrounding learning objects should refer to the work of Downes (2000), Wiley (2000), and Friesen (2003).

For the purposes of the current chapter, a DialogPLUS learning nugget in its simplest form must consist of the three core components: a learning objective, activity instructions, and supporting resources. Some nuggets might also include an assessment element, which can be either formative or summative. The cross-curricular mapping exercise provided the teaching staff with an opportunity to revisit and, if necessary, redesign their teaching practice and resources. The process was greatly assisted by the adoption of the DialogPLUS Toolkit, which was developed by project collaborators from education and computer sciences in order to support the design of learning nuggets. The toolkit maintained a nugget database, which embeds schema and metadata comparable with the emerging learning technology standards proposed by the IMS Global Learning Consortium (Bailey et al., 2006). Through the Toolkit, existing and new learning nuggets were arranged in terms of the three core and an optional assessment components. Learning nuggets were then matched and compared between institutions, based on examination of the actual learning activities rather than at the level of similarly titled materials. From the learning technology perspective, the process helped in populating metadata fields that in turn prepared each nugget to be more searchable and shareable.

In order to facilitate purposeful exchanges and sharing, the project had been organized around four sub-disciplinary themes: human geography, physical geography, earth observation, and geographical information systems (GIS). Each theme was labeled as a work package and led informally by the specialist teachers in the area across the collaborating institutions. These groupings emerged naturally from the team's subject specialties and research interests and facilitated focused communication between teaching staff in the different institutions. Resources such as an online census questionnaire, concept mapping tool, river catchment data, and remote sensing tutorials had all been identified for potential adaptation and reuse. An unintended consequence of the subject mapping exercise was a realization that all partners shared an interest in the promotion of academic integrity and referencing skills among their geography students. Penn State had already developed an academic integrity nugget (The Pennsylvania State University, 1999). The generic nature of the academic integrity nugget as well as the embedded quiz component made it an ideal candidate resource for experimentation in different models for content exchange. The nugget was thus seen to be of great interest to the other institutions and was added to the project as an additional work package, subsequently being repurposed and used at both Southampton and Leeds.

ISSUES IN EXCHANGING AND SHARING RESOURCES

Detailed accounts of how learning resources in each work package are developed, shared, and evaluated can be found in the other parts of this book. The foci of this chapter are the more general issues in the design, implementation, and repurposing of learning nuggets and the successes and shortcomings of particular chosen approaches.

Context Dependency and Content Portability

The content of learning material is often the starting point for any discussion of reuse (Neven & Duval, 2002; Nesbit et al., 2003; Krauss & Ally, 2005). Reinvention has long been at the heart of the teaching profession, but reuse can become an attractive option when a piece of learning material is deemed clearly relevant and useful (Gunn et al., 2007). Boyle and Cook (2001) asserted that the questionable quality and disjointed nature of the current online repositories for learning materials were the main reasons why educational practitioners were skeptical about their worthiness for reuse. While teachers are well placed to judge the quality and usefulness of materials in their own field, the structural deficiency of any repository or material bank could only be overcome by the agreement and enforcement of some forms of metadata standards. The need for such a systematic approach to archiving and classification will be discussed later in this section.

It is commonly agreed that highly contextualized content is less easily shared and reused than more generic material (Cisco, 2003; Duval & Hodgins, 2003). As noted by Rehak and Mason (2003), while it is mechanically possible to remove a component part of a larger resource collection, it can then be difficult to make use of the detached part out of its original context. Pedagogically sound learning resources tend to be bounded by a specific context that is critical for students to put their learning experience into perspective. Factors such as learning approach, learner profile, assessment regime, resources availability and accessibility, intellectual property rights, and so forth will all affect how the context is defined and formulated. Learning resources are therefore rarely both useful and reusable because of these local and institutional factors (Frisesn, 2003). The diverse academic settings among the DialogPLUS partners ensure that there are enough combinations for developing different approaches to context attachment and abstraction in order to maximize the chance of reuse.

Formatting and Standards

Technical aspects of exchanging and sharing learning resources include granularity, Web standards, and metadata. Closely affected by the dependency on a specific learning context, content reusability is also affected by the granularity with which the materials are sized and formatted. Granularity is not about the physical size but rather the amount of information communicated to the intended audience. Some have quantified size in terms of the number of learning objectives (Hamel & Ryan-Jones, 2002), while others base measurement on ideas or lesson time (Polsani, 2003). Whichever approach is chosen, the objective is to create online learning materials with an optimal granularity, both small enough to be portable but yet large enough to be independently meaningful. To achieve this, instructional designers often need to break up content objects into more granular items before each of them can be matched, remixed, and pieced together to form a useful learning resource (Duncan, 2003). It is indeed this flexibility to aggregate and disaggregate content that provide teachers the freedom to exchange and reuse others' learning resources (Koper et al., 2004; Purves et al., 2005).

Like any resource on the Web, the design of online learning materials should conform to the standards promoted by the World Wide Web Consortium (W3C at http://www.w3c.org) in order to ensure content interoperability and user accessibility across different browsing environments. As the majority of learning resources are presently text-based with still or animated pictures, the standards of hypertext markup language (HTML) and graphics are the most important aspects for content designers to consider. With the Web increasingly becoming more dynamic and sophisticated, the choice of scripting language will also affect the interoperability of any interactive content. Not only will the decision affect the interactive functionalities in different browsers and operating platforms, it will also dictate content continuity when newer Web technologies emerge.

Metadata is another item level consideration and concerns the standard description of learning materials. The concept owes its origin from library and information science whereby materials must first be described and classified in a commonly agreed framework before any of them can be searched and retrieved. As the information on the Web grows exponentially, resource discovery and retrieval are increasingly dependent on the availability and quality of metadata. Not surprisingly, metadata is identified as a key architectural component of the Semantic Web (Berners-Lee et al., 2001; Duval et al., 2002; van Ossenbruggen et al., 2002). Extending from single items to collections of aggregated learning resources, metadata remains relevant and is often considered together with ordering and location of granular content items in content packaging specifications. The issues of metadata and content packaging standards will be discussed fully in the following "System interoperability" section.

System Interoperability

All endeavors to create and reuse learning resources are situated within a set of broader technical and organizational systems. So far we have highlighted learning context, resource format, and Web standards, which are all item level concerns. At the system level, where individual units of activity are aggregated and delivered, the move to using a virtual learning environment (VLE) as an institutional learning environment has created new challenges for teachers who find themselves moving beyond use of familiar office software and venturing into the more open Web authoring environment. Hypertext documents on the Web are compact, portable, and designed to be viewed universally by all kinds of computers irrespective of their make (e.g., PC or Apple Macintosh) or operating system (e.g., Windows or Linux). From the students' perspective, the Web also alleviates the burden of acquiring any specific software or tool for them to read and download online learning materials.

While the Web has enhanced availability and accessibility of resources and spared users from the battle between computer and software vendors, the issue of system compatibility is transferred to the information provider, that is, the hosting institution as a whole and the content author as an individual. The Web in its bare form does not possess all the features, such as user interaction and security, which form part of a robust online learning environment. Most institutions have come to see the answer as use of a VLE, either a commercial system (e.g., Blackboard), open-source (e.g., Bodington, Moodle), or collaborative development (e.g., ANGEL), which can be relied on for teaching delivery and student management. While some may contradict this view and suggest that VLEs have not lived up to initial expectations and their future will soon be in doubt (Stiles, 2007; Weller, 2007), it is undeniable that educational establishments are experimenting and implementing learning strategies in which VLE use continues to play a key role (Jenkins et al., 2005; Iredale, 2006). The unplanned coincidence of working with three different VLEs within the DialogPLUS partnership initially seemed a technical inconvenience but has in fact become a catalyst for project partners to consider the importance of

interoperable content and open format (Martin & Treves, 2007), which are needless to say the two most important preconditions for reusability.

A lot of work has been done in defining and agreeing learning technology standards with respect to content packaging, learning design, and metadata description within the community. The IMS standards (http://www.imsglobal. org/specifications.html) and SCORM (Shareable Content Object Reference Model) specification (http://www.adlnet.gov/scorm/index.aspx) are the two most well known examples. Like any developing prototype, it takes several iterations before standards become stable and relevant to a critical mass of teachers. It was exactly this uncertainty that provided the DialogPLUS content developers with a window of opportunity to explore different avenues: for example, multi-versioned content for specific VLE upload; interoperable tutors delivering content materials to student groups using different VLEs; free-roaming students with full access to VLEs outside their native institution. A more detailed account of each approach will be discussed in the section "Exchanging e-learning resources."

A feature inherent in VLEs is the requirement for registered users to be able to log in to the system. Different types of users (students, tutors, guest users, etc.) are accorded varying levels of permission to access, upload, and modify material, based on their user credentials. However, user registration is usually carried out on a per-institution basis, with guest access for external users – where available – having only the lowest levels of access permission. This is problematic when cross-institutional teaching delivery is considered; either students and tutors need to have multiple accounts on VLEs at different sites, or there needs to be a mechanism for them to access material at a "foreign" VLE using an existing account. The Shibboleth project (http://shibboleth. internet2.edu/) is an open-source software layer that provides single sign on for Web services both within and across institutional boundaries, and can therefore potentially be used in conjunction with VLEs at different sites. One of the key features of Shibboleth, which is described more fully in the "Exchanging e-learning resources" section, is that users are authenticated using their home institutional username and password, regardless of the location of the servers delivering resources.

Human Factors

Technology on its own cannot improve teaching and learning. Its potential benefits can only be realized if teachers and learners are fully convinced, while skepticism and misunderstanding can easily lead to strong resistance to change. In the early days when content and learning management systems were piloted in many USA and UK institutions, the process was sometimes unfortunately linked with saving cost instead of enhancing educational quality (Graham, 2004; Sharpe et al., 2006). Setting a rigid "digital switch over" date whereby all taught courses must have a Web presence did little to encourage academics to voluntarily and "leisurely" examine the opportunities offered by e-learning. At the other end of the spectrum, many pioneering teachers, who spent a great deal of time developing innovative e-learning approaches and resources, can be let down if their institution decides to change the learning management system, facing the consequences of no further support and even an abrupt end to their developed work. It is against this backdrop of uncertainties about extra time and effort, system instability, and the complexity of accommodating different learning styles in the online environment that academics hesitate to develop and share online learning resources.

The realization that financial saving is not and should never be a main drive for e-learning (HEFEC, 1996) is underlined in the 2003 VLE survey conducted by the Universities and Colleges Information Systems Association (UCISA). The survey reports that there has been an "increas-

ing recognition that online learning is not a cost saver, which it was seen to be in the early stages of web-based developments" (Browne & Jenkins, 2003). The authors of the report stress that the provision of support staff and the relevance to career development have become the key incentives for e-learning uptake. It is therefore not a coincidence that higher and further education institutions are creating positions of instructional designers (in the USA) or learning technologists (in the UK) to bridge the gulf between learning and technology (Bricheno et al., 2004). Some of these learning technology professionals are employed in the central computing service and others attached to a teaching school. Each case has its own merits and drawbacks. School-based support may guarantee direct service accessibility and subject knowledge while the central provision tends to offer continuity and better system integration. The key is that academics would be more likely to consider "unproven" approaches and technologies in their teaching when there is a regular and reliable support framework in place.

Given keen academics and helpful support staff, the remaining piece of the jigsaw of developing online learning resources is the learner. While it is safe to assert that the current generation is highly computer literate, it will be premature to conclude that the Web will automatically become a fruitful environment for them to learn. Many of the them may well use the Web for social networking happily but whether such enthusiasm could be extended to educational activities remains unclear (Zemsky & Massy, 2004; Bond et al., 2007). Online learning has proved to be advantageous with respect to flexibility, convenience, and learner-centeredness (Metros & Bennett, 2004), but at the same time the independent element of online learning can frustrate learners with the sense of isolation, lack of immediate response, mismatched education needs, and learning style (Bricheno et al., 2004; Song et al., 2004). It is thus important that any shareable learning resources developed are not merely technology-proof but must be pedagogically sound in order to address and resolve learners' anxiety and needs.

Academic Governance

Where institutions elect to reuse learning materials created externally within their own courses, approval of the resulting program specification will remain with the teaching institution. Although there are international and inter-institutional differences in the required procedures, the syllabus is effectively under the control of the qualification-awarding institution, which is required to follow whatever regulations apply in its own context. If external learning materials form part of a local teaching program, rights management issues will generally have been addressed at the stage when the content was deposited or retrieved from a learning repository or received from a collaborating institution. Responsibility for all the course components remains unambiguously with the one teaching institution. A rather different situation prevails where collaboration results in actual teaching and delivery of course materials by another institution. In these circumstances, there will necessarily be a need for formal collaborative agreements between the institutions, which may incorporate provisions for the approval of prerequisites, credit points, assessment regulations and timetable, teaching staff, grievance procedures, external examining arrangements, and so forth. In the UK particularly, these will all be considered within the framework set by the Quality Assurance Agency for Higher Education (http://www.qaa.ac.uk/), whereas in the USA such matters are more likely to be regulated at the institutional level.

In the following sections we consider the exchange of e-learning materials at multiple levels, ranging from the reuse of individual learning nuggets from partner institutions to the inclusion of entire taught modules from another institution's degree programs. Each of the examples cited here

is set within a context of an appropriate framework of academic governance and regulation. Our more complex exchanges have been developed within the context of collaborative agreements at institutional level, which specify all the necessary elements identified above. Our overall experience is that the development of these governance arrangements has been at least as complex as the technical issues surrounding exchange, primarily because the online nature of our collaboration has presented the participating institutions with many new scenarios, inadequately addressed by existing procedures.

EXCHANGING E-LEARNING RESOURCES

Against all the potential barriers and resistance mentioned above, is it really worthwhile for academics to start sharing and reusing learning resources? Instead of killing the conversation with a simple "yes" or "no," the following paragraphs review how e-learning materials have been exchanged and repurposed among the DialogPLUS institutions. By assessing the issues and impacts of each type of exchange, readers are invited to arrive at their own conclusion of the opportunities and values that could be offered by sharing and reusing online resources.

"The Content Exchanges"

Photocopying textbook chapters within the terms and references of fair use (Casey, 2006) has long been a way for teachers to adapt and reuse learning materials. The arrival of scanners, digital cameras, and the widespread use of word processing software all make this routine content reproduction easier and quicker. These technological advances offer the opportunity for teachers to integrate smaller tracts of text and images as opposed to a whole page into their teaching handouts and slides. The use of the Web injects more dynamism into the process as the hypermedia is built to handle materials and activities of varying sizes, types, and sources. The main challenge confronting educators is no longer one of content digitization as learning resources increasingly co-exist in both paper and digital formats; educators instead have the time and tools at their disposal to focus on better content designs and alternative modes of delivery.

A well-designed learning nugget should enhance the student learning experience and easily adapt to changes in audience, curricula and learning environment. The first criterion is mostly pedagogical and was addressed by the learning design toolkit introduced in the DialogPLUS project. The adaptability part is about making a learning nugget sustainable beyond its first circle of use. Adaptable learning resources should be amenable to remolding while remaining meaningful. Many DialogPLUS learning nuggets were created with this adaptability for future reuse in mind. As discussed above, size and content type will affect resource reusability. The issue of granularity is however a complex one as there is no standard size and length for learning nuggets. Different considerations apply according to the resource type. In DialogPLUS, the common ones are course notes, laboratory activities and resource libraries. Courses or units of learning which have a defined set of learning objectives are divided into smaller learning sessions, each with one key learning objective. For example, "Drainage Basin Geomorphology," an undergraduate unit in physical geography at Southampton, is organized into ten topics each supported by a learning nugget. The topics are interdependent and arranged in a progressive order but their associated learning nuggets are self-contained. They could therefore be delivered in isolation to any learner provided that they have sufficient prior knowledge to follow the topic. A multiple-choice quiz is provided to help learners assess their mastery of the topic. The "Earth Observation of the Physical Environment" course at Leeds has a similar granular structure

comprising of ten introductory lectures on earth observation. Again, learners are given self-contained reading materials to follow and equipped with a multiple-choice quiz at the end to check their understanding. Below the course level, smaller activities such as laboratory experiments and questionnaires may be reassembled and become structurally detached from the original course. This reorganization helps facilitate reuse in a new context or environment. An example is the repurposing of the academic integrity materials. Online collections of reference resources are another form of DialogPLUS nugget. Their size tends to fall somewhere between a course and an activity. The Tyne Digital Library (http://www2.geog. soton.ac.uk/users/leungs/tdlsite/) and the Physical Geography Digital Library (http://www2.geog. soton.ac.uk/blackboard/units/year2/geog2016/ globaldlhome.asp) are two exemplar resources. Both were originally created for a specific course and have subsequently been adopted elsewhere.

Regardless of their granularities, the static component of all DialogPLUS learning nuggets is always written in compliance with existing hypertext (XHTML) and cascading style sheet (CSS) design standards to ensure cross-platform compatibility. The interactive elements such as surveys and quizzes demand more sophisticated coding beyond plain HTML. JavaScript, a lightweight client-side scripting language, can be used to create formative assessments that do not require tracking and recording. Flash, developed by Macromedia (now Adobe) to handle animated graphics on the Web, is most suited to produce media-rich content. Flash and JavaScript can also be combined together to create more complex multimedia resources such as drag-and-drop interaction. The concept mapping exercises in "Drainage Basin Geomorphology" are a product of these two technologies. For summative tests and exams where tracking and security are significant, the resources can be formatted based on the SCORM or IMS QTI (Question and Test Interoperability) standards, which are widely adopted by learning management systems and VLEs. Although many of these systems now offer an attractive integrated range of authoring and quiz content creation, this instant convenience can easily result in long-term reliance on one particular system. In our experience, which is shared by many such as Currier and Campell (2005) in their evaluation of the JISC Distributed National Electronic Resource (DNER) 5/99 projects and Clark's (2005) technical report from the SURF X4L Project, most VLEs do not offer adequate, standards-compliant export facilities for these interactive components. It is thus very hard to transfer content created natively within a specific VLE without extensive recoding. This restriction applies both to open source

as well as proprietary commercial packages. To avoid becoming trapped in a particular system, the DialogPLUS approach has been to create all content in HTML and interactive elements in Flash or JavaScript. The files are then uploaded to a VLE either individually or as a collection of resources. Loose files can now be assembled and organized using many freely available content and metadata editors such as Reload (http://www. reload.ac.uk/), elearning XHTML editor (http:// exelearning.org/), and Content Packager for IMS for Dreamweaver (http://www.adobe.com/cfusion/exchange/). In addition to reducing multiple files to one single zip file for ease of uploading, the editor creates a manifest of the content so

Table 1. A typical procedure to repurpose a mixture of static HTML pages and interactive quiz

	Academic Integrity Guidelines
Origin and destination	Penn State to Southampton and Leeds
Content type	Webpage (HTML) and quiz (HTML with server-side script)
Original source	Web quiz fully embedded in the ANGEL VLE supported by online reading materials related to referencing and plagiarism.
Transferral & decomposition	Angel does not offer content extraction or export facility. The quiz content is largely extracted via the "Cut and Paste" process.
Adaptation	The supporting reading documents are not transferred but replaced by the local documentation. The quiz elements are largely retained whereby the questions and answers are suitably adapted to make them in tune with the local context.
Content authoring	Supporting documents are created or modified in word processors such as Microsoft Word.
Web authoring	The text documents are either converted into PDF (for printing and download) or webpage (for viewing and linking). The quiz content go through the similar stage but the server-side functionality are provided by the use of a SCORM Runtime Wrapper. The wrapper adds the necessary JavaScript commands to the quiz content to enable this otherwise static content to interact with the VLE to which the guideline is destined.
Content packaging	Reload or the Content Packager extension for Dreamweaver is used to assemble every page and quiz into one compressed package. Because of the quiz element, the SCORM format is more favorable than IMS as the latter does not possess the communicative elements between the content package and the VLE. It is also desirable to add metadata at this stage.
Deposition	The package is imported to a chosen VLE. For deposition to Jorum or other content repositories, the IMS format is more appropriate as grading and recording are not required. Metadata should be added if not already present.
Delivery	Once uploaded successfully, the package will be able to run in a SCORM-compliant VLE provided that the suitable Java runtime environment has been installed beforehand. While not possible in VLEs, packages in Jorum are catalogued and thus can be searched, commented on and downloaded by the registered users of the service.
Future update and exchange	Updates are made to the individual HTML and quiz page which are the "highest common factor" of a learning resource package for different VLEs and repositories. Exchange, on the other hand, can take place at the package level.

Table 2. A typical procedure to repurpose Flash animations which are then delivered in a VLE

	Concept Mapping for Drainage Basin Geomorphology
Origin and destination	Penn State to Southampton
Content type	Flash animation
Original source	A series of drag and drop animation
Transferral & decomposition	Not applicable
Adaptation	The content is repurposed with six new concept maps that are used as part of the summative assessment.
Content authoring	The Flash content is edited and repurposed by the author of the original resource who is based at Penn State.
Web authoring	Minimal amount of authoring and editing such as changing titles and other descriptive text. The most challenging task is to integrate tracking and recording components in the resource which will run outside VLEs. (Flash offers an unrivalled level of sophistication compared with the simple drag-and-drop facility
Content packaging	Each Flash animation file is embedded in a webpage. All the web pages are then packaged using Reload or Content Packager in Dreamweaver.
Deposition	The package is uploaded to a chosen VLE or Jorum (or indeed any other learning object repository). In both cases, the IMS format is chosen. Metadata should also be added at this stage.
Delivery	Once uploaded successfully, the package will be able to run in an IMS-compliant VLE provided that the suitable Java runtime environment has been installed beforehand. While not possible in VLEs, packages in Jorum are catalogued and thus can be searched, commented on and downloaded by the registered users of the service.
Future update and exchange	Updates are made to the individual Flash files. Exchange, on the other hand, takes place at the overall package level.

Table 3. A typical procedure to build a digital library of learning resources aimed to support a project assignment

	The Tyne Digital Library
Origin and destination	Southampton to Southampton
Content type	Digital library type website
Original source	Paper-based materials, maps, river gauge website, digital photos
Transferral & decomposition	Reports, theses and maps are scanned. Maps are re-created digitally using data downloaded from the Edina Digimap service (http://www.edina.ac.uk/digimap). Many photos of the rivers with the Tyne catchment area are extracted from the CD of the River Habitat Survey conducted by the Environment Agency in the UK (http://www.environment-agency.gov.uk/subjects/conservation/840884/208785/?lang=_e).
Adaptation	Both maps and photos require copyright clearance. Maps of different scales are resized to ensure correct overlay while photos are all standardized to the same screen size.
Content authoring	As many of the existing files are predominantly from Microsoft Word, new files are also created in Word to allow ease of exchange and transfer.
Web authoring	While the resource uses HTML much more extensively with other D+ nuggets, it does not possess any interactive element so every page can be created using plain HTML.
Content packaging	Files are categorized into Project, Maps, Bibliographies and Data sections before assembled as a IMS learning package.
Deposition	The package is uploaded to a chosen VLE or Jorum (or indeed any other learning object repository). In both cases, the IMS format is chosen. Metadata should also be added at this stage.
Delivery	Once uploaded successfully, the package will be able to run in an IMS-compliant VLE provided that the suitable Java runtime environment has been installed beforehand. While not possible in VLEs, packages in Jorum are catalogued and thus can be searched, commented on and downloaded by the registered users of the service.
Future update and exchange	Updates are made to the individual page. Exchange, on the other hand, takes place at the overall package level.

that the collection can be easily understood and unpacked after it is imported into a standards-compliant system. Tables 1-3 give a summary of the key steps and required tools for authoring and packaging different resource types that are either subsequently exchanged among partner institutions or uploaded to the Jorum learning object repository. Readers who are interested in more details can refer to Lisa Rogers' Content Packaging Guide available at http://www.icbl.hw.ac.uk/learnem/outputs.html.

One of the key objectives for DialogPLUS has been to share the project deliverables and outcomes. By making sure that our learning nuggets conform to the established interoperability standards, the nuggets can be deposited in national learning object repositories such as Jorum (http://www.jorum.ac.uk) for reuse beyond the project. Jorum has been developed in the UK for service subscribers within the higher and further education communities to store and share their work. There are many similar learning object repositories developed in other countries, for example, Multimedia Education Resource for Online Learning and Teaching (Merlot: http://www.merlot.org) and Digital Library for Earth System Education (DLESE: http://www.dlese.org/library/) in the USA and eduSource (http://www.edusource.ca/) in Canada. Some have been launched as a result of international co-operations such as Global Learning Objects Brokered Exchange (GLOBE http://globe.edna.edu.au/globe/go). Common to all these repositories is a standard framework for resource description, which normally takes the form of metadata, that is, information about information. In the past, metadata has tended to be overlooked as learning resources are traditionally used within the classroom. There was little need to add textual explanation, as the face-to-face presentation should have provided all necessary contextual information. The convergence of the Semantic Web and learning repositories sees the growing wealth of distributed knowledge but also an ever-greater need for metadata. Without

metadata, online learning resources will be much harder to find, with the chances of discovery highly dependent on indexing by search engines. All DialogPLUS nuggets uploaded to Jorum are therefore tagged with metadata that conform to the UK Learning Object Metadata Core Element Set (UK LOM Core). The UK LOM Core is primarily for use within the UK academic community (Stevenson, 2005), but because it is derived from the globally accepted IEEE Learning Object Metadata (IEEE LOM), it is also compatible with other popular data models such as CanCore and Vetadata (Barker, 2005). It is thus possible for DialogPLUS resources to be adopted by users outside the UK domain.

Where it is desired to exchange not just static content, but actual teaching (incorporating content and tutor-led interaction) between institutions, rather different learning delivery models are required and these are discussed in the following section.

The "People" Exchanges

There are at least three different ways of configuring the delivery environment within which collaborating institutions can share online learning resources, in addition to the more common notion of deposition and sharing through a learning object repository.

Considered at a high level, if two institutions wish to share learning resources, either the resources must "move" to the students or the students must "move" to the resources. Under either scenario the student may in fact be sitting at the same PC and not be required to physically move at all, but the distinction is a very important one for the deployment of tutors, course materials, and VLEs.

In a conventional e-learning setup, each institution provides online learning content and interaction to its own students via a VLE (or some other content management system). Both the content (documents, tutorial materials, activities, datasets) and interaction (discussions, communication with tutor, feedback, student portfolios) are integral parts of the entire learning, but, they are treated differently under the different delivery models. In the following discussion, we explore delivery models where two or more institutions have agreed to share an online course module originating from one of the partners. These delivery models must all be set within the context of course approval, assessment and regulation considerations discussed above.

One option is to treat students from the other partner in much the same way as is conventional with physical student exchanges. This requires the incoming student to complete registration documents for the teaching institution and to become registered, albeit temporarily, to use that institution's VLE and learning resources. The students then login and take part in the module as part of the same class group as the "home" students. The advantage of this approach is that it presents a relatively familiar administrative workflow to the collaborating institutions and the whole class will be enriched by the interaction between one tutor and students from multiple institutions. An important disadvantage however is that students are faced with the potential confusion of multiple sequential registrations and unfamiliar interfaces. Whereas a student going on a physical exchange visit to another university will generally be clear about their institutional location, the "virtual exchange student" can readily become confused regarding tutorial responsibilities, coursework deadlines, submission procedures, and so forth.

An alternative model, which has been implemented by Southampton, Leeds, and Penn State, is to allow the student to remain within their own institutional VLE but to deliver a course from the partner institution. Under this arrangement, the entire course content must be either repackaged for the external VLE or placed in a location where it can be accessed from both VLEs simultaneously. The tutor is required to be registered in both VLEs. This approach offers the important advantage

that students are able to take their entire program of study from within the same VLE and are not required to undertake any additional registration procedures. However, this model places much greater demands on the tutor, who effectively teaches the course module to two separate classes simultaneously. Although the same course materials can be used the interaction aspects of the module are entirely separate so there is no direct communication between students from different institutions. We believe this to be an educational missed opportunity, but it also significantly increases the work of the tutor.

A third approach to the movement of students between collaborating teaching partners is to take advantage of new technologies for inter-institutional user authentication. These are considered in more detail below.

The "Access Management" Exchanges

Most VLE software provides some level of "guest access" whereby an external user can be granted access to one or more courses and may even be able to view all the learning content associated with a particular course. However, the interaction elements are not generally available to unregistered users, which means that granting guest VLE access does not provide an appropriate mechanism for allowing students from multiple institutions to be taught the same course module together. The underlying reason for this complexity is that the provision of static learning content does not require the system to have any knowledge of the user's characteristics; essentially, content pages are simply served up to the guest user as if they were on the open Web. However, course management functions require a personal identity in order to handle, for example, information on the users' membership of groups or discussion lists, marks achieved in assessment exercises and contact details for communication between the student and tutor. These functions cannot effectively be provided to an anonymous user, who would not be recognized as the same person returning in subsequent sessions, hence the need for the scenarios presented above whereby either the visiting student must complete standard registration processes and be granted a temporary login username by the teaching institution, or the tutor and the content must be made available within the students' local VLE. This general situation is not conducive to the most flexible sharing of students or learning resources between institutions.

A rather different alternative, with which the DialogPLUS project has engaged in some experimentation, is that of federated access management (http://www.ukfederation.org.uk/), or "single sign on." The key concept of federated access management is that a user has an institutional identity, which is accepted by other institutions within the same "federation." The user can gain access to a range of services within the federation of which their institution is a member, using their own institutional login details – hence they are required to sign on only once. Federated access management is set to become very familiar to academic users in the UK as this is the approach being adopted to supersede the Eduserve Athens authentication system, which has been widely used for access to national data services, online journals, and other similar resources such as those provided by the Economic and Social Research Council's Census Programme (http://census.ac.uk/). Athens requires each user to have an Athens username and password in addition to their own institutional login. The central Athens service checks the credentials of a user seeking to gain access to a resource and determines whether or not they should be granted access. Under federated management, the user has no separate login username for access to external services, but is prompted to enter their own institutional login details. These are referred by the service provider to the home institution and, if approved, access is granted. It should be apparent from this overview why the concept of the federation is so important: a user is being granted access to a service on the strength of approval by a trusted partner (a member of the same federation). This approach offers various advantages to users, most visible of which is the ability to access multiple services using only their own institutional login username and password. The establishment of access federations involves participating institutions signing up to the same terms and conditions covering system security, protocols for creating and managing user accounts, and so forth. It is thus possible for an institution to be a member of multiple federations and for different federations to have somewhat different terms and conditions. There are already established federations between some universities in the USA and within Europe. In the UK, the Joint Information Systems Committee's (JISC) access federation is expanding rapidly in the period 2006-08. Each federation member is required to run an "identity provider" service and a resource provider needs also to run a "where are you from" (WAYF) service, which is able to interrogate an incoming user and seek confirmation of their details from their home institution's identity provider. Shibboleth (http://shibboleth.internet2.edu/) is a set of software tools for implementing these exchanges, but the name has informally been used to refer to the entire system in recent UK discussions.

The figure below (Figure 1) illustrates a typical use case of Shibboleth. Firstly, the user attempts to access (1) a resource on the Web site of the service provider (SP). The resource in question is protected using Shibboleth, and so the Web server intercepts the action and creates an authentication request. A response is sent to the user that redirects them automatically to either their home Identity Provider (IdP) (2a) or to a Where Are You From (WAYF) service (2b). In the latter case, the user will select his or her own institution from a list, and then be subsequently forwarded to the correct identity provider. The use of a WAYF service permits users from a wide range of institutions to access the resource: if the resource is only ever intended to be used by users from a single institution, or if the user's institution has already been determined by other means, they can be sent directly to the IdP.

The IdP issues an authentication request to the user (3), who will respond (4) with appropriate credentials – typically a username and password pair, although other mechanisms could also be used. Assuming that the authentication is successful, a temporary response (5) is sent to the SP indicating that the access attempt is valid. The SP can then issue an attribute request (6) to the IdP, to discover more information about the user making the request. An important feature of Shibboleth is maximization of privacy, and thus the SP only requests as much information as is required to satisfy the resource restrictions. This does not necessarily include the user's name or any other individual identifying characteristics. In the case of course materials, it may only be necessary to confirm that the user is a registered student or member of staff at their home institution, or that they are registered for a particular course. In other cases, for example when the

resource provides access to data sets for which the licensing conditions dictate that users must be individually registered, then more attributes are required. The IdP responds to the attribute request (7), sending such information as has been requested, subject to any local privacy policies. Having received the attribute response, the SP can determine whether or not the user is permitted to access the resource, and then finally send the requested Web page to the user (8). The SP can also establish a unique session for that user, so that requests for further Web pages with the same access restrictions can be honored without further user interaction.

The concept of federated access management clearly offers some important potential benefits to e-learning exchanges between collaborating partner institutions, whereby students from one institution can gain full access to courses within the VLE of another partner using only their own institutional login details. As part of DialogPLUS we have engaged in an initial trial of this arrangement, with the University of Leeds and Penn State University establishing an experimental

Figure 1. The service request and user authentication process in Shibboleth

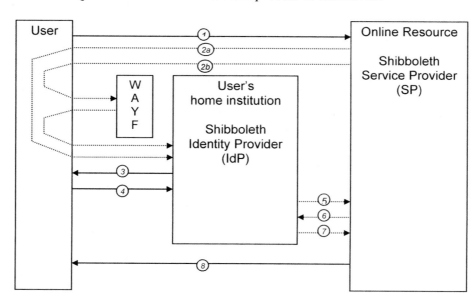

two-way "federation" and students on the Leeds online GIS program being able to gain direct access to Penn State materials within ANGEL. The University of Southampton and Penn State University are currently engaged in a further scoping study to permit Penn State students to gain access to Southampton's Blackboard service. These implementations present various challenges as there is relatively little relevant experience in the community and issues such as the selection of an appropriate federation and integration of student details for both local and external users can be tackled in a variety of ways. It seems likely that the use of federated access management to facilitate student e-learning exchanges will grow, especially as the use of access federation is becoming widespread for other types of shared resources.

LESSONS LEARNT AND CONCLUSION

The Web has undoubtedly provided educators with a largely free and extremely rich channel unrivalled by paper or other media for the distribution and sharing of knowledge. To date, with more than 500 million Web sites (Internet Systems Consortium, 2008), there is a common expectation that everyone can find something useful on the Web. The arrival of the Semantic Web not only helps shift the emphasis from quantity to quality but also reinforces the idea that the Web can be a productive educational medium. The additional effort involved in resizing, coding and reformatting learning resources into granular standards-compliant content objects should enhance resources discovery and reduce reliance on search engines. The growing acceptance of interoperability standards also paves the way for learning content to be moved in and out of any closed system such as a VLE or an institutional learning repository. Learners are becoming more aware of national resource sites such as Intute

(http://www.intute.ac.uk/), which provides support not only to academics but also to librarians, researchers, and students. With growing maturity of Internet technologies and increased confidence on the part of tutors and learners to explore the educational uses of the Web, it will be a great missed opportunity if digital learning resources remain merely downloadable but do not become truly reusable.

The focus of this chapter has been on exchanging e-learning materials, modules and students. It is initially tempting to consider that the key issues and solutions in such exchanges will be technical ones. But while it is possible to impose a specific technical format on an object, being in compliance with a standard and catalogued in a digital library do not, in themselves, make a resource more useful. A resource can only be useful for the other user if it is ever adopted and used again in a different context. Although there are many relevant technical considerations, as have been summarized here, our experience is that the governance and regulatory hurdles are actually the most important. There is also a fine balance between promoting open access academic resources, for example, MIT's OpenCourseWare (http://ocw.mit.edu/), and preserving individual or institutional intellectual property rights on the Web (Hoon & Van der Graaf, 2006). The issues are further complicated when resources are locked within a VLE or an internal institutional repository. While there is no magic solution to all these issues, the DialogPLUS project sees the way forward to lie in building trusted partnerships between institutions (DiBiase & Kidwai, 2007). Among all the factors including content, users, and systems, it is the preparedness of institutions to be interoperable that will have the most decisive influence.

REFERENCES

Bailey, C., Zalfan, M. T., Davis, H. C., Fill, K., & Conole, G. (2006). Panning for gold: Designing pedagogically inspired learning nuggets. *Educational Technology & Society, 9*(1), 113-122.

Barker, P. (2005). *What is IEEE learning object metadata/ IMS learning resource metadata?* CETIS Standards briefing series. Retrieved February 18, 2008, from http://wiki.cetis.ac.uk/What_is_IEEE_LOM/IMS_LRM

Berners-Lee, T., Hendler, J., & Lassila, O. (2001). The Semantic Web. *Scientific American.* Retrieved March 27, 2008, from http://www.sciam.com/article.cfm?id=the-semantic-web

Bond, S., Wodehouse, A., Leung, S., & Wallace, I. (2007). It takes a bit of imagination.... In *Institutional Transformation: The Proceedings of the JISC Innovating E-Learning 2007 Online Conference* (pp. 18-21). Retrieved February 18, 2008, from http://www.jisc.org.uk/media/documents/programmes/elearningpedagogy/ebookone2007.pdf

Boyle, T., & Cook, J. (2001). Towards a pedagogically sound basis for learning object portability and reuse. In G. Kennedy, M. Keppell, C. McNaught, & T. Petrovic (Eds.), *Meeting at the crossroads*: Proceedings of the 18th Annual Conference of the Australasian Society for Computers in Learning in Tertiary Education (ASCILITE 2001) (pp. 101-109). Melbourne: The University of Melbourne. Retrieved October 05, 2007, from http://www.ascilite.org.au/conferences/melbourne01/pdf/papers/boylet.pdf

Bricheno, P., Higgison, C., & Weedon, E. (2004). *The impact of networked learning on education institutions.* Bristol, UK: Joint Information Systems Committee (JISC). Retrieved March 17, 2008, from http://www.sfeuprojects.org.uk/inlei/

Browne, T., & Jenkins, M. (2003). *VLE surveys: A longitudinal perspective between March 2001 and March 2003 for higher education in the United Kingdom.* UCISA. Retrieved on March 27, 2008, from http://www.immagic.com/eLibrary/UNPROCESSED/Unprocessed%20eLibrary/eLibrary_uprocessed/JISC_Briefing_Papers/vle2003.pdf

Casey, J. (2006). *Intellectual property rights (IPR) in networked e-learning: A beginner's guide for content developers.* Glasgow: JISC Legal. Retrieved March 28, 2008, from http://www.jisclegal.ac.uk/pdfs/johncasey.pdf

Cisco Systems. (2003). *Reusable learning object strategy: Designing and developing learning objects for multiple learning approaches* (White paper). Cisco Systems Inc.

Clark, E. A. (2005). *SURF X4L - Reuse and repurposing of resources for content exchange, including technical considerations on interoperability.* A report from the SURF X4L Project, a JISC Exchange for Learning Project. Retrieved April 25, 2008, from http://www.staffs.ac.uk/COSE/X4L/X4Ltechnical.pdf

Currier, S., & Campbell, L. M. (2005). Evaluating 5/99 content for reusability as learning objects. *VINE: The Journal of Information and Knowledge Management Systems, 35*(1/2), 85-96.

Davis, H. C., & Fill, K. (2007). Embedding blended learning in a university's teaching culture: Experiences and reflections. *British Journal of Education Technology, 38*(5), 817-828.

DiBiase, D., & Kidwai, K. (2007). Wasted on the young? Comparing the efficacy of instructor-led online education in GIScience for post-adolescent undergraduates and adult professionals. In *Proceedings of the Association of American Geographers 2007 Annual Meeting*, San Francisco.

Downes, S. (2000). *Learning objects* (Essay). Retrieved October 23, 2007, from http://www.downes.ca/files/Learning_Objects.htm

Duncan, C. (2003). Granularization. In A. Littlejohn (Ed.), *Reusing online resources: A sustainable approach to e-learning* (pp. 12-19). London: Kogan.

Duval, E., & Hodgins, W. (2003). A LOM research agenda. In *Proceedings of the 12th International World Wide Web Conference*, Budapest, Hungary. Retrieved October 23, 2007, from http://www2003.org/cdrom/papers/alternate/P659/p659-duval.html.html

Duval, E., Hodgins, W., Sutton, S., & Weibel, S. L. (2002). Metadata principles and practicalities. *D-Lib Magazine, 8*(4). Retrieved March 27, 2008, from http://www.dlib.org/dlib/april02/weibel/04weibel.html

Friesen, N. (2003). Three objections to learning objects and e-learning standards. In R. McGreal (Ed.), *Online education using learning objects* (pp. 59-70). London: Routledge.

Graham, C. R. (2004). Blended learning systems: Definition, current trends, and future directions. In C. J. Bonk & C. R. Graham (Eds.), *Handbook of blending learning: Global perspectives, local designs* (pp. 3-11). San Francisco: Pfeiffer Publishing.

Gunn, C., Woodgate, S., & O'Grady, W. (2007). Repurposing learning objects: A sustainable alternative? *Association for Learning Technology Journal, 13*(3), 189-200.

Hamal, C. J., & Ryan-Jones, D. (2002). Designing instruction with learning objects. *International Journal of Educational Technology, 3*(1). Retrieved October 23, 2007, from http://www.ed.uiuc.edu/ijet/v3n1/hamel/index.html

Higher Education Funding Council for England (HEFEC). (1996). *Evaluation of the Teaching and Learning Technology Programme*. Bristol, UK: Higher Education Funding Council for England (HEFEC).

Hoon, E., & Van der Graaf, M. (2006). Copyright issues in open access research journal: The authors' perspective. *D-Lib Magazine, 12*(2). Retrieved February 18, 2008, from http://dlib.org/dlib/february06/vandergraaf/02vandergraaf.html

Internet Systems Consortium. (2008, January). *ISC Internet Domain Survey*. Retrieved on February 19, 2008, from http://www.isc.org/index.pl?/ops/ds/

Iredale, A. (2006). Successful learning or failing premise? A situated evaluation of a virtual learning environment. In D. Whitelock & S. Wheeler (Ed.), *Research proceedings: ALT-C 2006 The next generation*, Edinburgh, UK (pp. 1-10).

Jenkins, M., Browne, T., & Walker, R. (2005). *VLE surveys: A longitudinal perspective between March 2001, March 2003 and March 2005 for higher education in the United Kingdom*. Oxford: Universities and Colleges Information Systems Association.

Koper, R., Pannekeet, K., Hendriks, M., & Hummel, H. (2004). Building communities for the exchange of learning objects: Theoretical foundations and requirements. *ALT-J Research in Learning Technology, 12*(1), 21-35.

Krauss, F., & Ally, M. (2005). A study of the design and evaluation of a learning object and implication for content development. *Interdisciplinary Journal of Knowledge and Learning Objects, 1*, 1-22.

Martin, D., & Treves, R. (2007). DialogPLUS: Embedding eLearning in geographical practice. *British Journal of Education Technology, 38*(5), 773-783.

Metros, S. E., & Bennett, K. A. (2004). Learning objects in higher education: The sequel. *ECAR Research Bulletin, 2004*(11), 1-13.

Nesbit, J., Belfer, K., & Leacock, T. (2003). Learning Object Review Instrument (LORI)

user manual. *E-Learning Research and Assessment Network*. Retrieved October 23, 2007, from http://www.elera.net/eLera/Home/Articles/LORI%201.5pdf

Neven, F., & Duval, E. (2002). Reusable learning objects: A survey of LOM-based repositories. In *Proceedings of the 10th ACM International Conference on Multimedia*, Juan-les-Pins, France (pp. 291-294). New York: ACM.

Polsani, P. R. (2003). Use and abuse of reusable learning objects. *Journal of Digital Information*, *3*(4). Retrieved October 23, 2007, from http://journals.tdl.org/jodi/article/view/jodi-105/88

Purves, R. S., Medyckyi-Scott, D. J., & MacKaness, W. A. (2005). The e-MapScholar project – an example of interoperability in GIScience education. *Computers & Geosciences*, *31*(2), 189-198.

Rehak, D. R., & Mason, R. (2003). Keeping the learning in learning objects. In A. Littlejohn (Ed.), *Reusing online resources: A sustainable approach to e-learning* (pp. 20-34). London: Kogan.

Song, L., Singleton, E. S., Hill, J. R., & Koh, M. H. (2004). Improving online learning: Student perceptions of useful and challenging characteristics. *The Internet and Higher Education*, *7*(1), 59-70.

Sharpe, R., Benfield, G., Roberts, G., & Francis, R. (2006). *The undergraduate experience of blended e-learning: A review of UK literature and practice*. Oxford: The Higher Education Academy. Retrieved March 27, 2008, from http://www.heacademy.ac.uk/resources/detail/ourwork/research/Undergraduate_Experience

Stevenson, A. (2005). *Jorum application profile v1.0 draft*. Retrieved February 18, 2008, from http://www.jorum.ac.uk/docs/pdf/japv1p0.pdf

Stiles, M. (2007). Death of the VLE?: A challenge to a new orthodoxy. *Serials*, *20*(1), 31-36.

The Pennsylvania State University. (1999). *Course policies for GEOG482*. Retrieved October 23, 2007, from https://courseware.e-education.psu.edu/courses/geog482/policies.shtml

van Ossenbruggen, J., Hardman, L., & Rutledge, L. (2002). Hypermedia and the Semantic Web: A research agenda. *Journal of Digital Information*, *3*(1). Retrieved October 23, 2007, from http://journals.tdl.org/jodi/article/view/jodi-61/77

Weller, M. (2007). The VLE/LMS is dead. *The Ed Techie blog*. Retrieved February 18, 2008, from http://nogoodreason.typepad.co.uk/no_good_reason/2007/11/the-vlelms-is-d.html

Wiley, D. A. (2000). Connecting learning objects to instructional design theory: A definition, a metaphor, and a taxonomy. In D. A. Wiley (Ed.), *The instructional use of learning objects*. Retrieved October 23, 2007, from http://reusability.org/read/chapters/wiley.doc

Zemsky, R., & Massy, W. F. (2004). Why the e-learning boom went bust. *The Chronicle Review*, *50*(4), B6.

Chapter III
Collaborative Learning Activity Design:
Learning about the Global Positioning System

Helen Durham
University of Leeds, UK

Katherine Arrell
University of Leeds, UK

David DiBiase
The Pennsylvania State University, USA

ABSTRACT

Collaborative learning activity design (CLAD) is a multi-institution approach to the creation of e-learning material from the design phase through the development stage and onto the embedding of learning activities into existing modules at higher education institutions on both sides of the Atlantic. This was the approach taken by a group of academic and e-learning material developers at the Pennsylvania State University and the University of Leeds to develop a series of learning activities to support the use and understanding of the global positioning system (GPS). Aided by concept mapping, a Guidance Toolkit and Web conferencing facilities, the group worked seamlessly at producing a series of e-learning resources, including the basics of turning on a GPS unit and obtaining a spatial location, GPS data properties and GPS components, differential correction, and sources of GPS error and error correction. This chapter reflects on the success of this project, which, authors believe, hinged on the following: a clear vision in defining the learning outcomes of the collaborative resources; appropriate tools and technologies to support and facilitate the collaboration; excellent communication and a high level of trust between collaborators; and the identification of a robust iterative methodology to produce reusable e-learning resources.

INTRODUCTION

This chapter examines the practicalities of collaboration in the design and development of e-learning materials. In particular we discuss methodologies, benefits, and difficulties of producing reusable learning resources. International and national collaboration is increasingly common between academic institutions in research and teaching arenas, as evidenced by the Worldwide Universities Network (WUN), an alliance between 16 research-led universities throughout the world that fosters and supports interdisciplinary collaboration, faculty and student exchange, and e-learning, on a global scale (WUN, 2005). Investment in the development of high quality and innovative online resources can be high and, for a return on this investment, the sharing and reuse of resources should be encouraged. Any lack of reusability in resources is potentially caused by a misalignment of learning objectives and differences in the nature and depth of learning materials between user groups. A potential solution could be multiple authors (who have no geographical restrictions – they can be academics from the same institution or different institutions) sharing their knowledge and working as a team to produce resources. Educational institutions participating in collaborative courses, or creators of e-learning materials, need to consider new approaches to developing resources to guarantee reusability and adaptability for use by others; once developed, the resources can be uploaded to learning object repositories, for example, Intute and Jorum (United Kingdom [UK]), MERLOT and DLESE (United States of America [USA]), Ariadne (Europe), eduSource (Canada), and EdNA (Australia). To ensure that these resources can be re-deployed easily and integrated into appropriate courses there is an argument for collaboration in both the design and development stages of this material.

When defining the requirements of e-learning architecture in the development of reusable learning activities, Laurillard (2002) identifies learning design as a major recent development. Dalziel (2003, p. 594) describes learning design as concentrating on the context rather than just the content of e-learning, being activity-based rather than merely absorbing information, and having the requirement to meet multi-user environments, stating: "Much of the focus on learning design arises from a desire for reuse and adaptation at a level above simply reusing and adapting content objects."

Interest in learning design is motivated, in part, by the desire to make resources reusable but to successfully achieve this goal, an appropriate methodology is required. Brooks (1975) observed that, as in software engineering more generally, producing software that can be used by others is an order of magnitude more difficult than making software one can use oneself. Could this observation equally apply to producing courseware? One way to increase the transferability of learning objects between different institutional settings might be to involve authors from multiple institutions. With this in mind, the authors of this chapter devised a collaborative approach to learning design, called Collaborative Learning Activity Design (CLAD). By using this methodology to design and develop exemplar materials that can be disseminated to a larger audience in a more applied and reusable form, the success of collaboration can be considered. Will the integration of knowledge from multiple authors produce the sought-after goal of reusable materials or was Brook (1975) accurate in his belief that adding manpower [sic] to a software engineering project (or shared courseware as discussed in this chapter) does not make the project easier or likely to finish sooner?

Focusing on these issues, this chapter will describe the implementation of an inter-continental collaborative learning activity design by a consortium of academic and e-learning technologists who undertook this work as part of the Joint Information System Committee (JISC) and the National Science Foundation (NSF) funded

Digital Libraries in Support of Innovative Approaches to Learning and Teaching in Geography (DialogPLUS) project. E-learning material suitable for reuse and adaptation was designed and produced by teaching staff in both UK and USA academic institutions. Affiliated to geography departments, these project members applied the methodology to a teaching exemplar area familiar to them - the principles, use and applications of the global positioning system (GPS), with the aim of producing reuseable teaching and reference material suitable for use by staff and students (at all levels of study) requiring knowledge on how GPS can support them in their fieldwork.

The next sections will describe the background to CLAD and to the exemplar material, giving context to the choice of collaborative interaction and application. A more detailed section on the collaboration process including the methodology and tools implemented will follow. A series of e-learning resources were developed as a result of this collaborative process and are now integrated into modules hosted by consortium universities. Finally in this chapter we will reflect upon the successes and evaluate our understanding of the collaborative process.

BACKGROUND TO CLAD

Initially collaboration was defined by the consortium as "working together." Others have more complex interpretations of this term, for example Mattessich and Monsey (1992) suggested that there are three levels of "working together" defined as cooperation, coordination, and collaboration. Cooperation occurs when information is shared as needed, to design and develop independent resources without a commonly defined mission. The next level of working together is coordination, which requires greater sharing and communication, where common objectives are established and allocated to all stakeholders. Consequently resources and rewards are shared between all.

Collaboration is seen to extend this further, embedding goals and mechanisms for, and rewards of, collaboration within all stakeholders' institutions. Mattessich and Monsey (1992) use the quantity of risk to differentiate between coordination and collaboration, where the latter requires the greater risk. It is the authors' belief that boundaries between coordination and collaboration are fuzzy where measures of "risk" are hard to define. This is also noted by Fischer et al. (1992), who suggest coordination extends to collaboration but stress the need for innovative technologies to promote and maintain effective communication.

Given these definitions of collaboration we can now explore the phrase "Collaborative Learning Activity Design." CLAD was adopted in this instance to embrace the idea of creating teaching materials that could be used by multiple learning and teaching programs, not necessarily within the same educational institution or country. If we examine the component terms of CLAD, or a combination of some of these terms, we will see that there are a number of recurring themes. Saad and Maher (1996, p. 183) state that: "Design can be viewed as an activity in which teams of designers work together towards a final solution. The activity of designing through the interaction of designers and the environment is what we refer to as collaborative design." In their paper they investigate the use of computers for supporting collaborative design and stress the importance of communication in successful collaboration. To represent the types of interactions in collaborative work, they describe and illustrate a two-by-two time/space matrix (Figure 1) based on the location of those involved in the work and the timing of working together: face-to-face interaction, asynchronous interaction, synchronous distributed, and asynchronous distributed. All four methods of interaction were integral to the CLAD methodology presented here and are referred to throughout this chapter.

In a similar vein, Kvan (2000, p. 409) examined a new approach to learning design by concentrating on collaborative systems, which he states, "describe the computer systems which support distal communication between designers." He differentiates between loose-coupled collaborative design, where participants contribute their particular expertise at appropriate points in the design process, or a close-coupled process where participants work closely together to realize a design. The CLAD framework described within this chapter can be adopted as either a loose or close-coupled process.

In this study project members envisaged building upon existing experiences of cooperation to create resources that were truly collaborative. It was recognized that collaborating over distance and different time zones would introduce new hurdles in the exchange of ideas and working practices. Methods of interaction were identified as being critical to the success of working together and a combination of synchronous, asynchronous, distributed, and face-to-face meetings were pledged at the outset, using technology and tools to allow frequent communication. Some of these technologies and tools are described in greater detail later in this chapter.

BACKGROUND TO EXEMPLAR MATERIAL

There is an increasing need for the integration of GPS teaching and applications into higher education curricula, from geography-based studies such as fieldwork (Stoltman & Fraser, 2000) and geographical information systems courses (Dana, 1997, 1999) through to studies in criminology (Althausen & Mieczkowski, 2001). This is supported by Wikle and Lambert (1996) who propose that GPS technology is important to anyone in the spatially oriented disciplines. As a method for data collection to feed into a geographical information system (GIS) or for tracking criminals and creating maps of crime scenes, the application of GPS in a wide variety of higher education curricula and their use within work-related situations is increasing. However, users of GPS may require training in the use of a GPS data collection device (commonly known as a receiver) to make the most of its functionality and, in order to use the data captured by the technology, they need to appreciate the limitations of the system.

Discussions between geographers in the DialogPLUS consortium universities identified a common need to develop learning materials to

Figure 1. Time/space matrix to illustrate collaborative interactions (Adapted from Saad and Mayer, 1996)

	Same Time	Different Time
Same Place	Face to Face Interaction (e.g. Project meetings)	Asynchronous Interaction (e.g. File management)
Different Place	Synchronous Distributed (e.g. Access Grid, HorizonWimba)	Asynchronous Distributed (e.g.Email, Toolkit)

support students and staff in their understanding and use of GPS. Existing material at the Pennsylvania State University (Penn State) on sources of error, GPS components, GPS data properties, and differential correction was about to be reviewed and updated and could be supplemented by learning materials in using and understanding GPS receivers. Instead of working independently, a collaborative approach to the design of new and updated learning and teaching resources was taken by geographers at the University of Leeds and Penn State resulting in the development of a series of digital resources on GPS.

Here follows a brief introduction to GPS, for more detailed information on GPS, Wikle and Lambert (1997) and Dana (1999) provide a useful reference.

Navigation Systems in Use

Dana (1997, paragraph 1) describes GPS as "an earth-orbiting-satellite based system that provides signals available anywhere on or above the earth, twenty-four hours a day, that can be used to determine precise time and the position of a GPS receiver in three dimensions." Owned and operated by USA's Department of Defense (DOD), the GPS gives free access to users and can work in any weather conditions. Earth-orbiting satellites transmit navigation signals, which GPS receivers decode to provide a variety of information. The first GPS satellite was launched in the early 1970s and the navigation system became fully operational in 1993 after the establishment of a 24-satellite constellation. It was originally launched to provide navigational support to the USA military but is now also used by the general public all over the world, though at a lower level of accuracy than that available to the military (Hoffman-Wellenhof et al., 2001; Ordnance Survey, 2007a). The configuration of the system constellation ensures that there are at least four satellites above the horizon thus enabling the triangulation of any position on the earth's surface (Wikle &

Lambert, 1996). Members of the public using this system can expect to obtain an instantaneous, real-time position to an accuracy of approximately 10m. The accuracy of GPS can be improved by the differential mode whereby positioning errors can be determined and corrected by comparing the measured and known coordinates at a base point (Dana, 1999).

The future value of location finding systems is highlighted by the investment by European Union member states in the Galileo project (Lindström & Gasparini, 2003), which promises to deliver a positioning system of greater accuracy to European users and one not subject to the security or defense policy of the USA. In addition to this forthcoming European satellite navigation system, the continuing modernization of GPS and the long-awaited improvements to GLONASS, the Russian Global Navigation Satellite System (Russian Space Agency, 2006; Ordnance Survey, 2007b) can only mean that the use of GPS receivers for positioning and navigational purposes becomes more mainstream for both work and leisure time.

The use of such position location and navigation systems by geographers and other field scientists will start to have a long-term effect on the technological and cultural attitude at higher education level. GPS receivers are becoming affordable solutions for students to collect their spatial data, and results from a survey, carried out by the Fieldwork Education and Technology group of the LTSN-GEES pedagogic research and fieldwork program (France et al., 2002), highlights the importance of GPS hardware in the fieldwork element of undergraduate modules from geography, earth, and environmental science programs in higher education institutions.

THE COLLABORATION PROCESS

For the project members involved in the design and development of these international collaborative learning activities, the creation of reusable learning material for embedding in more than one institution was a new experience. Hitherto, learning materials had been developed to suit the learning objectives and aims of a particular module rather than the wider approach required in this instance. This section will examine the methodology of collaborative design of the series of e-learning resources. Reference to the time/space matrix as defined and illustrated by Saad and Maher (1996) will be made when considering how remote and disparate teams worked together, evaluating the methods used to share workspaces over distance. The elements of CLAD will be described followed by a look at how the technology and pedagogical approach to the design and development process was used by the team.

Methodology of CLAD

At the outset of the collaborative process; a face-to face meeting was held. This was the only synchronous interaction between all project members for the duration of the case study. This initial meeting allowed stakeholders to discuss potential topics that could be used for the e-learning "nuggets", re-usable, exchangeable, and scaleable e-learning objects as described in greater detail in Chapter I. In order to test the robustness of CLAD and produce collaborative learning activities, the methodology was applied to a topic in which project members were knowledgeable and for which there was a common learning requirement. Following agreement by consortium members that the collaborative process was to be tested on the design and development of e-learning resources that would support the use and understanding of GPS, the first stage of the CLAD methodology was instigated.

CLAD consists of a number of iterative processes as illustrated in Figure 2. There were two cycles in the process: the design cycle and the development cycle. To initiate the design cycle, a collaborative process throughout, the

identification of personal and common teaching requirements were needed. This was, to a large extent, completed in the face-to face meeting with discussions on desired learning objectives and outcomes, topics to be covered by a series of nuggets, and the relationships within and between nuggets. Timescales and responsibilities were a feature of these discussions and appropriate methods of communication were established.

Subsequent stages in the collaboration process would need to take into consideration the geographical dispersion of the collaborators so technologies and tools were identified to facilitate asynchronous interaction, asynchronous distributed, and synchronous distributed communication, as defined in the Saad and Maher (1996) time/space matrix. Five time zones and several thousand miles separate the collaborators in this study, which made real-time discussions more challenging. E-mail allowed for asynchronous exchange, but for synchronous communication, instant chat services such as Windows Messenger/MSNTM were used. More sophisticated technological tools, such as the Web conferencing tool HorizonWimbaTM, were also identified and used to help synchronous communication; as a licensee of HorizonWimbaTM, Penn State was able to host two-way audio, slide shows, application sharing, whiteboard, and document sharing. This provided the means for collaborators to meet in a virtual room during the design phase to share requirements, ideas, and knowledge to create a concept map of the e-learning resources. Concept mapping and the Guidance Toolkit (as described in greater detail in the next section of this chapter) supported asynchronous distributed communication throughout the design phase.

The design cycle was iterated several times before all members of the collaboration team were happy that the resources would meet all requirements and maximize the reusability potential. Once collaborators were satisfied with the design phase then the second iterative cycle was entered, whereby the material was developed by individu-

Figure 2. Flowchart of CLAD process (Adapted from Durham and Arrell, 2007)

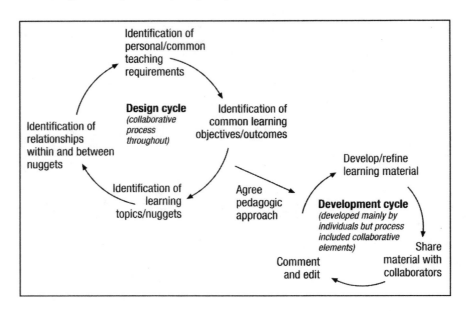

als or small groups, reviewed by the collaborative team, and then edited by the developers, repeating this development process until satisfactory e-learning resources were produced. As in the design phase, the use of technologies and tools supported and facilitated the collaborative nature of the development phase.

THE COLLABORATION TOOLS

There were two tools employed in the CLAD case study: the concept mapping tool, which was used to facilitate the e-learning resources planning, and the Guidance Toolkit, which supported the development of the nuggets. We will look at each of these tools in turn to find out more about their origin and application and will then examine how they were used in the methodology to create the resources.

Concept Mapping

Concept mapping was integral to this CLAD project, a tool for facilitating and guiding the learning material developers through the visualization and allocation of topics into a series of nuggets. Use of concept mapping was not new to some members of the team, as it had been successfully used previously in the design of reusable learning material for e-education in cartography and GIS at Penn State (DiBiase, 2005). Chapter X of this book describes applications of concept mapping in greater detail, so in this chapter we will mainly concentrate on justifying the adoption of the tool in the context of CLAD.

Concept mapping has been recognized as a useful tool for science education and has its origins in a longitudinal study, which started in the 1970s to assess school children's understanding of science concepts (Novak, 1990). Through that study, it was identified that concept mapping was a useful tool for students to learn how to learn and to help teachers be more effective in their teaching and curriculum planning. It was this second application, facilitating the production of more effective learning materials, that lead the team to adopt this tool.

Concept maps are organized in such a way that they parallel human cognitive structure, and are therefore hierarchical in structure (Wandersee, 1990). Key concepts are identified and arranged in an order (general to specific) and then the concepts are linked in a meaningful way by establishing a relationship. The team extended the technique further into curriculum planning by packaging, or parsing, inter-related concepts into a unit of learning or a nugget, but those units were themselves concepts, with relationships with other units. This

proved successful in facilitating the identification of a series of learning units.

Figure 3 shows the concept map agreed as part of a consensus-building process. The small boxes in the diagram represent the concepts, such as "Data Point" and "Coordinate Systems." Other boxes contain concepts that relate to the accuracy of the data obtained by GPS, using a GPS receiver, introduction to the concept of differential correction, and the various positioning systems in place, including GPS, GLONASS, and the European Galileo system. Links, or relationships between the concepts, can be identified and are clearly marked on the diagram, for example the relationship between "Data Point" and "Coordinate System" is "georeferenced to." By recording these concepts and relationships

Figure 3. The process of concept mapping

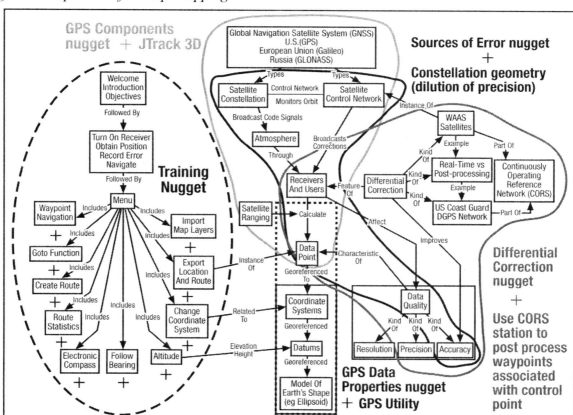

in this way, patterns emerge allowing nuggets to be visualized. This is followed by the packaging (or parsing) of the concepts into multiple units of study, as nuggets become concepts in their own right. In Figure 3, concepts have been packaged into broader concepts such as "Training" or "GPS Data Properties," and links, for example, "instance of" can be established between several nuggets to create a series of inter-related learning resources

Consensus among multiple authors in different institutional contexts is thought to increase reusability - or perhaps more accurately, "institutional interoperability," as discussed in Chapter II: Exchanging e-learning materials, modules, and students. In addition, the ability to parse concept maps in different ways, and to accommodate different locally applicable learning activities (symbolized by the cross or plus sign in Figure 3), should increase the adaptability of the resources to different institutional contexts. This is illustrated by some of the components that can be packaged into several different nuggets, such as the component "Data Point," which appears in four parsed units in our example.

To concur with the findings of DiBiase (2005), the authors believe that concept mapping has the advantage of helping foster collaboration better than text outlines. A subject topic can be broken down into inter-related key concepts and relationships can be identified to link within and between these concepts. The resulting structure and content constitute a concept map that is simple and visually easy to interpret. It should be noted however, that the process of producing a concept map may require several revisions but, with appropriate communication facilities, they can easily be shared synchronously in a face-to-face or distributed meeting or shared asynchronously among collaborators by, for example, e-mail.

Guidance Toolkit

The Dialog-Plus Nugget Developer Guidance Toolkit was used to facilitate the seamless production of the GPS nuggets (http://www.nettle.soton.ac.uk/toolkit/) and is described in greater detail in Chapter IX: A toolkit to guide the design of effective learning activities. The Toolkit, developed by

Figure 4. Extract from the Guidance Toolkit: Summary page of "GPS Training" nugget

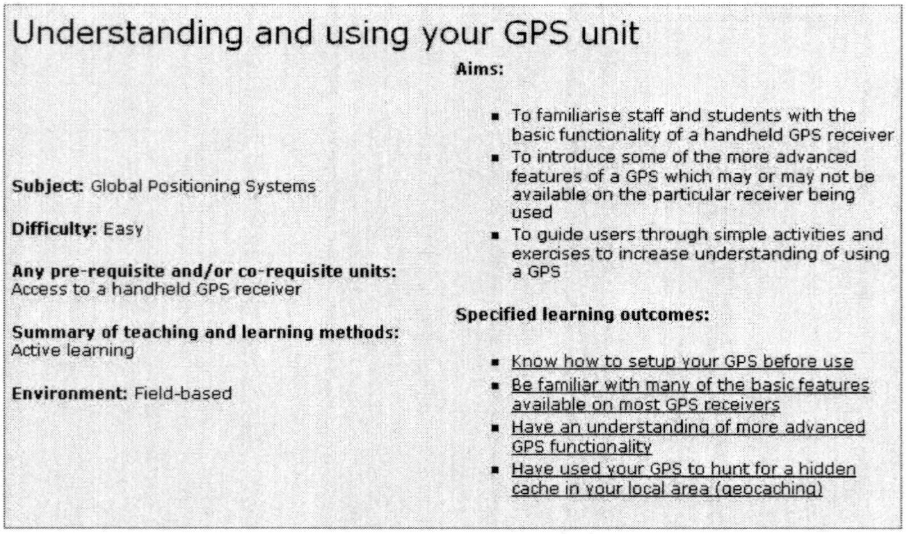

Understanding and using your GPS unit

Aims:

- To familiarise staff and students with the basic functionality of a handheld GPS receiver
- To introduce some of the more advanced features of a GPS which may or may not be available on the particular receiver being used
- To guide users through simple activities and exercises to increase understanding of using a GPS

Subject: Global Positioning Systems

Difficulty: Easy

Any pre-requisite and/or co-requisite units: Access to a handheld GPS receiver

Summary of teaching and learning methods: Active learning

Environment: Field-based

Specified learning outcomes:

- Know how to setup your GPS before use
- Be familiar with many of the basic features available on most GPS receivers
- Have an understanding of more advanced GPS functionality
- Have used your GPS to hunt for a hidden cache in your local area (geocaching)

project members at the University of Southampton, guides and supports teachers and learning material developers in the modification, creation, and sharing of learning activities and resources. Using this toolkit, developers at Penn State and the University of Leeds could record and formalize the nugget design by defining aims and learning outcomes, describing tasks to be undertaken by nugget users and identifying the learning and teaching approaches to be adopted.

Figure 4 shows the aims and learning outcomes and active learning method of the GPS training unit as articulated in the Toolkit. This introductory page of an entry summarizes the difficulty, pre-requisites for undertaking the activity, the learning and teaching methods, and a description of the environment in which the activity is carried out. Aims and learning outcomes can be readily viewed and links exist to the tasks associated with the learning outcomes.

DEVELOPMENT OF COLLABORATIVE RESOURCES

Following on from the concept mapping and working in parallel with the Guidance Toolkit, the team developed a series of nuggets. Collaboration and communication (both synchronous and asynchronous) continued to be a vital element throughout this phase of the project, requiring a combined effort to ensure that the evolving material continued to meet all stakeholder requirements.

The outcome from this exercise was the design and creation of e-learning resources on the use and applications of GPS, but of particular note was the production of the training unit on using a GPS, which had been designed and developed from first principles using the CLAD methodology. These resources are in active use within modules hosted by consortium universities (as illustrated in Figure 5), and the training unit is available to the wider academic community via the Jorum repository.

Figure 5. Screenshot of GPS resource within Bodington (University of Leeds VLE)

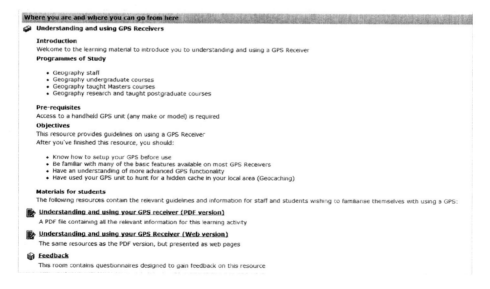

REFLECTION AND EVALUATION

Upon completion of the exemplar unit and its dissemination we can reflect on the successes and evaluate our understanding of the collaborative process. We can consider whether we believe the team truly worked in a collaborative way or was it, in reality, coordination or even cooperation? Many learning units developed as part of the DialogPLUS project, prior to this application of a collaborative methodology, fell under the cooperation grouping. Although project members attempted to share resources, these were always designed and disseminated independently without learning objectives being agreed by all stakeholders. This may potentially explain the minimal use of resources by institutions other than the author institution. In the exemplar material described in this chapter, we built on existing experiences of both cooperation and coordination to produce truly collaborative materials, characterized by common goals for all stakeholders, a firm team framework, open communication channels, and, as identified by Mattessich and Monsey (1992), a high level of trust between all stakeholders.

We can reflect on the nature of the collaborative design as outlined by Kvan (2000). It is the authors' belief that successful learning design can be achieved using either a close or loose-coupled process. At the outset, when the CLAD methodology was devised, the intention of the collaborative design was to use a close-coupled process. Although participants did work very closely together to design the courseware, there was an element of individuals contributing their experiences and expertise in specific areas, and this experience should probably be described as a loose-coupled process. Maybe this would be true of any group of academics who are collaborating, as it would be unusual for academics to all have the same knowledge and experience? Examining the stages of the design process there was collaboration in the form of a brainstorming session at the beginning, the use of collaborative

systems to further develop the design and establish key concepts and relationships between elements of the nuggets and a level of individual work as pockets of expertise were built up into appropriate activities.

In the introduction we presented the views of Brooks (1975) who suggested that adding extra personnel to a project does not necessarily make the work easier or allow the project to be completed earlier. Brooks was referring to software engineering projects but is the same true in the design and development of collaborative learning material? Do multiple authors help or hinder the process? The application of CLAD to this single exemplar is not enough to categorically state that collaborative design and development produces better resources in a faster time. It could be argued by the authors that in this example of collaboration the design and development stages were more thorough, more considerate of other requirements beyond the aims and objectives of a single module or course of study, and therefore more reusable. Conversely, there was no evidence that the collaborative process made the creation of this unit any faster than if produced by a sole academic; synchronous collaboration relied heavily on the availability of technology and multiple authors with relevant expertise, although the use of asynchronous communication such as e-mail could be used for most consultations, which maintained the pace of working together.

What this exemplar unit did identify was that a team that works well together, can share knowledge, is prepared to try different approaches to the design and development of learning material, and ultimately trust their team members to play their part in the collaborative process results in a learning outcome that by the very nature of its collaborative beginnings must have some reusable qualities. A successful outcome of this project would be material that was easily adaptable and reusable within multiple courses at more than one academic institution. Evidence would suggest that this team was successful in producing such an outcome with the integration of the learning unit into several programs of study, at different levels, in the consortium universities. In addition, its availability to the wider academic community via the Jorum repository will allow quantitative analysis of the reuse of the material in the future.

Englebart (1995) and Arias et al. (2000) argue that complex design problems cannot, by definition, be achieved by an individual alone and require the aggregated knowledge of multiple designers or authors. Kvan (2000, p. 413) states that, " collaboration is time consuming and requires relationship building and is suited to very particular problems that require such close coupling of the design process and its participants." It has already been identified that this exemplar material was not produced from close-coupled collaborative design, yet this application of CLAD was successful as it was implemented within an existing stakeholders group, with a high level of trust, respect, and aggregated knowledge. It was this aggregated knowledge and common goals that initiated and drove the potency, or effectiveness, of the e-learning resources design.

An interesting reflection to make at this point would be whether the choice of a GPS exemplar was appropriate to showcase the process of producing reusable learning material. In fact, the choice was particularly relevant when we consider one of the most reused resources on the Web about GPS: Wikipedia's entry (http://en.wikipedia.org/wiki/Global_Positioning_System). This resource is accessible to everyone, multilingual, community-authored, and amenable to local activities. With the increasing success of such reusable resources and with the advent of Web 2.0, is there the possibility that educators' emphasis could shift away from content development to (local) activity development, supported by community resources like Wikipedia?

CONCLUSION

This chapter has considered the practicalities of applying a collaborative methodology to successfully produce reusable learning materials. Perhaps the single factor identified as contributing to the success was communication. Face-to-face meetings, the low-level asynchronous technology such as e-mail, and the higher-level synchronous technology such as Web conferencing via HorizonWimbaTM supported the existing trust and good working relationships between team members. Successful collaboration requires interaction between participants, that is, participants need to share data as well as communicate with one another. Successful collaboration results in accomplishing something together that may not have been as successful if undertaken by an individual. This formed the motivation for using CLAD in the design and development of the learning material.

Equally important to successful collaboration was a clear vision of the learning outcomes. With an agreed consensus-built design agenda, team members worked together to produce materials with wider-reaching learning objectives, appropriate for adapting to a greater range of study level and the ability to be reused in more than one program of study in more than one educational establishment. Outcomes from this project are available via institution virtual learning environment or content management systems, in addition to learning object repositories.

The experiences of this consortium of practitioners have identified how new technologies and different cultures need to be accepted and embedded within courses to encourage global learning. Concept mapping, the Toolkit, and Web conferencing tools have introduced a new approach in learning material design. The methodology of identifying personal and common teaching requirements, identifying learning objects and topics and the relationships both within and between the elements, parsing elements into units of learning, and the iterative process of refining the overall concepts into reusable learning activities can be applied to independent or collaborative learning material development in any discipline and topic where e-learning material is to be reused or adapted.

The enthusiasm by members of the project to overcome all difficulties and produce a series of learning activities suitable for deployment in any of the member institutions was essential to successful collaboration. This was supported by the identification of a robust, iterative methodology to ensure the design and development of reusable resources for straightforward embedding in geography programs in both the UK and USA.

REFERENCES

Althausen, J. D., & Mieczkowski, T. M. (2001). The merging of criminology and geography into a course on spatial crime analysis. *Journal of Criminal Justice Education, 12*(2), 367-383.

Arias, E., Eden, H., Fischer, G., Gorman, A., & Scharff, E. (2000). Transcending the individual human mind – creating shared understanding through collaborative design. *ACM Transactions on Computer-Human Interaction, 7*(1), 84-113.

Brooks, Jr, F. P. (1975). *The mythical man-month: Essays on software engineering.* Reading, MA: Addison-Wesley Pub. Co.

Dalziel, J. (2003). Implementing learning design: The learning activity management system (LAMS). In G. Crisp, D. Thiele, I. Scholten, S. Barker, & J. Baron (Eds.), *Interact, Integrate, Impact: Proceedings of the 20th Annual Conference of the Australasian Society for Computers in Learning in Tertiary Education, Adelaide.*

Dana, P. H. (1997). *Global positioning system overview.* NCGIA Core Curriculum in GIScience. Retrieved April 11, 2008, from http://www.ncgia.ucsb.edu/giscc/units/u017/u017.html,

Dana, P. H. (1999). *Global positioning system overview.* The Geographer's Craft Project, Department of Geography, The University of Colorado at Boulder. Retrieved March 1, 2008, from http://www.colorado.edu/geography/gcraft/notes/gps/gps_f.html

DiBiase, D. (2005). Using concept mapping to design reusable learning objects for e-education in cartography and GIS. In *Proceedings of the XXII International Cartographic Association* (ICC2005), A Coruña, Spain.

Durham, H., & Arrell, K. (2007). Introducing new cultural and technological approaches into institutional practice: An experience from geography. *British Journal of Educational Technology, 38*(5), 795-804.

Englebart, D. C. (1995). Towards augmenting the human intellect and boosting our collective IQ. *Communications of the ACM, 38*(8), 30-32.

Fischer, G., Grudin, J., Lemke, A., McCall, R., Ostwald, J., Reeves, B., & Shipman, F. (1992). Supporting indirect collaborative design with integrated knowledge-based design environments. *Human-Computer Interaction, 7,* 281-314.

France, D., Fletcher, S., Moore, K., & Robinson, G. (2002). *Fieldwork education and technology.* The GEES Subject Centre. Retrieved May 6, 2008, from http://www.gees.ac.uk/pedresfw/pedrcit.htm

Hoffmann-Wellenhof, B., Lichtenegger, H., & Collins, J. (2001). Global positioning system: Theory and practice (5th rev. ed.). Austria: Springer-Verlag.

Kvan, T. (2000). Collaborative design: What is it? *Automation in Construction, 9*(2000), 409-415.

Laurillard, D. (2002). Design tools for e-learning. Keynote presentation for ASCILITE 2002. Winds of change in the sea of learning: Charting the course of Digital education, Auckland, New Zealand, 8-11 December 2002. Retrieved May

6, 2008, from http://www.ascilite.org.au/conferences/auckland02/proceedings/papers/key_laurillard.pdf

Lindström, G., & Gasparini, G. (2003). *The Galileo satellite system and its security implications.* Occasional Paper No. 44. Paris: European Union Institute for Security Studies. Retrieved March 1, 2008, from http://aei.pitt.edu/682/01/occ44.pdf

Mattesssich, P. W., & Monsey, B. R. (1992). Collaboration: What makes it work? St. Paul, MN: Amherst H. Wilder Foundation.

Novak, J. D. (1990). Concept mapping: A useful tool for science education. *Journal of Research in Science Teaching, 10,* 923-949

Ordnance Survey. (2007a). *Beginners guide to GPS – accuracies (single GPS receiver).* Retrieved July 23, 2007, from http://www.ordnancesurvey.co.uk/oswebsite/gps/information/gpsbackground/beginnersguidetogps/whatisgps_08.html

Ordnance Survey. (2007b). *GPS background – emerging satellite navigation systems.* Retrieved July 23, 2007, from http://www.ordnancesurvey.co.uk/oswebsite/gps/information/gpsbackground/satnavsystems.html

Russian Space Agency. (2006). *Information-analytical centre.* Retrieved March 1, 2008, from http://www.glonass-ianc.rsa.ru/pls/htmldb/f?p=202:1:2558501722024360015::NO

Saad, M., & Maher, M. L. (1996). Shared understanding in computer-supported collaborative design. *Computer-Aided Design, 28*(3), 183-192.

Stoltman J. P., & Fraser, R. (2000). Geography fieldwork: Tradition and technology meet. In R. Gerber, & G. K. Chuan (Eds.), *Fieldwork in geography: Reflections, perspectives, and actions.* Kluwer Academic Publishers.

Wandersee, J. H. (1990). Concept mapping and the cartography of cognition. *Journal of Research in Science Teaching, 27*(10), 923-936.

Wikle, T. A., & Lambert, D. P. (1996). The global positioning system and its integration into college geography curricula. *Journal of Geography, 95*(5), 186-193.

Worldwide Universities Network (WUN). (2005). Worldwide Universities Network. Retrieved March 1, 2008, from http://www.wun.ac.uk/index.php

USEFUL WEBLINKS

Intute: http://www.intute.ac.uk

Jorum: http://www.jorum.ac.uk

DLESE: http://www.dlese.org

MERLOT: http://merlot.org

ARIADNE: http://www.ariadne-eu.org

eduSource: http://edusource.licef.teluq.uquebec.ca

EdNA: http://www.edna.edu.au

Horizon Wimba: http://www.wimba.com/

Penn State Module: http://natureofgeoinfo.com (Chapter 5)

Windows Messenger: http://get.live.com/messenger/overview

Section II
Geography Exemplars

Chapter IV
Census and Population Analysis

David Martin
University of Southampton, UK

Philip Rees
University of Leeds, UK

Helen Durham
University of Leeds, UK

Stephen A. Matthews
The Pennsylvania State University, USA

ABSTRACT

This chapter presents the development of a series of shared learning materials prepared to facilitate teaching in human geography. The principal focus of this work has been on how to use census data to understand socio-demographic phenomena such as ethnic segregation or neighborhood profiles. In this area, students are required to address a combination of substantive and methodological issues that are particularly well suited to blended learning. Much of the information describing population characteristics is itself published online and it is therefore necessary to engage with external online data resources in order to obtain and analyze information for specific study areas. Our teaching exemplars include those designed to develop students' understanding of the data collection process, for example through the use of an online census questionnaire; analysis methods, through the provision of visualization tools to show demographic trends through time; and substantive examples, by comparison between urban social geographies in the USA and UK. Particular challenges are presented by the different nature (format, content, detail) and licensing arrangements for the census data available for student use in the UK and USA. In the UK students and researchers access census data via a research council funded program of data support units, which provides access to data from four successive censuses. In the USA open access to extensive data holdings is provided by the national statistical agency, the U.S. Census Bureau. However, the UK National Statistics offices are providing an ever-larger portfolio of datasets online and available to all, facilitating international collaboration and the types of data being provided are developing rapidly to fill the gaps between censuses.

INTRODUCTION

This chapter deals with the theme of the Dialog-PLUS project concerned with teaching in human geography. Our aim is to describe the ways in which e-learning have been used in a variety of geography degree programs across three universities. This will provide readers with examples of how the e-learning materials were used in specific courses, which may help them in using similar approaches and of the issues that were faced in incorporating these new materials. Table 1 shows the relationship of the courses the authors have delivered to the sections of the chapter in which different aspects are discussed. The courses range from first year modules taught to 150 undergraduate students to master's modules taught to five distance learning students. We hope that readers will find the chapter useful on several levels: as a guide to the range of ways in which to introduce e-learning materials into a human geography course, as a guide to recent UK and U.S. census data and their utility in thematic courses in human geography, and as an inspiration to students to explore the wealth of electronic information now available on our human world.

The exemplar material incorporated into the courses that is discussed in this chapter largely derives from the periodic censuses of population. These statistical instruments deliver vital population and housing information on small geographic areas. Census datasets were the first to be comprehensively made available online in the UK and USA. However, one major drawback of the census is its infrequency, as it is administered every ten or five years. Hence the currency of outputs declines over time for up to ten years. This data gap has been filled by use of vital statistics (on births and deaths) and of migration statistics (still poorly developed at small area scale) and

Table 1. Summary of the courses described in this chapter

Program Level	Module topic (University)	Activities delivered by e-learning	Chapter sections
Undergraduate			
Level 1 (campus module)	Geography of the UK (Leeds)	A practical on using the Online Census Atlas; a practical using the 2001 Census Output Area Classification and the current Neighbourhood Statistics web resources for England and Wales	Data exploration: an online census atlas
Level 3 (campus module)	Census and Neighborhood Analysis (Southampton)	All lecture materials online; practicals involve online and web based activities	Data collection: an online census form. Data analysis: geodemographic and multivariate indicator demonstrators. Engaging with real places: using students' own neighborhoods
Masters			
Module (campus module)	Spatial demography (Penn State)	Online materials on spatial analysis of population groups	Data analysis: demographic modelling
Module (campus module)	Census Analysis and Geographical Information Systems (Leeds)	All module materials including practicals are available online. Combined with lectures and individual consultations	Using census data to explore ethnic segregation: comparing US and UK cities
Module (distance module)	Census Analysis and Geographical Information Systems (Leeds)	All units of module delivered online with intensive electronic communication and planned telephone tutorials	Using census data to explore ethnic segregation: comparing US and UK cities

hence population estimates. National statistical offices have recently invested in the development of small area statistics based on population registers (available in many European countries), vital statistics registers, on administrative registers, and on state benefit records. In the UK there has been a major effort to deliver these inter-censal statistics for neighborhoods (National Statistics, 2007a). As these new statistical systems develop and improve we will need to move our focus from mainly using the census to using more current sources of information, following the lead of many Scandinavian countries, such as Finland, in which the census is just a set of tables produced annually from an integrated database of population, housing, address, and geographical data.

The scope of the authors' collaboration within human geography has been very specific, involving the teaching of census and population analysis. Across the participating institutions this was one of the earliest common areas to be identified and it is a widespread element in undergraduate geography degree programs. Teaching in this area supports many areas of the UK's Quality Assurance Agency (2000) subject benchmark statement for geography, contributing specific knowledge and understanding as well as to a wide range of student skills and abilities. A common approach among the project partners was the teaching of demographic structures and principles through student use of census information, and some of the authors had previously collaborated on the creation of a research-level handbook (Rees et al., 2002) and as part of a JISC-funded collection of online tutorial materials, discussed further below and described in See et al. (2004). This area particularly lends itself to the use of blended learning approaches in which face-to-face teaching of concepts and principles is supported by student activities, which engage them with real data. Census and population data are increasingly published online and students who go on to use such information in professional employment will need to understand the data structures and engage with retrieval interfaces. There are thus strong arguments in support of exposing students to the types of data retrieval and analysis scenarios that would be faced, for example, by an employee in local government, a health care planner, or retail analyst. Analysis of these types of data is also inherently spatial, involving an appreciation of neighborhood definition and identity, and benefiting from the use of appropriate visualization tools. These concepts and skills are often taught to student groups, particularly in the UK, who are well motivated in terms of the social and policy aspects of the subject but whose quantitative analysis skills are relatively weak (Keylock & Dorling, 2004). Geographic information systems (GIS) offer important data handling and analysis tools for census analysis, but these are most commonly taught as separate modules within undergraduate degree programs and GIS teaching is not the principal focus of the materials discussed here.

In Southampton, the DialogPLUS development work took the form of an entirely new optional module Census and Neighborhood Analysis aimed at second and third year undergraduates. Leeds had well-established census and demographic teaching within the undergraduate program. A long running second year course Research Methods in Human Geography (Gould et al., 2007-2008) provides students with an introduction to census area statistics, census interaction data and computer mapping methods. This material and associated techniques are taught in the first semester. In the second semester the students learn about qualitative methods. The School of Geography (Gould, See, Durham, & Rees) led one of the projects in a JISC-funded program, the Collection of Historical and Contemporary Censuses (CHCC, 2003), which included the commissioning of a series of online tutorial materials designed to introduce students to major themes in the use and analysis of UK census data. Further, during this period Leeds developed a module Census Analysis and GIS as part of a wholly online distance learning master's program in GIS, delivered collaboratively with Southampton (University of Leeds et al. 2008). At Penn State, learning materials were developed to support a module on Spatial Demography.

Our course mapping exercise revealed separate stages in the teaching of this subject area. Our teaching typically commences with an attempt to explain the data collection process in order that students can fully appreciate the scope and limitations of the information sources being used. Teaching then generally proceeds through a series of techniques for the exploration and analysis of published data and focuses on selected substantive issues, often promoted through student engagement with the in-depth study of particular localities. There is also a wide range of supporting material referenced by courses in this area, dealing with the organization of government statistical agencies, the documentation and publication of census results, and the policy uses of these data.

The organization of this chapter reflects this progression. The following section provides a brief overview of the subject area with which we are concerned, highlighting key learning objectives and common features. We then present a series of teaching exemplars, drawing on development work across the DialogPLUS partnership, which implement each of the stages of teaching identified here. Following this, we deal specifically with some issues of online access to census data in an international context and the chapter concludes by drawing out general principles and guidance from our collaborative activities. Many of the educational issues discussed here are truly international in nature but it should be noted that the specific datasets and examples cited relate only to the UK and USA, reflecting the transatlantic nature of the DialogPLUS collaboration.

TOPIC OVERVIEW

Most undergraduate geography degree programs include modules on population geography, sometimes taking a global or historical perspective but frequently focusing on demographic processes within the national context in which the course is being delivered. These often support other areas of teaching on urban development and spatial policy. In practice, much geographically informed government policy is developed on the basis of small-area geographical data derived from geo-referenced censuses and administrative data sources. An understanding of these data sources, their potential, limitations, and policy relevance is of value to human geography students and those concerned with environmental management. The collaborations described in this chapter focus on our teaching in this area. Although none of the DialogPLUS partners had exactly matching module courses, instructors found little difficulty in identifying significant blocks of teaching with common interest and development potential. UK subject benchmark statements require the specification of learning outcomes for taught program in terms of "knowledge and understanding" and "skills and abilities." Teaching in this area supports both strands and the learning outcomes defined for one of the DialogPLUS modules, the Southampton Census and Neighborhood Analysis module, are shown in Table 2.

One of the principles we have applied in our course design incorporating e-learning is that we expose our students to concepts and analysis methods using the full data sources available to the research community, and not just to 'toy' examples developed by the lecturer. So in the UK the practical exercises which our students undertake involve accessing Economic and Social Research Council (ESRC) Census Programme resources or National Statistics resources that university, research institute, and government researchers are themselves using. This gives our students an advantage in the labor market where

Table 2. Example of learning outcomes addressed through DialogPLUS learning activities ("Census and Neighborhood Analysis" module, University of Southampton, 2006-7)

Outcome Type	Specific Learning outcomes
Knowledge and understanding	Understand the influence of spatial and temporal scale on socioeconomic processes Understand the distinctiveness of social neighborhoods within the global mosaic Understand the various approaches available for representing the human world Understand the use of concepts of space and spatial variation in geographical analysis
Subject specific intellectual skills	Assess the merits of contrasting geographical explanations and policies Abstract and synthesise information from a range of different geographical sources Understand the importance of the spatial characteristics of geographical data Analyse and critically interpret primary and secondary geographical data
Subject specific practical skills	Collect, analyse and understand data in human geography, using computer techniques Understand the ways in which geographical data of various types can be combined, interpreted and modelled
Transferable/general (key) skills	Use your computational skills and ability in the use of statistical software Confidently use a range of relevant forms of IT software Marshall and retrieve data from library and internet resources. Understand the importance of data integrity, quality assurance and archiving in field and laboratory contexts

the skill to access and analyze quantitative statistics of a geographical nature relevant to business and government missions is highly prized. We avoid therefore what is known as the "Blue Peter" syndrome: in a popular British television magazine program for children, the presenters demonstrate some creative activity, but because of the time pressures of real-time TV, switch rapidly to the finished product ("here is a cake I baked earlier").

We also aim to nurture the ability in students to find new resources for themselves and to be able to adapt to re-organizations of Web resources in a resilient fashion. Despite a growing level of information technology (IT) literacy among human geography students, the ability to think laterally in such situations does not come naturally.

One interesting by-product of this approach is some frustration felt by students from developing countries when studying on our courses in the UK or USA. They wish to research key questions about the human geography of their own countries, but then find that the wealth of data available in the UK and USA is not replicated in their own. Our response to this frustration is to say to the students "when you become Director of National Statistics in your country, then you have a model of how to develop the accessibility of population statistics for all citizens for their benefit." In fact, rapid progress is being made by many countries and international bodies. Canada provides a wealth of statistics from Canadian censuses and other sources (Statistics Canada, 2008), comparable to those provided by UK National Statistics and U.S. Census Bureau. France also provides extensive information on recent censuses (INSEE, 2008), much of it translated into English. South Africa has developed a Web site that provides access to a selection of key statistics from the 2001 Census (Statistics South Africa, 2008a), together with an innovative online atlas (Statistics South Africa, 2008b).

International organizations also provide a growing range of country statistics useful for teachers and students of human geography. Examples include the world population estimates and projections prepared by the UN Population Division (United Nations, 2008), which are used extensively by demographers. The World Bank (2008) provides a superb overview of the level and changes in the development of countries in its World Development Reports. An example of the information provided by a university research center for worldwide topics is the CIESIN resource base delivered by Columbia University (2008). Eurostat (2008a) now provides extensive databases on economic, social, and environmental topics across not only the member states of the European Union but also across the three level a hierarchical system known as the nomenclature of territorial units for statistics (NUTS), with some additional information for Local Administrative Units (Eurostat, 2008b). ESPON (2006) delivers a sophisticated atlas that maps a rich set of economic, social, technological and environmental variables at NUTS2 region scale across Europe (EU members plus EEA countries).

TEACHING EXEMPLARS

Data Collection: An Online Census Form

One of the immediately apparent features of teaching about censuses is that many university students will generally never have taken part in one. Students entering direct from school or after a gap year in the UK or USA will only have personal experience of completing a census form if a decennial census enumeration (2001 in the UK and 2000 in the USA) took place between their 18th birthdays and taking the census module. They thus bring a range of misconceptions with regard to census content and public interpretation of the form, which have the potential to significantly impair their critical use of reported census data. In response to this, lecturers have sometimes designed a student group exercise in which each group conducted a census of the class, beginning with design of the questions, delivery, and collection of the questionnaire, design of a coding scheme for the data, data entry into a statistical package such as Statistical Package for the Social Sciences™ (SPSS, 2008), generation of a set of

tables and graphs, and production of a report on the census. This exercise worked well in a face-to-face class setting but was not transferred to distance learners because of the difficulties in creating student groups and moderating their progress. Given the recent development of social networking sites such as Facebook (2008) this possibility could be revisited.

A logical development has been to develop a Web-based version of the census questionnaire, which captures responses to provide a local dataset for teaching and a catalyst for class discussion. This was first implemented in Southampton and accompanied by a discussion forum in the module's Blackboard virtual learning environment (VLE) site. Completion of the form was voluntary and no student names and addresses were captured, but the resulting micro-dataset has then been used with each student group in order to demonstrate further data processing principles such as tabulation and disclosure control methods. Inclusion of the tutor's own responses in the class dataset serves to demonstrate clearly the issues associated with protection of a potentially "risky record," being the only member of the group married, in employment and aged over 40! The online questionnaire and resulting dataset has proved to be a rich learning resource, which can be drawn on at intervals throughout the module. Students consistently express difficulty interpreting the census instructions with regard to their own place of residence and their economic status—the majority of them being at a term-time address and having some part-time employment to support their full time studies. These insights are a very powerful stimulus to more critical use of published census results and a sound understanding of the data creation process. They also provide a discussion-starter for current consideration of online enumeration as a mainstream census data collection exercise, as used in the 2006 Canadian census (Statistics Canada, 2005). The specific practical activity and discussions are not formally assessed although instructor comment on the discussion forum provides an opportunity for formative feedback.

It became clear that the original Southampton implementation using .asp scripting and running on the School of Geography server was not generally transferable between institutions, with differences in IT security standards and policies particularly affected server-side scripting. A more portable version of the online form for Leeds use was subsequently developed using PHP scripting. An excellent source of guidance on the construction of online questionnaires, which could be readily adapted to this purpose, is available at University of Leicester (2008) (see Madge & O'Connor, 2004).

Data Exploration: An Online Census Atlas

The JISC-funded CHCC project, which ran until 2003, had initiated the development of an interactive Web-based visualization and exploratory census data analysis tool but the release dates of the 2001 Census data did not permit the completion of this resource within that project. One of the main responsibilities for the University of Leeds at the start of the DialogPLUS project was to integrate 2001 Census data with just under one hundred key variables from 1971, 1981, and 1991 Censuses and produce the Online Census Atlas. The resource delivers a simple mechanism for mapping UK census statistics using two forms of map boundaries to allow universal access to the Atlas. This section will outline the background to the creation of this resource and will briefly describe the map boundaries used to display the data.

The CHCC project was a collaborative venture involving four UK universities, the Universities of Leeds, Manchester, Essex, and Glasgow. The development of the Online Census Atlas was part of the Leeds team's remit to create online learn-

ing and teaching material, which would support and enhance student learning about the census area statistics from 1971 to 2001. Census area statistics are detailed aggregate data published for small geographical areas. The accessibility of the Online Census Atlas is based on the use of Java-based mapping software called CommonGIS. The original, full version of CommonGIS was adapted by the software developers to remove features not required for the Atlas and to create a focused tool for simple exploration and visualization. CommonGIS developers Andrienko & Andrienko (2006) have described their mapping software in more detail and have expanded on the approaches, techniques, and methods used for exploring spatial and temporal data in a recent publication. Project members at Leeds created an interface for this adapted tool, which allows users to select geography, variables, and census years of interest and to visualize and explore patterns in human geography.

Access to the full version of the Online Census Atlas is via an Athens-authenticated Web address (Durham et al., 2003a), which restricts users to those registered within UK academic institutions. Non-UK or non-academic users may use a reduced version of the Atlas available via a University of Leeds server (Durham et al., 2003b). To use the Atlas the user does not need any knowledge of visualization or mapping packages because the census data have already been combined with the map boundaries. The interface to both versions is identical and all features of visualization and exploration are replicated but, in the full version, users can view the census data as both conventional boundaries and a population-weighted cartogram, whereas in the reduced version only the cartogram boundaries are available. The more experienced user may wish to download boundaries and data for use in their own GIS, but the resource is equally suitable for the those with more limited applications who just wish to use the mapping and exploratory tools within the

Atlas, supported by the save and print capabilities of the software.

The Atlas allows users to select from six geographical scales: Government Office Regions, Local Authorities, Historical Counties, County and Unitary Authorities, European Constituencies, and Westminster Parliamentary Constituency. Maps of 88 key variables from the Census Area Statistics collated from the four most recent UK Censuses can be displayed for any geography and users can select either an individual census year to explore or two sequential/non-sequential census years may be selected to analyze change over time. Visualization of census variables are provided by means of conventional, generalized boundaries and, independently, by population-weighted cartograms, called Universal Data Maps (UDM). An example of a pair of maps produced using the online census atlas is shown in Figure 1.The conventional boundaries were created, in-house, by aggregating simplified 1981 ward boundaries to 2001 higher geographies using

look-up-tables. The origin of these boundaries enforced access protection via Athens. The UDMs were created by the University of Leeds, using Microsoft Excel™ and are a representation of the geographies, based on population, so they are not restricted in the same way and are available for use and download by the general public.

The starting point for the construction of the UDMs was creating the parliamentary constituencies as a cartogram. Parliamentary constituencies have an approximate population of 90,000 so each constituency was given an area of equal weight. Dorling (1995) describes the circular cartogram algorithm with which these constituencies were built initially. The constituencies were then transferred into a spreadsheet format where each constituency was represented by two cells and this lower level geography was then aggregated to five higher geographies, using look-up-tables, taking care to maintain topology wherever possible. Because each area is proportional to its population, the resulting shape of Great Britain is distorted,

Figure 1. Illustrative maps from the Online Census Atlas: the percentage of households with no car access in 2001

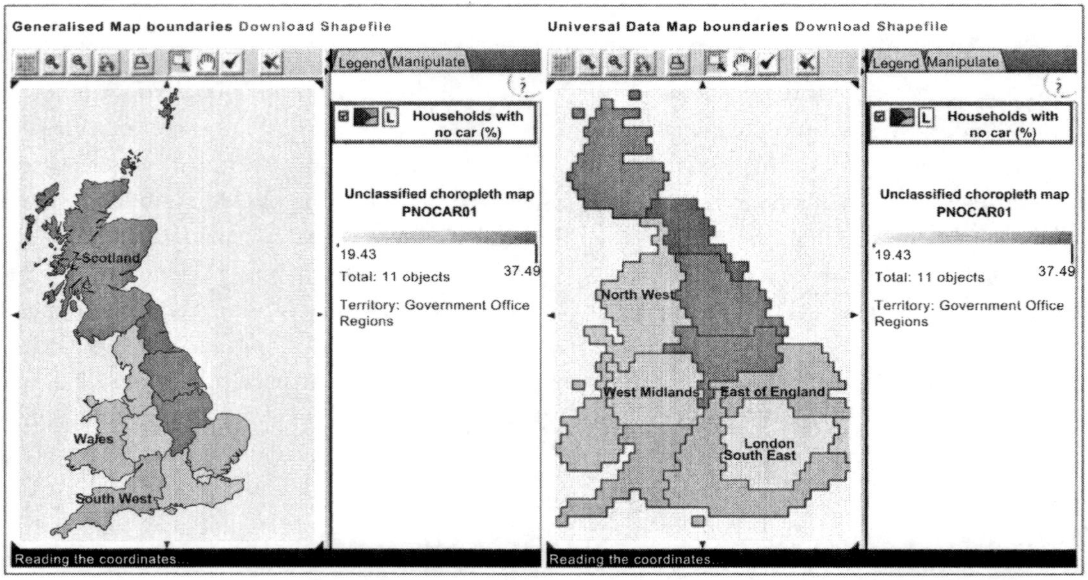

but what it does ensure is that densely populated area, such as cities, are fairly represented on the map, whereas in conventional maps it is often the sparsely populated rural areas that are emphasized. UDMs can be downloaded from the Atlas in spreadsheet format where simple mapping can be applied; or vector-format versions of the UDMs have been produced to allow users to use more advanced mapping and analysis techniques within their own GIS.

Detailed descriptions of the creation of the conventional boundaries and UDM, and instructions of how to use and interpret maps displayed within the Atlas, are given in Durham et al. (2006). The output from this teaching exemplar is a streamlined access to census data from the last four UK Censuses, across various geographical scales, permitting visualization, exploration, and download of census data for application by users. The Atlas uses easily recognizable conventional maps but also introduces a population-weighted cartogram as an effective method of displaying data; and because these boundary representations do not have copyright issues, census data exploration is accessible by all.

Data Analysis: Demographic Modeling

At Penn State, specific learning objectives were being developed to support the teaching of a course on "spatial demography" offered by Stephen Matthews. The spatial demography course focuses on substantive demographic research topics while exposing students to the challenges in, and opportunities for, using GIS and spatial analysis in exploring these topics. A core substantive topic in the spatial demography courses is race/ethnic segregation. Specifically, race/ethnic segregation is the focus of a lab exercise that includes a general introduction to the extraction of U.S. Census data and boundary files via the American Factfinder (U.S. Census Bureau, 2008), the mapping of individual race/ethnic groups, and finally the

calculation and mapping of summary measures of race/ethnic segregation. Students are encouraged to select their own study areas, typically a single county or aggregation of counties. Other useful Web sites that provide access to race/ethnic data include the Lewis Mumford Center (2008). This exercise and related discussion section occurs relatively early in the semester. There are two key reasons for this. First, as the U.S. Census Bureau is a key resource for much socio-economic and demographic data and one that students are likely to draw on, it is important that students become familiar with the geography of the U.S. Census (the levels of analysis), data availability (which varies by geographic level), and extraction procedures for both tabular and geographic boundary files. Second, race/ethnic residential segregation is an important backdrop for the study of other substantive topics introduced later in the semester, including but not limited to the spatial mismatch hypothesis, spatial entrapment, poverty, crime/delinquency, health inequalities, neighborhood resources, food environments, and environmental justice.

The race/ethnic segregation exercise is introduced in the context of several salient policy and social science themes. First, there is a general discussion to better understand why the U.S. Census Bureau asks questions on race and ethnicity. Specifically, the questions are asked to fulfill a variety of legislative and program requirements (e.g., data on race are used in the legislative redistricting process carried out by the states and in monitoring local jurisdictions' compliance with the Voting Rights Act, and these data are also essential for evaluating federal programs that promote equal access to employment, education, and housing and for assessing racial disparities in health and exposure to environmental risk). This is followed by a discussion of national and sub-national trends in race/ethnic distributions in the USA, which provides a convenient jumping off point for discussing the changes in how race/ethnicity is classified in the U.S. Census. The

2000 U.S. Census was the first to allow persons to affiliate with multiple races. The twin problems of changes in both the definitions of race/ethnicity over time and the boundaries of census units (typically for small geographical units relevant for a study in a single metropolitan area) leads in to a discussion of how results could be artifacts of the data rather than true change.

It is important that students also better understand how residential race/ethnic segregation is measured (Penn State Population Research Institute, 2005-2008). Following Massey & Denton (1988) and Iceland et al. (2002), there are up to seventeen measures of segregation; these can be grouped into five key dimensions of segregation namely: evenness (diversity), exposure, concentration, centralization, and clustering. The main focus is on the 'evenness' dimension and specifically the measures based on the dissimilarity and entropy indices. In the data extraction exercise students calculate the standardized entropy score: the relative evenness of groups within the population where if all "n" groups are represented in equal numbers the entropy value would be one, while at the other extreme in cases where the population was homogenous (i.e., all in one group) the value would be zero. The students extract and map individual race/ethnic data for selected metropolitan areas (typically at the census tract level), as well as produce a map of the race/ethnic entropy scores using a four group measure.

The race/ethnic segregation exercise and discussion ends with a focus on extending the aspatial nature of segregation measures to better incorporate broader spatial contexts and spatial segregation measures. These include a discussion of the work of Wong (2002, 2003, 2004, 2005) and his Arcview™ 3.x segregation program and the more recent work at Penn State on measuring spatial segregation by Reardon et al. (2008) and Lee et al. (2008).

Later in the semester the students are exposed to the freeware package GeoDa (see Anselin et al., 2006a, 2006b). Many use the Geoda package to generate both global Moran's I values of spatial autocorrelation and Local Indicators of Spatial Association (Anselin, 1995) based on race/ethnic variables and measures such as the entropy index. This helps them to better understand the spatial concentration of race/ethnic groups and possible causes and consequences of residential segregation in metropolitan areas of the USA. Elsewhere during the course the students learn about geocoding and the use of GIS to facilitate the creation of contextual databases (for example, in one exercise the students geocode school locations within Allegheny County, Pennsylvania [Pittsburgh] and link the school addresses to both data on the race/ethnic composition of the school and the race/ethnic composition of the census tract of the school). The use of geocoding and extraction of census data (beyond race/ethnicity variables) are strongly encouraged as part of building contextual databases to be analysed using packages like GeoDa and/or a standard statistical package in the final project of the spatial demography course.

Data Analysis: Demonstrators for Geodemographics and Multivariate Indicators

The GeoDemo and IndDemo Excel demonstrators were created to allow students on the Southampton Census and Neighborhood Analysis module to see for themselves how geodemographic classification and multivariate deprivation indicators are constructed (for general references on the issues covered, see Senior [2002] and Rees et al. [2002]). The blended learning approach taken for this course means that students will be working with these data independently and not necessarily within a university setting where software such as SPSS will be readily available to them. Further, these students will not all have encountered the same statistical analysis packages in previous modules, making it difficult to find a common software platform for practical assignments. These

Excel-based demonstrators allow these students to work in a familiar environment so that they gain understanding of the entire processes, from data selection to the implications of choosing weights, naming classes, and so forth, which is very hard to achieve without working with real data.

These Excel demonstrators have been programmed within Excel worksheets using Visual Basic for Applications (VBA), which allows the students to fill in design options for a multivariate indicator or hierarchical classification and then see the results applied in a series of worksheets on census data they have previously retrieved as simple .csv format files. Excel, although widely available to the students, is not directly suited to iterative processing tasks such as hierarchical classification through its spreadsheet functions but the embedding of program code and the necessary control forms allows these operations to be included simply. These applications have the disadvantage of using a proprietary software solution, yet this must be balanced against the many benefits of the self-contained package, which allows students to experiment with the effects of, for example, building deprivation indicators from normalized compared to un-normalized variables, or choosing different numbers of output clusters in a classification algorithm. The effect of the familiar software tool is to focus attention on the underlying concepts rather than the software environment.

In a year one module on the Geography of the UK at the University of Leeds students are asked to construct a profile of their parental home neighborhood. They first have to link their neighborhood to census geography by inputting their postcode into the GeoConvert resource (Census Dissemination Unit, 2008a) to obtain their 2001 Census Output Area (OA) code. This is then used to access the 2001 Census Output Area classification (National Statistics, 2007b; National Statistics, 2007c; National Statistics, 2007d; SASI, 2007; Vickers, 2006; Vickers et al. 2005, Vickers & Rees 2006; 2007) to determine

the Super-Group their neighborhood belongs to and to map the classification of OAs on a back-cloth of a roads and settlement map (Milton, 2007) using Google Maps™. The students then access the Neighbourhood Statistics resource of National Statistics (2007a) and extract post-census indicators for their neighborhood. These Web resources hold census data in a highly processed form suitable for providing context in a variety of situations, including in this case identifying where students came from.

Engaging with Real Places: Using Students' Own Neighborhoods

The Southampton Census and Neighborhood Analysis module has been developed to involve students in the study of a specific neighborhood throughout the duration of the module. This begins with an initial review of academic and policy-based conceptions of neighborhood, particularly drawing on Galster (2001), and an overview of the tensions between such conceptualizations and the geographical definitions and data available to researchers in practical settings. Students select their own neighborhood and describe its distinctive features and boundaries through a discussion list within Blackboard. Successive practical exercises involve students in the reconciliation of their initial reasoning with available statistical and administrative boundary systems, census and neighborhood statistics, and area profiles. These ideas are further developed through exploratory spatial data analysis, and the placing of the neighborhood in a regional and national context through construction of their own deprivation indicators and geodemographic classifications. The module closes with a reconsideration of the practical challenges to the implementation of area-based policies and students develop a personalized understanding of the challenge of developing and implementing evidence-based, theoretically informed policy. The goal here has not been to provide an entirely online version of a previous lecture-based module but to develop a new module in which learning objectives, teaching methods, and assessment are well-aligned. Lectures provide a weekly commentary and major on the delivery of conceptual and policy-oriented perspectives and which are primarily assessed through a conventional unseen examination at the end of the module. These are accompanied week-by-week with self-paced (i.e., not formally timetabled) online learning activities, which utilize practical skills and real data to exemplify the issues encountered at each stage. These are assessed entirely through the coursework submissions.

Practical assignments are set throughout the course and build up cumulatively to form two assessed coursework submissions, involving the students in a sequence of structured problem-solving while developing their understanding of a specific neighborhood in depth. An added advantage of this approach is that all students necessarily work on different neighborhoods and this facilitates online discussion of generic issues such as data availability and neighborhood size while promoting academic integrity as the concepts and techniques must be appropriately interpreted and applied in the context of each unique neighborhood. This approach has much in common with the problem-based learning approach to GIS teaching described by Drennon (2005), which recognizes that many of the students taught will not eventually be professional geographers but will nevertheless find themselves evaluating data and making complex decisions in ways akin to the learning undertaken here. This does not however mean that students will only make generic use of these materials: the instructor was recently contacted by a student who, having gained first exposure to these issues through the first delivery of this module in 2003-2004, has gone on to work for a leading commercial geodemographic supplier.

Using Census Data to Explore Ethnic Segregation: Comparing Cities in the USA and UK

At the University of Leeds a master's level module is taught on Census Analysis and GIS within two master's in GIS programs, one a face-to-face course and the second a distance learning course. Table 3 shows the structure of the latest offering of the face-to-face module (Rees, 2000-2008). To access the materials of either course requires registration for the relevant degree program, of course, but Unit 4 from the module has been made available via the Jorum repository for learning materials in the UK (Durham & Rees, 2006). Jorum is an online repository of e-learning materials, supported by the Joint Information Systems Committee of the UK's Higher Education Funding Councils. Several of the units in the Census Analysis module incorporate materials from the teaching materials developed under the Collection of Historical and Contemporary Censuses project (CHCC, 2003).

The module aims to show the rich resources on the human geography of the UK provided by successive censuses and to use these resources to explore the ethnic geography of an exemplar city in the UK, Birmingham and a parallel city, Birmingham, USA. Units 2 through 6 and 9 contain material from both the UK and USA so that students can appreciate the differences in available data and social situations in two countries. The many overseas students who take this module are eager to put the skills they learn to work on the human geography of their own countries but this is almost impossible, as few countries provide such rich socio-economic and demographic data for small areas online.

Unit 1, Introduction to the Census, explains the rationale for a census and the basic characteristics from administration through collection, data capture, editing, imputation, database creation, output creation, and output dissemination. This is delivered via a lecture in the face-to-face mod-

Table 3. The syllabus of a Masters module incorporating e-learning materials: GEOG5100 Census analysis and GIS, University of Leeds

Unit	Description
1	Unit 1: Introduction to the Census *Practical Exercise 1: Exploring the ESRC/JISC and ONS/GROS/NISRA census web pages* Activity 1: Filling in the 2001 Census Form; Evaluation of the Questions
2	Unit 2: The Census Data System: procedures, outputs and geography. *Practical Exercise 2: Extraction of ethnic data for Birmingham, UK from the 2001 and 1991 Censuses, issues of comparability over time, Extraction of racial data for Birmingham US from the 2000 and 1990 Censuses* Activity 2: Project ideas and brainstorming [Practical Exercise 1 due as Assessment 1]
3	Unit 3: The Census Data Matrix: distribution, composition and change percentages, location quotients, indicators of change, indicators of diversity (Diversity and Entropy indexes). *Practical Exercise 3: Analysis of ethnic data for Birmingham, UK and analysis of racial data for Birmingham, US.* Activity 3: Discussion of the nature of ethnic and racial groups in the UK and the US.
4	Unit 4: Mapping census data for large areas, map resources and techniques *Practical Exercise 4: Geographies and Map Resources for Use with Census Data (CHCC Unit 12); Universal Census Data Mapping (CHCC Unit 13); Map Resources for Use with US Census Data.* Activity 4: Maps for your project. [Practical Exercises 2 and 3 due as Assessment 2]
5	Unit 5: Understanding Output Areas for the 2001 UK Census (CHCC Unit 8); Mapping census data for small areas, map resources and techniques. *Practical Exercise 5: Output Areas from the 2001 UK Census (CHCC Unit 8). Mapping for OAs for Birmingham, UK; mapping Census Tracts from the 2000 US Census for Birmingham, US.* Activity 5: Project topics to be decided.
6	Unit 6: The measurement of spatial patterns using summary indexes, measures of evenness (Dissimilarity, Entropy), exposure (Isolation), Concentration (Delta), Centralization & Clustering. *Practical Exercise 6: Computation of Indexes and their interpretation for Birmingham UK or Computation of Indexes and their interpretation for Birmingham US.* Activity 6: Stages in project planning and execution. [Practical Exercises 5 & 6 due as Assessment 3]
7	Unit 7: Deprivation variables and indicators from the Census and other sources (Indices of Deprivation 2000, 2004); analysis of long term illness using deprivation and other determinants; use of SPSS for means analysis, regression analysis. *Practical: Exercise 7: Explaining the Distribution of Limiting Long Term Illness.* Activity 7: Ongoing project work.
8	Unit 8: Understanding geographical conversion and lookup tables; problems with projects *Practical: Using CHCC Unit 11 to understand Geographical Conversion* Activity 8: Ongoing project work [Practical Exercise 7 due as Assessment 4]
9	Unit 9: Spatial dimensions of race/ethnicity in the United Kingdom and in the United States. Activity 9: Ongoing project work

ule but via a detailed document in the distance learning, both made available via the institution's VLE or course management system. The unit document includes full instructions for the practical exercise. The computer-based practical asks the students, for example, to build up their own archive of census documentation from national statistics and academic program Web sites, for future reference in the module.

Unit 2, The Census Data System, explains in more detail the outputs and geography of the Census Data System (Rees et al., 2002). Practical exercise 2 takes students through procedures to extract ward level ethnic data for Birmingham UK from both the academic resource (ESRC, 2006) and one of the National Statistics resources (NOMIS, 2008). These resources offer Web-based extraction engines of different design and deliver data in different formats. The students also used the American Fact Finder to extract tract level census data on races and ethnicities (U.S. Census Bureau, 2008). Students are provided with a generic skill for extracting online census data, to which they can return when researching the module project, when working on their master's dissertations or in future careers. Use of the data extraction engine CASWEB (Census Dissemination Unit, 2008b) becomes so embedded in their learning that most come to think it is the producer of the data! Unit 3, The Census Data Matrix, explains the properties of the matrix, which is not just any table of census statistics but one with particular and useful properties. The columns hold group populations, the rows hold area populations, and

there is a summary column of area population totals and a summary row of group populations over all areas. Embedded aggregations of groups or areas, which are usual in census tables, are removed for ease of analysis.

Units 4, Mapping of census for large areas, and Unit 5, Understanding output areas, show students how census statistics can be mapped at different scales. The module also introduces students to alternative mapping techniques, such as the population cartogram. Figure 2 shows one example of the cartograms used as illustrations in the module. This cartogram was developed by Durham, Dorling, and Rees (see the section on the Online Census Atlas) and then used by Rees & Butt (2004) to map ethnic population change at local authority scale in England. The map depicts the 2001 spatial distribution of the Black African population using location quotients. Students normally report considerable resistance to using

Figure 2. Example of a population cartogram introduced in a Leeds Masters module (Source: Appendix to Rees and Butt, 2004)

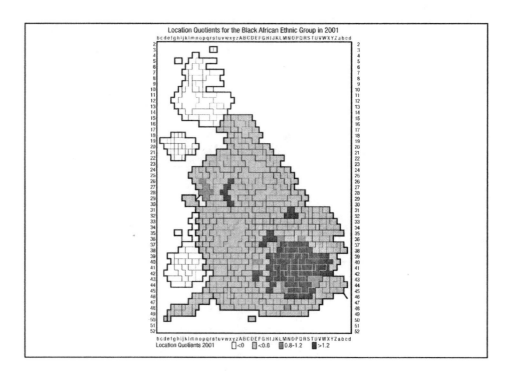

Figure 3. Illustrative ethnic group maps for Birmingham UK and Birmingham US produced by students in the Census Analysis and GIS module, University of Leeds:Location quotient maps for an ethnic minority group in each city

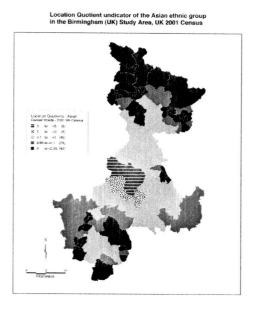

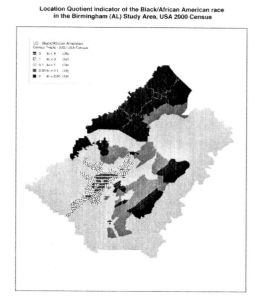

population cartograms, which are unfamiliar distortions of their "standard" view of geographical space. Students who wish to explore these techniques further are directed to the Web pages of the Social and Spatial Inequalities research group at the University of Sheffield (SASI, 2008). Figure 3 shows an example of small area mapping produced by students for their Unit 5 work for the two cities studied. Unit 6, The measurement of spatial patterns using summary indexes, uses the Unit 3 derived indicators to compute indexes that summarize the spatial distributions of ethnic groups for comparison purposes. Earlier, parallel work in the Penn State course on data analysis and demographic modeling was discussed.

Unit 7, Deprivation variables and indicators, takes students through the methods used to construct summary deprivation indicators and how they can be used as explanatory variables in an analysis of the spatial distribution of limiting

long-term illness, a self-reported health variable. Unit 8, Understanding geographical conversion and lookup tables, tackles the tangled web of spatial harmonization. Administrative areas, postal zones, and service regions undergo continual change. To analyze change in populations over time, the geographical units must be defined in exactly the same way. National statistical offices rarely give much priority to the conversion of old, out-of-date geographical areas to new current ones. Geographical conversion techniques need to be used. These involve the aggregation or disaggregation of source geographical areas to target geographical areas. Geographical conversion is also increasing needed for contemporary datasets because national statistical agencies are becoming increasingly reluctant to publish census statistics for more than one set of areas, possibly because of work on the risks of disclosure from overlapping geographies carried out by Duke-Williams & Rees (1998). Unit 9, Spatial dimensions of race/ethnicity in the UK and U.S., completes the syllabus by providing two exemplar spatial analyses to stimulate student ideas for their own projects.

Our discussion so far has centered on the logic behind the syllabus of modules, which is what students are interested in. However, it is useful to reflect as well on the development of e-learning materials from the instructor's point of view.

The Census Analysis and GIS module began life in the 1990s as a face-to-face component of a master's degree in GIS with extensive materials prepared to support computer-based practicals. When the distance learning version of the master's degree was developed early in the 2000s, the previous face-to-face materials were re-written for fully online use, with support from JISC via the CHCC and DialogPLUS projects. This conversion acted as a rigorous quality control for the learning activities because distance learners will report faults and demand solutions very quickly compared with campus students. They are more mature and have the self-confidence to demand quality instruction, which they are usually paying for personally. Campus students can often find the

work-around through their peer support network. Having developed the higher quality e-learning materials for distance learners, the materials were re-employed in the face-to-face module to create a blended learning experience, which is a combination of lectures, seminars, and individual consultations combined with do-it-yourself practical exercises. In particular, instructor attendance at practical lab sessions was no longer necessary. The time could be reallocated to individual discussions with students about their project work.

Table 4, showing the offerings of Census Analysis at the University of Leeds in the period 2003-2007, illustrates another consequence of the development of modules based on e-learning (Rees, 2003-2008). They can be offered more frequently. So in the five teaching years 2003-2004 to 2005-2006, the face-to-face and distance versions of Census Analysis were both offered three times. In 2006-2007 and 2007-2008 the face-to-face module has been run twice but the distance learning version six times. Distance learning degrees are characterized by small cohorts who join the programs each semester or each quarter, necessitating more frequent offerings. Also, distance students have to take breaks

Table 4. Offerings of GEOG5100 and GEOG5101 at the University of Leeds, 2003-2008

GEOG5100 Face to Face module	GEOG5101 Distance Learning module
Academic year	Academic year
2003-4	2003-4
2004-5	2004-5
2005-6	2005-6
2006-7	2006-7 (1)
	2006-7 (2)
	2006-7 (3)
2007-8	2007-8 (1)
	2007-8 (2)
	2007-8 (3)

in their programs of study because of work or family commitments and so they develop their own individual schedule for module completion. Frequently extensions are requested and granted and the schedules need to be very flexible. From the point of view of the institution, with e-learning delivery of this kind it is economically viable to offer modules to smaller classes than in a face-to-face context. For the instructor it means accepting a different distribution of tutoring effort, which continues across traditional university inter-term breaks. Combining both traditional and distance modes of learning is difficult for instructors from a time commitment point of view. Most academic teachers are very reluctant to take on this dual commitment. We may see specialization into face-to-face and online instructors as a result.

CENSUS DATA ACCESS IN THE USA AND UK

Internationally, it is a feature of censuses conducted in the 2000-2001 round that Web dissemination has been seen as a principal means of data dissemination. This marks a major change from the 1990-1991 round of censuses, which effectively took place before the Web. Tools for online census access are thus still essentially in their first generation (Martin et al., 1998). Extensive publicly available census data are accessible from the Web sites of the U.S. Census Bureau and the UK statistical agencies. In the USA, the American FactFinder (U.S. Census Bureau, 2008) offers an extensive and freely-accessible interface to census datasets that is similar in concept to sites such as the Neighborhood Statistics Service (National Statistics, 2007a) for England and Wales, Scottish Census Results Online (GROS, 2008a; 2008b), and Northern Ireland Census Access (NISRA, 2008a; 2008b), although these (and other) sites offer only a limited subset of the extensive census datasets produced in the UK. Although a major innovation in public access to census

data, searchable by postcode or place name, the Neighborhood Statistics service contains census data only for 2001 and only univariate tables, thus it is not possible to retrieve even the basic age/sex structure necessary to create a population pyramid of a specified locality. The teaching potential for these data is thus severely limited. Full access to the very extensive census datasets for the UK is available online to registered UK academic users through the ESRC's Census Programme (ESRC, 2006) and the registration system encompasses all students in UK higher and further education institutions. Whereas the U.S. Census Bureau is effectively supporting a global community of users, it is outside the funding scope of the UK ESRC to support global access to the full range of publicly produced census datasets. This situation presents difficulties for the delivery of truly internationally transferable teaching materials in the field of census and demographic analysis.

The project team has been able to exchange teaching materials authored within the project without difficulty, but are limited in their ability to embed census data within these materials or to assume student access to UK census resources from outside the UK. Census datasets are an important element of our conceptualization of digital library resources relevant to human geography, being extensive catalogued online repositories of materials valuable to teaching and learning, and research. Further, JISC-funded census teaching materials created as part of the CHCC project are also a valuable resource in this area comprising online tutorials and exercises that make reference to the UK datasets. Unfortunately, the datasets available to the UK academic community are in several cases restricted by license to use by UK-registered academics and accessible only through the Athens authentication system.

This creates an imbalance whereby UK students could use materials and data from the USA without obstacles, but students in the USA would be restricted to use of the somewhat limited public datasets in the UK's Neighborhood Statistics systems. Although the UK census organizations will provide more detailed data on CD/DVD, the products and delivery formats are complex and not a viable means of obtaining UK census data for small contributions to taught courses in the USA. Changes to the UK's academic authentication system during 2006-2008, involving the adoption of federated access management, will remove some of the technical obstacles to truly international data sharing but will not resolve the licensing and funding constraints on the fuller international dissemination of UK census datasets.

CONCLUSION

This chapter has presented a number of examples of e-learning in human geography. Common themes across all of the exemplars include the objective of helping students to navigate the Web and find data resources about people that can be used to research particular human geographic topics. The theme of ethnic spatial distribution recurs in the Penn State and Leeds master's courses, along with a focus on teaching the methods of analysis of population count data. The theme of neighborhood in its different guises is at the heart of the Southampton course, providing students with tools and data with which to explore different definitions. The neighborhood theme also appears in the description of an exercise carried out by Leeds first year geographers. In all of these applications of e-learning the aim is to provide students with generic skills so that they can easily move from the examples and exercises of the courses to using equivalent materials and techniques in their own project or dissertation work. Also common across all the e-learning activities described above is the aim of improving the student's skills in quantitative analysis and geographical information handling. In the course outline in Table 3, for example, in successive units students learn to implement index formulae using spreadsheet functions (unit 6), employ regression analysis to explain the spatial distribution of a dependent variable that has been age-standardized (unit 7), and develop an understanding of the principles and techniques behind geographical harmonization (unit 8).

An article in The Economist reporting on technology and government (The Economist, 2008) introduces a crucial distinction between i-government and e-government. I-government is concerned with delivering relevant information to the citizen to help them in their discourse and interaction with government. E-government means converting the processes of government, which are currently inefficient, into easy-to-use and very cheap Web processes. The article rightly says that i-government has made much progress; e-government is still in its infancy and has a lot to achieve. This typology is quite useful in reflecting on the experience reported in this chapter in using electronic resources to teach human geography. We have been quite successful in information-learning and in technique-learning (e.g., mapping, statistical analysis). The courses have been focused on these two aspects of information and techniques for the exploration of human geographic issues. However, we have still to achieve as much in addressing the challenge of using electronic learning to help students achieve understanding of the processes that are changing contemporary populations and societies.

NOTE

A note on the varying and sometimes confusing nomenclature of higher education learning programs is given here. The terms program/course, course/module and unit/week have different meaning, depending on the context and the institution. At the University of Leeds, for example, "program (spelled programme)" refers to the sequence of learning components that, if successfully navigated by a student, lead to the award of a particular degree. "Module" at Leeds refers to one of the components of the program. "Unit" is a sub-component with a module, although in traditional face-to-face modules, the term "week" is often used, reflecting the timing of delivery of the syllabus. Other institutions use terms such as "course" or "unit" for what Leeds call a module, and "course" is sometimes used to refer to "program." Despite the differences in nomenclature, most higher education institutions deliver a three tier hierarchy of programs containing modules containing units and assign weights ("credits," "hours") to each element.

REFERENCES

Andrienko, G., & Andrienko, N. (2006). *Exploratory analysis of spatial and temporal data: A systematic approach.* London: Springer.

Anselin, L. (1995). Local Indicators of Spatial Association—LISA. *Geographical Analysis, 27*(2), 93-115.

Anselin, L., Syabri, I., & Kho, Y. (2006a). GeoDa: An introduction to spatial analysis. *Geographical Analysis, 38*(1), 5-22.

Anselin, L. et al. (2006b). *GeoDa: An introduction to spatial analysis.* Champaign-Urbana: Spatial Analysis Lab, University of Illinois. Retrieved February 20, 2008, from https://www.geoda.uiuc.edu/

CHCC. (2003). *Developing the collection of historical and contemporary census data and related materials (CHCC) into a major learning and teaching resource.* Retrieved February 14, 2008, from http://www.chcc.ac.uk

Census Dissemination Unit. (2008a) *GeoConvert: A resource for geographical lookups and conversion.* Manchester: University of Manchester. Retrieved February 20, 2008, from http://geoconvert.mimas.ac.uk/

Census Dissemination Unit. (2008b). *CASWEB: Web interface to census aggregate outputs and digital boundary data.* Retrieved February 20, 2008, from http://casweb.mimas.ac.uk/

Columbia University. (2008). *Centre for International Earth Science Information Network (CIESIN).* New York: Earth Institute, Columbia University. Retrieved March 23, 2008, from http://www.ciesin.org/

Dorling, D. (1995). Visualising changing social-structures from a census. *Environment and Planning A, 27*, 353-78.

Drennon, C. (2005). Teaching geographic information systems in a problem-based learning environment. *Journal of Geography in Higher Education, 29*(3), 385-402.

Duke-Williams, O., & Rees, P. H. (1998). Can Census Offices publish statistics for more than one small area geography? An analysis of the differencing problem in statistical disclosure. *International Journal of Geographical Information Systems, 12*(6), 579-605.

Durham, H., Dorling, D., & Rees, P. (2003a). *Online census atlas.* Retrieved February 16, 2008, from http://www.ccg.leeds.ac.uk/teaching/chcc/; http://devchcc.mimas.ac.uk/cgi-bin/CAS/atlas/showdata.cgi

Durham, H., Dorling, D., & Rees, P. (2003b). *Online census atlas.* Retrieved February 16, 2008, from http://www.chcc.ac.uk/atlas/index.html

Durham, H., Dorling, D., & Rees, P. (2006). An online census atlas for everyone. *Area, 38*, 336-341.

Durham, H., & Rees, P. (2006) *Census analysis and GIS – unit 4.* Retrieved February 6, 2008, from http://repository.jorum.ac.uk/intralibrary/IntraLibrary?command=preview&learning_object_id=3645

ESPON. (2006, October). *ESPON ATLAS: Mapping the structure of the European territory. The European Spatial Planning Observation Network and the partners of the ESPON programme.* Retrieved March 8, 2008, from http://www.espon.eu/mmp/online/website/content/publications/98/1235/file_2489/final-atlas_web.pdf

ESRC. (2006). *Moving you closer to the data.* Economic and Social Research Council. Retrieved March 26, 2008, from http://census.ac.uk/

Eurostat. (2008a). *Regions: Databases and publications.* Luxembourg: Statistical Office of the European Communities. Retrieved March 23, 2008, from http://epp.eurostat.ec.europa.eu/portal/page?_pageid=1335,47078146&_dad=portal&_schema=PORTAL

Eurostat. (2008b). *Nomenclature of territorial units for statistics—NUTS statistical regions of Europe.* Retrieved March 25, 2008, from http://ec.europa.eu/comm/eurostat/ramon/nuts/home_regions_en.html

Facebook. (2008) *Facebook is a social utility that connects you with the people around you.* Retrieved February 14, 2008, from http://www.facebook.com/

Galster, G. (2001). On the nature of neighborhood. *Urban Studies, 38*(12), 2111-2124.

Gould, M., Stillwell, J., & Vanderbeck, R. (2007-2008). *Research methods in human geography.* University of Leeds Module GEOG2680. Module outline and materials. Retrieved February 16, 2008, from http://webprod1.leeds.ac.uk/catalogue/dynmodules.asp?Y=200708&M=GEOG-2680; http://vle.leeds.ac.uk/site/nbodington/geography/geoglev2/geog2680/sem1/

GROS. (2008a). *Welcome to SCROL: Scottish Census Results Online.* Edinburgh: General Register Office. Retrieved February 20, 2008, from http://www.scrol.gov.uk/scrol/common/home.jsp

GROS. (2008b). *Scottish neighbourhood statistics.* Edinburgh: General Register Office. Retrieved February 20, 2008, from http://www.sns.gov.uk/

Iceland, J., Weinberg, D. H., & Steinmetz, E. (2002). *Racial and ethnic residential segregation in the United States: 1980-2000.* U.S. Census Bureau, Series CENSR-3. Washington DC: U.S. Government Printing Office.

INSEE. (2008). *Portail INSEE.* Paris: Institut National de la Statistique er des Etudes Economiques. Retrieved March 23, 2008, from http://www.insee.fr/en/home/home_page.asp

Keylock, C. J., & Dorling, D. (2004). What kind of quantitative methods for what kind of geography? *Area, 36*(4), 358–366.

Lee, B. A., Reardon, S. F., Firebaugh, G., Farrell, C. R., Matthews, S. A., O'Sullivan, D. (2008). Beyond the census tract: Patterns and determinants of racial segregation at multiple geographic scales. *American Sociological Review, 73,* 766-791.

Lewis Mumford Center. (2008). *Lewis Mumford Center for Comparative Urban and Regional Research.* Retrieved February 20, 2008, from http://www.albany.edu/mumford/

Madge, C., & O'Connor, H. (2004). Online methods in geography educational research. *Journal of Geography in Higher Education, 28*(1), 143-152.

Martin, D., Harris, J., Sadler, J., & Tate, N. J. (1998). Putting the census on the Web: Lessons from two case studies. *Area, 30*(4), 311-320.

Massey, D. S., & Denton, N. A. (1988). The dimensions of residential segregation. *Social Forces, 67,* 281-315.

Milton, R. (2007). *Dan Vickers' output area classification.* London: Centre for Advanced Spatial Analysis, University College London. Retrieved February 20, 2008, from http://www.casa.ucl.

ac.uk/googlemaps/OAC-super-EngScotWales.html

National Statistics. (2007a). *Neighbourhood Statistics home page.* London, UK: Office for National Statistics. Retrieved February 20, 2008, from http://neighbourhood.statistics.gov.uk/

National Statistics. (2007b). *UK census based classification of output areas.* London: Office for National Statistics. Retrieved February 20, 2008, from http://www.statistics.gov.uk/census2001/cn_139.asp

National Statistics. (2007c). *Area classification for output areas.* London: Office for National Statistics. Retrieved February 20, 2008, from http://www.statistics.gov.uk/about/methodology_by_theme/area_classification/oa/default.asp

National Statistics. (2007d). *Understanding the 2001 census area classification for output areas.* London: Office for National Statistics. Retrieved February 20, 2008, from http://neighbourhood.statistics.gov.uk/dissemination/Info.do;jsessionid=ac1f930dce62afd8c80e527429aaefdb6cfe17d733c.e38OaNuRbNuSbi0LbhyNb3eOb3uLe6fznA5Pp7ftolbGmkTy?page=CaseStudies_Classification.htm&bhcp=1

NISRA. (2008a). *NICA: Northern Ireland census access.* Belfast: Northern Ireland Statistics and Research Agency. Retrieved February 20, 2008, from http://www.nicensus2001.gov.uk/nica/public/index.html

NISRA. (2008b). *NINIS: Northern Ireland neighbourhood information service.* Belfast: Northern Ireland Statistics and Research Agency. Retrieved February 20, 2008, from http://www.ninis.nisra.gov.uk/mapxtreme/default.asp

NOMIS. (2008). *Official labour market statistics.* Census 2001 on Nomis. Retrieved February 18, 2008, from https://www.nomisweb.co.uk/home/census2001.asp

Quality Assurance Agency. (2000). *Subject benchmark statements: Geography.* Retrieved February 6, 2008, from http://www.qaa.ac.uk/academicinfrastructure/benchmark/honours/geography.asp

Reardon, S. F., Matthews, S. A., O'Sullivan, D., Lee, B. A., Firebaugh, G., & Farrell, C. R. and Bischoff, K. (2008). The goegraphic scale of metropolitan segregation. *Demography, 45*(3), 489-514.

Rees, P. (2000-2008). GEOG5100: *Census analysis and GIS.* Masters Module. University of Leeds, Leeds, UK. Retrieved February 6, 2008, from http://vle.leeds.ac.uk/site/nbodington/geography/geoglevm/geog5100 (registered users only)

Rees, P. (2003-2008). *GEOG5101 census analysis and GIS.* Distance Learning Masters Module. University of Leeds, Leeds, UK. Retrieved February 6, 2008, from http://vle.leeds.ac.uk/site/nbodington/geography/geoglevm/geogodl/geog5101/. (registered users only)

Rees, P., & Butt, F. (2004). Ethnic change and diversity in England, 1981-2001. *Area, 36*(2), 174-186. Retrieved February 6, 2008, from http://0-www.blackwell-synergy.com.wam.leeds.ac.uk/doi/full/10.1111/j.0004-0894.2004.00214.x

Rees, P., Martin, D., & Williamson, P. (Eds.). (2002). *The census data system.* Chichester, UK: Wiley.

SASI. (2007). *The National classification of census output areas.* Sheffield: Sheffield Social and Spatial Inequalities Research Group, Department of Geography, University of Sheffield. Retrieved February 20, 2008, from http://www.sasi.group.shef.ac.uk/area_classification/index.html

SASI. (2008). *Social and Spatial Inequalities Research Group.* Sheffield: Geography Department, University of Sheffield. Retrieved February 6, 2008, from http://www.sasi.group.shef.ac.uk/

See, L., Gould, M. I., Carter, J., Durham, H., Brown, M., Russell, L., & Wathan, J. (2004). Learning and teaching online with the UK census. *Journal of Geography in Higher Education, 28*(2), 229-245.

Senior, M. (2002). Deprivation indicators. In P. Rees, D. Martin, & P. Williamson (Eds.), *The census data system* (pp. 123-138). Chichester, UK: Wiley.

SPSS. (2008). *Data analysis with comprehensive statistics software.* Retrieved March 25, 2008, from http://www.spss.com/spss/

Statistics Canada. (2005). *2006 Census.* It's not too late. About the online questionnaire. Ottawa: Statistics Canada/Statistique Canada. Retrieved February 17, 2008, from http://www12.statcan.ca/IRC/english/about_e.htm

Statistics Canada. (2008). *2006 Census, 2001 Census and 1996 Census.* Ottawa: Statistics Canada/Statistique Canada. Retrieved March 23, 2008, from http://www12.statcan.ca/english/census/index.cfm

Statistics South Africa. (2008a). *Census 2001: Interactive & electronic products.* Retrieved March 23, 2008, from http://www.statssa.gov.za/census01/html/C2001Interactive.asp

Statistics South Africa (2008b). *Census 2001: Digital census atlas.* Retrieved March 23, 2008, from http://www.statssa.gov.za/census2001/digiAtlas/index.html

The Economist. (2008, February 16). The electronic bureaucrat: A special report on technology and government. *The Economist.* Retrieved February 17, 2008, from http://www.economist.com/printedition/

The World Bank. (2008). *WDRs: World development reports.* Washington, DC: World Bank. Retrieved March 23, 2008, from http://econ.worldbank.org/

United Nations. (2008). *What's new.* New York: Population Division, Department of Economic and Social Affairs, United Nations. Retrieved March 23, 2008, from http://www.un.org/esa/population/unpop.htm

University of Leeds, University of Southampton, Penn State World Campus & Worldwide Universities. (2008). *GIS online learning.* Retrieved February 14, 2008, from http://www.gislearn.org/

University of Leicester. (2008). *Exploring online research methods incorporating TRI-ORM. ESRC Research Methods Programme 2004-6 and ESRC Researcher Development Initiative 2007-2009.* Retrieved February 17, 2008, from http://www.geog.le.ac.uk/orm/

U.S. Census Bureau. (2008). *American factfinder.* Retrieved February 14, 2008, from http://factfinder.census.gov

Vickers, D. (2006). *Multi-level integrated classifications based on the 2001 census.* Unpublished doctoral dissertation, University of Leeds. Retrieved February 20, 2008, from http://www.geog.leeds.ac.uk/people/old/d.vickers/thesis.html

Vickers, D., Rees P., & Birkin, M. (2005). *Creating the national classification of census output areas: Data, methods and results.* Working Paper 05/1. School of Geography, University of Leeds, Leeds. Retrieved February 20, 2008, from http://www.geog.leeds.ac.uk/wpapers/05-2.pdf

Vickers, D., & Rees, P. (2006). *Introducing the National Classification of Census Output Areas.* Population Trends, 125, 15-29. Retrieved February 20, 2008, from http://www.statistics.gov.uk/downloads/theme_population/PT125_main_part2.pdf

Vickers, D., & Rees, P. (2007). Creating the National Statistics 2001 Output Area Classification. *Journal of the Royal Statistical Society, Series A, 170*(2), 379-409.

Wong, D. W. S. (2002). Spatial measures of segregation and GIS. *Urban Geography, 23,* 85-92.

Wong, D. W. S. (2003). Implementing spatial segregation measures in GIS. *Computers, Environment and Urban Systems, 27,* 53-70.

Wong, D. W. S. (2004). Comparing traditional and spatial segregation measures: A spatial scale perspective. *Urban Geography, 25*(1), 66-82.

Wong, D. W. S. (2005). Formulating a general spatial segregation measure. *The Professional Geographer, 57*(2), 285-294.

Chapter V
Using Digital Libraries to Support Undergraduate Learning in Geomorphology

Stephen Darby
University of Southampton, UK

Sally J. Priest
Middlesex University, UK

Karen Fill
KataliSys Ltd, UK

Samuel Leung
University of Southampton, UK

ABSTRACT

In this chapter we outline the issues involved in developing, delivering, and evaluating a Level 2 undergraduate module in fluvial geomorphology. The central concept of the module, which was designed to be delivered in a "blended" mode, involving a combination of traditional lectures and online learning activities, was the use of online digital library resources, comprising both data and numerical models, to foster an appreciation of physical processes influencing the evolution of drainage basins. The aim of the module was to develop the learners' knowledge and understanding of drainage basin geomorphology, while simultaneously developing their abilities to (i) access spatial data resources and (ii) provide a focus for developing skills in scientific data analysis and modeling. The module adopts a global perspective, drawing on examples from around the world. We discuss the process of course and assessment design, explaining the pedagogy underlying the decision to adopt blended delivery. We share our teaching experiences, involving a particular combination of "face-to-face" lectures and online sessions, complemented by independent online learning, and supported by the associated virtual learning environment. Finally, we discuss the issues highlighted by a comprehensive module evaluation.

INTRODUCTION

In this chapter we describe the development, delivery, and evaluation of a Level 2 module designed to support undergraduate learning in fluvial geomorphology. The central concept of the module was the use of a suite of online digital library resources, comprising both data and numerical models, to foster an appreciation of the physical processes influencing the evolution of drainage basins. In the following sections we discuss the design of the module and its assessment, and share our experiences in delivering it, reflecting on the challenges and opportunities involved in blending the "face-to-face" lectures and online sessions. Finally, the module has been comprehensively evaluated, and we discuss the issues highlighted by this process.

DESIGNING THE MODULE

Pedagogic Rationale

The central subject concept of the digital library *Drainage Basin Geomorphology* module was to employ a suite of online resources as a means of fostering appreciation of drainage basins as fundamental environmental units. Specifically, the aims were to develop knowledge and understanding of drainage basin geomorphology, while simultaneously developing students' abilities to (i) access spatial data resources within the digital library, (ii) provide a focus for developing skills in scientific data analysis and modeling, and (iii) develop abilities to solve problems and think critically. A set of learning outcomes (Table 1) was devised around two key scientific questions: (i) how and why does the geomorphology of drainage basins vary between different geographical locations?, and (ii) the prediction, using process modeling, of drainage basin changes driven by natural (e.g., climate, tectonic) and anthropogenic (e.g., land cover, river engineering) forcings. Consistent with the nature of the DialogPLUS collaboration

itself, a key requirement of the learning approach was to enable a global perspective to be adopted, using the potential of digital library resources to provide examples and experiences of drainage basin geomorphology from around the world.

The starting point in the module design process was to consider how it might contribute to the broader undergraduate curriculum in geography, both within the specific context of the School of Geography at the University of Southampton, but bearing in mind the DialogPLUS philosophy that the learning materials would be available for use elsewhere. Focusing on the Southampton context, the general subject matter is integral to one of the main research themes, *Environmental Processes and Change* (EPC), within the School of Geography. In particular, the interdisciplinary nature of the subject matter, and its focus on predicting drainage basin responses to natural and anthropogenic drivers of local and global environmental change, offered an opportunity to add a coherent, and previously lacking, link between the diverse modules offered by academic staff members within the EPC theme. Developing the module was viewed as an opportunity to bridge a perceived gap in the curriculum between existing Level 2 and 3 modules concerned with the drivers of environmental change (e.g., *Hydrology, Quaternary Environmental Change, Global Climate Change*), and other Level 2 and 3 modules concerned with the impacts on fluvial process systems (e.g., *Fluvial Sedimentological Processes, River Channel Dynamics*). It was thought that this could be optimally achieved by designing the module primarily for Level 2 students. The timing of the module is coincident with students initiating their undergraduate research projects, a major piece of coursework that is submitted in the final year of the degree program. The module was also, therefore, intended to introduce students to a research-oriented approach, as discussed further below. Nevertheless, the particular nature of its structure and delivery retained the possibility of delivering tailored content and assessments to Level 3 students as well.

Table 1. Overview of learning outcomes for the drainage basin geomorphology module

Category	Learning Outcomes
Knowledge and understanding	Students will be able to display knowledge and understanding of: • the terminology, nomenclature & classification systems used in drainage basin geomorphology; • the nature of drainage basin evolution and change; • the relationships between physical and human processes in shaping terrestrial landscapes; • past, present and future variability in drainage basin geomorphology, with in-depth competence and detailed knowledge of specific local contexts; • the distinctiveness of particular physical environments within the global mosaic.
Subject-specific intellectual skills	Students will be able to display the following subject specific intellectual skills: • assess the merits of contrasting geomorphologic theories, explanations and policies; • abstract and synthesize information from a range of different sources • structure conceptual and empirical material into a reasoned argument.
Subject-specific practical skills	Students will be able to display a range of subject specific practical skills: • use appropriate techniques, including computer software, to produce clear diagrams and maps; • collect, analyze and understand data in physical geography using computer techniques; • understand the ways in which geographical data of various types can be combined, interpreted and modeled.
Transferable key skills	Students will have developed a range of transferable/general (key) skills: • pursue knowledge in an in-depth, ordered and motivated way; • use computational skills and ability in the use of modeling software; • marshall and retrieve data from library and internet resources; • be aware of the role and importance of evidence-based research.

However, the particular context for the creation of the module as part of the DialogPLUS project also meant that it was clear from the outset that the module would, and should, differ fundamentally from others in the existing EPC theme. Critical to this difference was the use of a suite of digital library resources that are accessed online and based around the study of drainage basins as fundamental environmental units. The pedagogic rationale for this had two key dimensions:

1. *Why drainage basins?* Drainage basins tend to be singular entities (and indeed are recognized as fundamental units of the terrestrial landscape) that are, therefore, "easy" to analyze and study. However, they have fundamental processes, both physical and human, that operate within them. Providing an educational experience of a range of different catchments for comparative context helps in understanding these issues. The use of digital library resources offers a means of providing this comparative context (e.g.,

Tooth, 2006). Not only could students adopt a global perspective by exploring drainage basins beyond their local vicinity, the study of drainage basin geomorphology requires analysis over broad spatial and temporal scales. Both these characteristics contrast with much of the existing geomorphology curriculum in that geomorphology options typically adopt a reductionist approach that tends to be narrowly focused in both scope and scale.

2. *Why modeling?* Traditional university training in physical geography adopts conventional teaching methods, perhaps supported by a limited number of field visits or laboratory practicals. Such methods are necessarily event-based and it is therefore difficult to promote understanding of either the role of sequences of events inherent in the long-term evolution of landforms and landscapes, or the long-term influence that contemporary processes have on landscape development. This is where the use of model-

ing and digital library resources was seen as having immense potential, with associated data visualization and animation tools allowing students to explore drainage basin evolution over long time-scales, and the response of drainage basins to multiple environmental change scenarios. It was hoped that this would give students the opportunity to think laterally, synthesize complex sources of information, and identify the nature of interactions between driving processes.

This dual focus on large-scale analysis and modeling was seen as essential in equipping physical geography undergraduates with the training and skills required in modern contemporary geomorphology. By coincidence, a geomorphologic luminary published a commentary article (Church, 2005) highlighting precisely this point, in the very week the module was launched. Church argues that a geographical education in geomorphology is becoming disconnected from the research discipline because it does not teach students the skills required to participate in research or professional application. Specifically, many geographical curricula do not include the mathematical and physical foundations necessary to understand the governing physical processes and to become confident in applying appropriate analytical (modeling) tools. The digital library concept provides an educational framework that might help address these issues.

Structure of the Module

If the potential for using digital library resources was both clear and exciting, it was recognized at the outset that the key to embedding them into effective student learning would lie in providing a coherent structure for this purpose. Careful consideration was, therefore, given to the design of the module, which was structured to allow students to progress along a cognitive "pathway" comprising five substantive learning stages (SLS,

Figure 1). The pathway has the potential to be customized (e.g., for students of different levels), but in all cases seeks to move from *conception* at the beginning of the unit towards the higher-order notion of *perception* by the end. At the outset of the module design process, the pathway was also structured in such a way that students would be required to reflect on the process of learning, as much as the learning itself (Schon, 1987; Seale & Cann, 2000). However, as we shall see, this aspect of the module proved difficult to execute successfully in practice and was subsequently significantly revised. The key steps in the cognitive pathway are as follows:

1. **Induction** in which students are introduced to the culture of using online learning resources (something that many have no prior experience of) while also establishing prior knowledge and assuring that they understand the expectations and objectives of the module. The induction stage in current versions of the module is focused on familiarizing students with the layout of the digital library resources, but it is recognized that there is the potential for them to use this phase of learning to identify a personal development program that can be tailored to their specific learning styles, backgrounds and prior knowledge.

2. **Conception** in which each student develops a conceptual framework that underpins the remainder of the module. This is achieved through a combination of introductory lectures and formative learning activities, culminating in an interactive simulation exercise in which students develop a "concept map" illustrating the generic functional structure of the drainage basin sediment transfer system.

3. **Knowledge development** in which the student's scientific knowledge is developed, based on the conceptual foundation developed in the previous phase. Specifically, lectures serve to introduce the principle that

drainage basins are subject to a generic set of physical and human processes of sediment transfer. During this phase, engagement with digital library resources is necessary to inculcate how the rates and relative dominance of these processes vary geographically in response to changing environmental boundary conditions.

4. **Perception/knowledge transfer** in which the student's ability to apply previously developed knowledge to new and broader contexts is developed. This phase relies less on supporting lectures but leans instead towards stronger and more independent engagement with digital library resources and associated modeling tools. In this phase, students undertake a research project in which landscape modeling tools are researched and then deployed to investigate how the role of drainage basin characteristics influence basin responses to drivers of environmental change.

5. **Reflection** in which students reflect on the materials covered in the module and their approaches to learning. Specifically, this stage seeks to engage students in thinking about the structure and content of the digital library resources, thereby encouraging their use to support learning elsewhere in the geography curriculum. It is also the intention that the process of reflection will help to promote a mature approach to learning in subsequent units.

Having identified the structure of the module, its content can now be reviewed.

Content of the Module

The *Drainage Basin Geomorphology* module comprises a set of lecture slides and an associated online learning activity (OLA) for each of ten topics (*see* Tables 2 and 3). Each OLA includes a statement of learning outcomes, its specific content and associated online data resources, an assessment activity (discussed below), and suggested further reading, where possible hyperlinked to the title pages of the specific journal issues. The resources are available to students at Southampton within the institutional Blackboard Virtual

Figure 1. Pedagogic structure of the module Drainage Basin Geomorphology indicating the relationship between the learning phases

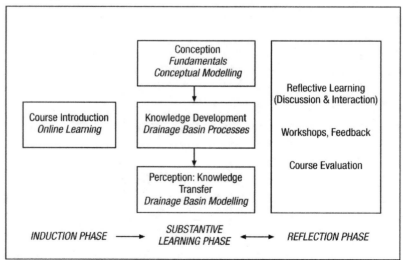

Learning Environment (VLE), the integration of online delivery and e-learning tools into the teaching and learning process is discussed in the next section. Readers who are interested in examining the online resources should refer to Table 2 where the public links to the OLAs, lecture slides, and assessment activities can be found. The OLAs are also all available within the Jorum repository (*see* Chapter II), from where registered users may freely download and use the resources. By agreeing to the terms of use of Jorum, users may repurpose the whole or part of the downloaded materials to suit their own specific teaching requirements (Jorum, 2007).

To a certain degree the syllabus content of the module corresponds to traditional curricula in geomorphology and physical geography. For example, the *conception phase* emphasizes the use of sediment budgeting as a core theoretical framework, as well as dealing with basic aspects such as how to characterize and compare the morphology of different drainage basins. Likewise, in the *knowledge development phase*, the lectures provide factual content regarding the basic physical, chemical, and biological processes that contribute to the catchment sediment budget. In this part of the course, the OLAs are used to reinforce these aspects, as well as to explore the notion of spatial and temporal variability in process rates as functions of changing environmental boundary conditions. This focus is common to many existing curricula, but it is the availability of the digital library resources that enables a broader (i.e., global) perspective to be taken. Where the content of the module differs from some curricula is perhaps most keenly expressed during the *perception phase*. This focuses on drainage basin models, and indeed the process of modeling, critically exploring how models may be used to address a range of scientific questions. The development of multiple modeling scenarios to emulate the use of multiple working hypotheses is explored, with an emphasis on how these scenarios must then be parameterized with geographic data in

order to address the postulates. Having inculcated the necessary modeling skills, the remainder of this phase is devoted to the students attempting to transfer their knowledge to new contexts, via their coursework project assignments, described in more detail below.

Although it is not the intention here to review each specific learning activity (interested readers can do so for themselves by downloading them as described above), one does require special mention. With much of the course content online, and with all the assessment by coursework, the potential for plagiarism and other forms of academic dishonesty (e.g., collusion) was perhaps higher than on more traditional courses. To address this issue, a dual-focused strategy was developed in which the aims were to (i) educate students about the educational value of upholding standards of academic integrity and (ii) emphasize that tools were in place to detect infringements, and that such infringements would not be tolerated. Thus, all students taking the module were initially required to take and pass an online formative exercise concerned with inculcating appropriate values of academic integrity (see Table 2). The principles underlying, and details of, this "academic integrity quiz" are detailed in full in Chapter VIII. In summary, the exercise is designed to ensure that all students taking the module demonstrate a complete understanding of why and how sources must be referenced. In addition, students were required to submit their written coursework assignment online using a plagiarism detection service provided to UK Higher Education Institutes by the Joint Information Systems Committee (JISC). Students were advised to submit drafts of their work to the plagiarism detection service software, enabling them to identify poorly referenced work and correct it prior to final submission. This constructive approach has proved highly effective in preventing instances of poor referencing and overt plagiarism. Admittedly, such cases are rare, but thus far no cases at all have occurred since the launch of the module in 2005.

Table 2. Overview of structure, content and digital library resources employed in the Drainage Basin Geomorphology module. Entries highlighted in bold represent learning activities that were authored by partners within the DialogPLUS collaboration (Pennsylvania State University in both cases) and repurposed for this module. Note that the acronyms used in this table are defined in Table 3, where full details of the digital library learning resources are provided

Week	Topic	Materials	OLA Learning Objectives	Assessment	Resources
Part 1: Induction					
1	Introduction/ Induction	PPT OLA	1. Provide an introduction to working with the Blackboard VLE 2. Understand the structure, rationale, aims and expectations of the module 3. Introduce use of online discussion forums 4. Explain how to cite academic references and online sources appropriately, to maintain high standards of academic integrity	MCQ(F) AI Quiz (F)	
Part 2: Conception (Fundamentals of drainage basin geomorphology)					
2	Fundamentals of Drainage Basin Geomorphology	PPT OLA	1. Understand (i) why a range of variables are used to describe and represent drainage basin morphology; (ii) how they are calculated, and (iii) how they vary across different geographical environments 2. Encourage discussion and reflection on the extent to which drainage basin morphology is linked to controlling variables in different environments around the world	MCQ (F)	PGDL
3	Conceptual Model of Drainage Basin Geomorphology	PPT OLA	1. Build and apply a conceptual model of drainage basin geomorphology founded on the principles of sediment budgeting 2. Understand that the development of the conceptual model – and the identification of constituent processes - underpins the remainder of the course	cMAP (FS 15%)	
Part 3: Knowledge Development (Drainage basin processes)					
4	Sediment Production: Crustal Deformation and Weathering	PPT OLA	1. Use the principles of isostasy to determine the extent to which post-glacial isostatic rebound is complete or ongoing 2. Determine contemporary rates of crustal deformation for selected drainage basin sites 3. Investigate how weathering rates and mechanisms are predicted to vary in different global climatic zones using Peltier's (1950) model 4. Reflect on the limitations of Peltier's model	MCQ (F)	NASA JPL ISOSTASY PELTIER DLESE NSDL PGDL
5	Water Mediated Sediment Transfer	PPT OLA	1. Learn to undertake basic terrain analysis (digitising drainage areas and extracting local slope gradient values) using MapInfo GIS software 2. Use MapInfo to estimate the local slopes and contributing drainage areas for a number of channel heads in a UK drainage basin using 1:10000 scale Ordnance Survey topographic map data 3. Use the data to construct a slope-area model for channel head locations 4. Compare the Lake District slope-area model with other published models in order to identify and compare the processes influencing channel head formation in different drainage basins located in varying environments	MCQ (F)	Digimap PGDL
6	Gravity Mediated Sediment Transfer	PPT OLA	1. Identify the relative importance of site-specific (i.e. geographical) factors as controls on (a) landsliding and (b) runoff-driven sediment transfer on slopes in drainage basins 2. Recognise the geographical circumstances under which landsliding or runoff processes dominate hillslope sediment transfer 3. Make predictions about the effects of future environmental change on sediment transfer dynamics for drainage basin hillslopes situated in discrete geographical locations.	MCQ (FS 25%)	USDAHEM ISLOPE PGDL

continued on following page

Table 2. continued

Week	Topic	Materials	OLA Learning Objectives	Assessment	Resources
Part 4: Perception (Drainage basin modelling and prediction)					
7	Workshop	PPT	1. Provide feedback and reflection on first two parts of the course 2. Initiate the individual project report. This involves students using the WILSIM landform simulator, appropriately parameterised with Digital Library data, to predict the establishment and evolution of landscape geomorphology in different global locations.		WILSIM PGDL
8	Linking Hill-slope Form and Process	PPT OLA	1. Understand the theoretical basis of a simple hillslope evolution model 2. Identify the relative importance of different environmental drivers and model parameters in controlling hillslope evolution 3. Apply the hillslope evolution model to investigate the environmental conditions associated with the formation of specific hillslope forms 4. Understand how interactions between different processes are expressed in geomorphological signatures in the landscape	MCQ (F)	NOHPM PGDL WILSIM
9	Landscape Evolution Models and Modelling	PPT OLA	1. Enable you to articulate how you might use a LEM to answer specific geomorphic questions 2. Understand how sensitivity analysis combined with LEM scenarios of geomorphic change can be used to explore multiple working hypotheses 3. Recognize how sensitivity analysis can help you identify and analyze uncertainties in a LEM 4. Encourage you to consider two somewhat contradictory consequences of LEM sensitivity analysis: first, how LEM uncertainties limit the geomorphic questions you can address with LEMs; secondly, how knowledge of LEM response can give valuable insight into dominant factors in landscape evolution.	MCQ (F)	WILSIM PGDL
10	Basin Response to Environmental Change	PPT OLA	1. Complete the individual project report, using the WILSIM landform simulator, appropriately parameterised with Digital Library data, to predict the establishment and evolution of landscape geomorphology in different global locations.	IPR (S 60%)	WILSIM PGDL JISC PDS
Part 5: Reflection and Conclusion					
11	Reflection and Conclusion	PPT OLA	1. To complete an online, anonymised, student evaluation of the course 2. To complete an online, anonymised, student evaluation of the OLAs	MCQ (F)	

Notes

Materials: PPT = Lecture slides, OLA = Online learning activity.

Assessment: F = formative assessment, S(x%) = summative assessment (with weighting), FS = assessment that is both formative and summative.

Accessing Resources: All the resources described in the table can be assessed from the physical geography home page of the DialogPLUS project website at http://www.dialogplus.soton.ac.uk/physical/index.htm. The left menu on the web page allows interested readers to navigate to all OLAs, the full set of lecture slides and any online assessment activity students took part in during the course.

Table 3. Key to acronyms used in Table 2. Items highlighted in bold text are JISC/NSF resources

Category	Acronym	Description (JISC/NSF resources are highlighted in bold text)	Used in OLAs
Digital Library	PGDL	Physical Geography Digital Library A collection of global and catchment data sets collated specifically for this module. See Table 5 for details of PGDL content. (Available at http://www2.geog.soton.ac.uk/blackboard/units/year2/geog2016/globaldlhome.asp)	2, 4, 5, 6, 7, 8, 9, 10
Digital Library	NASA JPL	An online database of GPS Time Series, hosted by NASA's Jet Propulsion Laboratory, from GPS monitoring stations located around the globe. Used in this course to explore how global tectonic and isostatic processes combine to determine rates of crustal uplift/subsidence (Available at http://sideshow.jpl.nasa.gov/mbh/series.html)	4
Simulation	ISOSTASY	ISOSTASY is an interactive Flash Animation, produced by Cornell University, that enables students to experiment with the principles of isostasy. Used in this course to demonstrate that post-glacial isostatic rebound can be a significant component of uplift in formerly glaciated regions. (Available at http://discoverourearth.org/student/isostasy.html)	4
Web Page	PELTIER	PELTIER is an interactive (ASP) web page, created by Jim Milne of the School of Geography at the University of Southampton for the specific purposes of this course. It enables students to explore how physical and chemical weathering processes vary in intensity and dominance as a function of varying climates around the globe. (Available at http://www2.geog.soton.ac.uk/blackboard/units/year2/geog2016/weathering.asp)	4
Digital Library	DLESE	The NSF funded Digital Library for Earth Science Education is introduced to students on this course as a potential resource that they might use to extend their learning. (Available at http://dlese.org/library/)	4
Digital Library	NSDL	The NSF funded National Science Digital Library is introduced to students on this course as a potential resource that they might use to extend their learning. (Available at http://nsdl.org/)	4
Digital Library	DIGIMAP	EDINA's Digimap is a JISC funded service that enables students to access UK Ordnance Survey mapping and mapping data at a range of spatial scales. Used in this course to facilitate geomorphic analysis of different regions of the UK. (Available at http://edina.ac.uk/digimap/)	5
Simulation model	USDAHEM	US Department of Agriculture Hillslope Erosion Model A web based modelling tool to predict the sediment yield from hillslopes as a function of varying slope gradients, slope shapes, runoff yields, land cover and land management practices. (Available at http://eisnr.tucson.ars.ag.gov/hillslopeerosionmodel/)	6
Simulation Model	ISLOPE	Infinite Slope Stability Analysis A standard approach in soil mechanics for predicting the stability of hillslopes with respect to shallow landsliding. The analysis was coded into an excel spreadsheet by Dr Steve Darby of the School of Geography at the University of Southampton for use by students on this course (NB: not fully documented) (Available at http://www.dialogplus.soton.ac.uk/physical/2016-ola6/data/Infinite Slope Analysis.xls)	6

continued on following page

Table 3. continued

Category	Acronym	Description (JISC/NSF resources are highlighted in bold text)	Used in OLAs
Simulation Model	WILSIM	WILSIM (Web-Based Interactive Landform Simulator) is a web based modelling tool, funded by the NSF, designed to predict the topographical evolution of a landscape as a function of varying tectonic, climatic and surface erodibility (which is a surrogate for geology, vegetation, etc) parameter values in different environments. (Available at http://www.niu.edu/landform/)	7, 8, 9, 10
Simulation Model	NOHPM	A slope-profile model developed by Nick Odoni (School of Geography, University of Southampton) that predicts the variation of hillslope profiles as a function of varying tectonic, climatic and erodibility parameters. Coded into excel spreadsheet form for use by students on this course (NB: not fully documented) (Available at http://www.dialogplus.soton.ac.uk/physical/2016-ola7/data/Odoni Hillslope Model.xls)	8
Learner Support	JISC PDS	An online Plagiarism Detection Service operated by JISC. Students were required to submit their 3000 word project reports to the PDS prior to final submission of their work. They received reports (also visible to the tutor) indicating the extent to which their text matched text in the PDS database (which includes journal articles as well as the other reports submitted by students). (Available at http://www.submit.ac.uk/static_jisc/ac_uk_index.html)	10

Underpinning the learning activities was the digital library. This was not accessed via a single portal, but consisted of three separate strands. First, students were encouraged to use traditional library resources (e.g., bibliographic search engines, journals, books, etc.), as well as Internet-based information resources and services, and so forth. This aspect was facilitated by efforts to include hyperlinked journal articles within the online reading lists provided. Second, a variety of extant data and modeling resources were embedded as hyperlinks within associated learning activities (see Tables 2 and 3 for details). However, it became clear that although there are a large number of extant online geomorphologic learning resources, many of which are accessible through existing digital libraries such as the USA National Science Foundation's (NSF) Digital Library for Earth Science Education (DLESE, see Tables 2 and 3), almost all of them focus on landform description, or understanding or simulating drainage basin processes. In contrast, the provision of *data* that students might use to parameterize, calibrate and

validate numerical modeling investigations was less common. In particular, there seemed to be a total absence of systematically structured sites that could provide data and information about specific drainage basins in different areas of the globe. This was problematic in that such site-specific information was necessary to support the type of exploratory modeling work that was integral to the pedagogic approach employed in this module. Consequently, effort was invested in constructing a specific digital library infrastructure suitable for the purposes of this module. The result (Table 4) is a Web site that comprises (i) global-scale, searchable data sets for a range of variables that control drainage basin geomorphology (e.g., global climate data for current and future scenarios, land cover data sets, remote sensing images, sediment yield, etc.), and (ii) catalogues of resources and data for seven drainage basins, covering a range of geological, topographical and climatological contexts, at sites in Europe and the USA. Each of the sites therefore contains resources such as aerial photographs, remotely sensed imagery, topographic maps, digital elevation models, and catalogues providing a range of morphometric data (e.g., stream order, drainage density, relief, hypsometry, etc.) for each basin. In principle, the digital library can be extended to include new drainage basins as data become available. In practice, the existing range of sites, though limited to two continents, covers a wide range of environmental contexts so that students can use the library to access and understand unfamiliar locations, as well as adding to their knowledge and understanding of databases.

Teaching and Learning Methods

In this section we summarize teaching and learning methods adopted on the module. The module was designed to be taught in a "blended" mode (Kerres & De Witt, 2003), that is to say involving a combination of face-to-face teaching (lectures) complemented by a series of self-paced online activities, though it should be noted that the latter are introduced in timetabled (i.e., face-to-face) practical sessions. A blended approach was adopted for a variety of reasons. It recognises that students learn best in different ways and therefore caters for a range of learning preferences (Honey & Mumford, 1992). Gold et al. (1991) argue that students benefit from a more personal approach to teaching, and the face-to-face contact within this approach supports this notion. The lectures and practicals allow staff to introduce theories, to enthuse students about the subject, and provide feedback, whilst the online activities provide

Table 4. Details of Learning Resources created and/or repurposed for use in the "Physical Geography Digital Library" (see Table 3 for full URL)

Type of Data/ Model	Specific Items	Description
Global Topography (Terrestrial)	Global Digital Elevation Model (DEM)	GTOPO30 (http://edc.usgs.gov/products/elevation/gtopo30/gtopo30.html) is a global DEM produced and distributed by the US Geological Survey's EROS Data Center
Global Climate (Terrestrial)	Mean Annual Temperature	It is important to note that data are provided for present day (1961-1990) and future (2050s and 2080s) conditions. For present day conditions, data are sourced from the Climatic Research Unit's (CRU at http://www.cru.uea.ac.uk/) global climatology (New et al., 1999). For future conditions, data are derived from a global climate model (the Hadley Centre's HadCM3) under one specific "greenhouse gas" emissions scenario (IPCC A2). IPCC A2 is a relatively high emissions scenario, representing a heterogeneous market-led world with high population growth (IPCC, 2000).
	Mean Annual Precipitation	
	Mean Annual Actual Evapo-ration	
	Mean An-nual Potential Evaporation	
Global Hydrology (Terrestrial)	Mean Annual Runoff	These data were provided by Professor Nigel Arnell of the School of Geography, University of Southampton. The data are outputs from a global terrestrial water balance model (Arnell, 2003) that is parameterised with a range of climate data sets (see above) and operating at 0.5 by 0.5 degree resolution. Importantly, the data are provided both for present day (1961-1990) and future climate conditions (2050s and 2080s). For present day conditions, Arnell's model uses climatic input data taken from the Climatic Research Unit's (CRU) global climatology (New et al., 1999). For future conditions, the water balance model is parameterised with data derived from a global climate model (the Hadley Centre's HadCM3) under one specific "greenhouse gas" emissions scenario (IPCC A2). IPCC A2 is a relatively high emissions scenario, represent-ing a heterogeneous market-led world with high population growth (IPCC, 2000).
	Soil Moisture Data	This archive (1992-2000) of soil moisture data (http://www.ipf.tuwien.ac.at/radar/index.php?go=ascat) estimated from ERS Scatterometer data is hosted by the Vienna University of Technology, Institute of Photogrammetry and Remote Sensing.

continued on following page

Table 4. continued

Global Imagery (Terrestrial)	Landsat Satellite Imagery	A global archive of high quality LANDSAT imagery (https://zulu.ssc.nasa.gov/mrsid/) is managed by NASA's John C Stennis Space Center
Global Land Cover (Terrestrial)		A global land cover map (http://maps.eogeo.org/GLC2000/), sourced from the Joint Research Centre of the European Commission
Global Rivers Sediment Yield	Online database	The FAO have commissioned a database (http://www.fao.org/ag/agl/aglw/sediment/default.asp) containing data on annual sediment yields in worldwide rivers and reservoirs, searchable by river, country and continent. Compiled from different sources by HR Wallingford, UK.
Watershed Geomorphology	1. Highland Water, Hampshire, UK 2. River Tyne, Northumberland, UK 3. River Zwalm, Belgium 4. Goodwin Creek, Mississippi, USA 5. Walnut Gulch, Arizona, USA 6. Smith River, Oregon, USA 7. Sulphur Creek, California, USA	The PGDL offers access to a library of specific drainage basins located in various locations around the world. For each drainage basin, a range of data resources are provided, including: DEM, aerial photos, topographic maps, and a summary of various physiographic indices (e.g. basin elevation, relief, ruggedness, hypsometry, stream order, drainage density, etc). This facilitates comparison of basin geomorphology in different locations, as well as offering opportunities for hypothesis generation, model parameterisation, calibration and validation.

students with the opportunity to develop skills and understanding within an independent-learning situation (Stefani et al., 2000; Laurillard, 2002; Biggs, 2003). In addition, the ability for students to work at their own pace allows them to study topics in more depth and to consolidate their learning, a recognised advantage of computer-based learning (Perkin, 1999). These aspects are now briefly discussed, but further details on specific elements of teaching practice are also reviewed in the sections on assessment, our reflections of the course, and the course evaluation.

- Lectures are used primarily to introduce each topic in the syllabus (see Table 2), provide a context for the associated online learning activities, and a framework for students' independent reading. Each topic is covered in a single coherent presentation, albeit timetabled into a double slot (two halves of 45 minutes each). This enables the lecture to cover a somewhat wider range of material than would otherwise be possible in single sessions, as well as creating "room" for discussion and questions during the presentation.

- Students develop their knowledge and understanding by interacting with the online resource-based learning activities available in the Blackboard VLE, and by undertaking further reading. Consistent with the pedagogic rationale articulated previously, suggested reading is deliberately focused on journal articles (where possible these are hyperlinked from reading lists made available in Blackboard). As previously discussed the online learning activities are self-paced, but are introduced in a face-to-face practical class. Each activity comprises an integral formative and/or summative assessment, so further discussion of online content is discussed below, and also in Table 2.

- An infrastructure to support student learning is created by timetabling three additional formal workshops and helpdesks, which tend to focus on problem-solving associated with the project work, but also provide additional opportunities for feedback, as well as tutor-to-student and student-to-student interaction, and through the infrastructure of the digital library itself. The latter involves the provision of online message boards to promote student interaction with peers and tutors, as well as the various document tracking facilities offered within the Blackboard VLE. The latter functionality is used to check the degree to which students with lower than average grades for their summative and formative assignments (see following section) are engaging with the course materials (e.g., reading lists, digital library resources, message boards) and intervening where necessary to either admonish or encourage. Students are also encouraged to become reflective learners by articulating what they think they have learned in each activity. Initially (in the 2004/2005 academic session) an online reflective learning diary was provided for this purpose, but this proved to be very unpopular (see below). In subsequent sessions the reflective learning diary has been abandoned, with students instead encouraged to use the online discussion forums for the same purpose.

Assessment

Specific issues considered in designing the assessment regime for *Drainage Basin Geomorphology* included (i) the balance between coursework and examinations within the physical geography curriculum as a whole; (ii) the need to ensure that where possible the assessments should serve a formative, as well as a summative, function, and;

(iii) the desire to reduce the amount of marking required by the teaching staff. This last point was not merely a matter of reducing staff time in a research-led institution. Rather, given the moderate to large cohort sizes anticipated in a Level 2 module, it was seen as fundamental to enabling the teaching staff to spend more time providing detailed feedback and contact with the students, which Lauillard (2002) and Mutch (2003) identify as being essential for student learning and improvement. Based on these considerations the decision was taken to continuously assess the module using coursework exclusively. The nature of the assessments, and their relationship to the learning outcomes, are now discussed in more detail.

Summative Assessments

The aggregate mark for the module as a whole was derived from the grades achieved in each of three formal coursework assignments, of increasing level of difficulty, timed to coincide with the end of the three stages of substantive learning (Figure 1). Although these assignments were designed to provide the necessary summative function, all include formative elements:

1. **Interactive concept mapping exercise.** In this exercise (see OLA3 in Table 2), students assemble a series of concept maps by connecting a set of concepts and their various inter-relationships. Individual concept maps are used to classify drainage basin processes, to build a conceptual catchment sediment budget, and then to use the sediment budget model to provide qualitative predictions of basin responses to environmental (tectonic, climatic, land cover) change (Figure 2). The concept maps comprise a series of Flash™ animations (drag and drop exercises), with individual marks awarded for placing the correct concept or relation into the appropriate area or areas of the concept maps. The marking process for this assignment

Figure 2. Example of a concept map used in Drainage Basin Geomorphology as accessed (a) prior to and (b) after the submission deadline. The latter version is a feedback version which informs students how to assemble the map correctly.

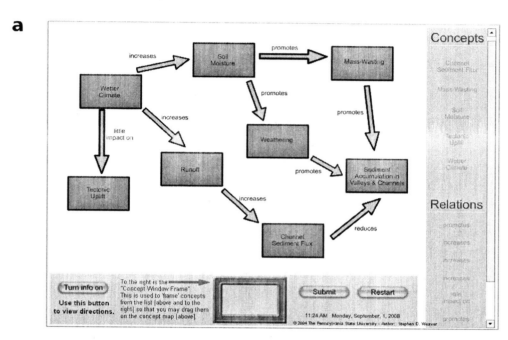

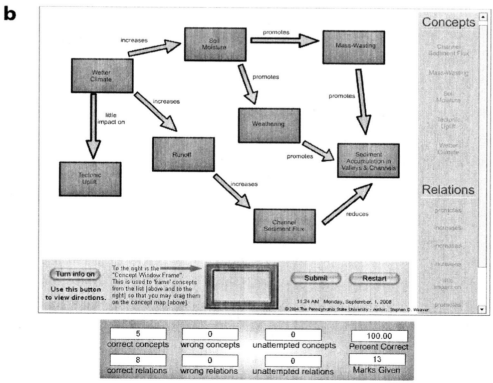

is, therefore, fully automated. After the submission deadline, feedback versions of the concept maps are made available within the Blackboard VLE (Figure 2). Further feedback is given in a dedicated workshop, where some of the more complex elements of the relationships are discussed further.

2. **Sediment transfer modelling.** In this exercise, students engage with hillslope process models to explore (a) the dominant factors controlling hillslope sediment transport by runoff and by landsliding, and to (b) predict the response to changes in environmental variables for specific drainage basins within the digital library (see OLA6 in Table 2). The students' ability to correctly parameterize the models, evaluate and analyze the model output, and to interpret key findings is assessed through two multiple choice quizzes (MCQs), delivered from within Blackboard. To avoid collusion, each individual student's MCQs are derived from large, randomized question banks. Two MCQs are required to assess different learning outcomes separately. MCQ-A addresses the issue of whether the modeling was conducted correctly. MCQ-B addresses the ability of students to interpret the resulting output -- these questions are posed in the form of a range of interpretations of given output data, only one of which is correct. In this way the possibility that the student's ability to interpret model output is hindered by a failure to obtain the correct output is avoided. As with the concept mapping, the marking of the MCQs is automated, and detailed, automated, feedback is released after the submission deadline. A second workshop is also held to discuss issues arising.

3. **Research project report.** This is a substantive piece of project work (a 3,000 word scientific report written in the style of a journal article) centered on the use of an online landscape evolution model (WIL-SIM, see Luo et al., 2004, 2005, 2006) and

digital library data resources to address a specific project title (see Table 2). The project involves undertaking a review of the relevant literature to ensure that the work is grounded within the appropriate disciplinary context, as well as learning how to use the landscape modeling tools and how to design a modeling-based project in a scientifically robust manner. Support for the project work is provided in a series of lectures and OLAs in Part 4 of the course (see Table 2). The project report is submitted online at the end of the academic session, with detailed written feedback released with the marks some weeks later. Marking criteria for the project report are identical to those of the final year undergraduate research project. Note that while the feedback on the project work is, therefore, not obtained until the end of the module, that feedback still serves a formative function in that the project work can be seen as a pre-cursor to the expectations of the final year dissertation.

The weightings for each of the three summative assignments reflect the varying levels of difficulty of each assignment. In the initial two years that the module was offered, the first two assignments were worth 25% each, while the individual project report contributed the remaining 50%. However, by the third year of the module (the current 2006/2007 academic session) it was clear (based on personal reflection, mark profiles, and feedback obtained from student evaluations, discussed below) that greater emphasis should be placed on the project report, and less on the concept mapping. The weightings of these two assignments have, therefore, now been adjusted by small amounts to 60% and 15%, respectively.

Formative Assessments

To provide learners with feedback on their progress and understanding of each topic, the decision was taken to create a formative assessment for each topic covered in the course syllabus (Table 2). The formative assessments also provided the students with the opportunity to become familiar with the format and approach used for the summative assessments (Pattinson, 2004). These assessments took the form of MCQs, intended to be taken after completing the learning activities within each OLA. Each MCQ was graded, but as entirely formative exercises the marks did not contribute to the aggregate marks for the module. Nevertheless, to provide some motivation for students to engage with the OLAs, students were required to take each assignment. Specifically, students were notified that failure (except in special circumstances) to complete all of the formative MCQs by the end of the module would automatically lead to them being awarded an overall mark of zero for the module.

Each of the MCQs was taken and assessed automatically online, with grades recorded in the student gradebook area of Blackboard. Detailed feedback on each individual question was also provided (again delivered automatically and immediately after submission). Feedback was provided in instances where questions were answered either correctly or incorrectly, to ensure that students who may have answered correctly but for the wrong reasons (e.g., by guessing) had the opportunity to have misconceptions disabused. Feedback on incorrectly answered questions was focused on explaining how to achieve the correct answer, rather than merely providing it.

Since each MCQ was associated with an online learning activity that was introduced in a face-to-face session (timetabled in a University of Southampton computer workstation cluster), academic staff were able to observe how students interacted with both the OLA and the associated MCQ. Further evidence about these aspects was also available through student comments posted in reflective learning diaries and on discussion boards, as well as in frank interviews undertaken as part of the evaluation process (see the discussion in the sixth section). For all these reasons, it became

clear (perhaps unsurprisingly) that a substantial number of students were engaging with the formative assessments in a way that was neither intended, nor which was likely to enhance their learning experience. Instead, such students merely "guess" the answers to complete the module requirements as quickly as possible. Many students provided feedback in their evaluations that suggested that credit (i.e., summative marks) should be awarded for these MCQs, though it must be emphasised that students also clearly value the learning resources and activities for their own sake, not to mention the associated feedback that is provided. The latter is consistently highlighted in student evaluations as a positive element of the module, and clearly is an aspect where online delivery has helped in providing timely and substantive feedback.

The real question, therefore, is whether "disengaged" behaviour is promoted or inculcated by the nature of the learning activity and designs, or whether it is merely reflecting a broader and well-documented tendency for students to focus their attention on summative assessment (Brown et al., 1999; Rust, 2001; Biggs, 2003). Academic staff take the view that it is better to provide the opportunity for students to engage voluntarily in substantive learning than it is to "coerce" them into it by setting incentives, such as awarding grades that contribute to the overall marks for the module. Indeed, taking the latter path seems to us to be surrendering the notion of learning for the sake of learning itself and pandering to an assessment culture that is itself damaging to learning. For this reason we have thus far resisted such calls. What we have done is to modify our teaching practice, both in the face-to-face lectures and OLA sessions, and in our contributions and interventions on the VLE discussion boards, to encourage students that engaging with the formative assessments is ultimately rewarded in an improved understanding that is likely to be reflected in the summative work.

EXPERIENCES OF DELIVERING THE MODULE

The first main point to note is that the amount of staff time invested in designing and authoring the learning materials and associated digital library was very significant. However, this investment has been recouped subsequently through time savings offered by automated marking and feedback strategies, as well as more effective time management strategies (e.g., use of online discussion boards vs. personal e-mails) offered by the VLE. Although it is difficult to estimate precisely, we believe that the project was, over this three-year time horizon, approximately time neutral. If a module of this nature is unlikely to offer staff time savings, then it is clearly important that other advantages (e.g., related to the potential enrichment of the curriculum offered by the use of digital library resources) accrue instead.

In fact, our experiences have highlighted a number of pedagogic issues that warrant mention here. It is not our intention to offer definitive answers to these questions, merely to highlight them to interested readers so that they may better decide for themselves if they wish to adopt such approaches to their teaching. Specific issues to consider are as follows:

- **Technical issues.** In the first session that the module was taught, a significant number of technical problems were encountered, despite extensive testing of the electronic resources prior to their launch. These problems were reflected in the student evaluations (below), though to a lesser extent than anticipated. What is clear is that no amount of testing by academics prior to the launch of new materials can replicate the ways in which electronic resources are used by students. Such "teething problems" (all were corrected for the second cohort) should therefore be accepted as a necessary evil in which the initial cohort of students are, to a certain

degree, treated as guinea pigs. This need not be problematic if academics are honest with students about this and are prepared to offer the necessary support during ramp-up periods.

- **Managed versus unmanaged learning.** It very rapidly became clear that the provision of a rich array of digital library and associated learning resources was highly valued by the students, and that the resources met their goal of offering broad horizons for exploration. However, in some sense, we became victims of our own success, experiencing some difficulties in managing the enthusiastic response of learners. By this we mean that the provision of digital library resources can lead to a "snowballing" effect in which students spend a great deal of time exploring them, and seeking additional materials. This is not in itself problematic, but we noted that there is a tendency for students to sometimes enter "blind alleys" as they travel beyond the resources managed within the structure of the provided OLAs and digital library.
- **Practical versus analytical skills.** Despite the clear focus of this module on critical analysis and research skills, it became apparent that it was often easier for us to design and resource learning materials that focused on practical skills. Consequently, we sometimes had to redraft our materials as it became clear that corrective action was required. This raised the question of whether e-learning inherently favours the development of practical skills rather than critical thinking and analytical skills (interested readers are referred to Fox & MacKeogh [2003]).
- **Customised learning.** Given its broader context within the DialogPLUS project, one of our initial aims was to explore the possibility of students from different institutions sharing the learning materials. Although this fascinating possibility was ultimately not explored further, we sense that the use of e-

learning and online learning resources offers potential to customise learning to match the differing levels and cultures associated with different cohorts of students across a range of institutional contexts.

- **Do the online learning activities developed for this module favour a particular type of learning style?** The activities developed for this course are essentially "linear" in structure (students enter them at the beginning and progress through them sequentially) and it is recognised that not all students learn most effectively in this manner (Honey & Mumford, 1992).

- **Individualised learning versus the student dynamic.** The learning activities discussed here have been embedded into a module essentially as pieces of coursework that are additional to, rather than replacements for, lectures. Furthermore, the activities were supported by timetabled "computer practical" sessions. The structure and environment of these sessions implies that students engaged with the activities in a particular way. More often than not, small teams of students spontaneously developed and worked through the materials collaboratively. Again, this was not in itself seen as a disadvantage, but we merely note that if the learning activities had been delivered as a module offered in a *distance* learning programme, then such collaboration would likely not have taken place.

MODULE EVALUATION

Chapter XIII of this book describes and discusses the DialogPLUS evaluation strategy, processes, and outcomes. In the first academic session (2004/2005) in which the *Drainage Basin Geomorphology* module ran, the specific activities undertaken to evaluate the innovations in this module involved (1) observations of the introductory lecture and two face-to-face workshops; (2)

monitoring student use of the course virtual learning environment (Blackboard), online discussion boards, and reflective diaries; (3) interviews with individual students; and (4) analysis of responses to an online questionnaire. This initial cohort comprised a total of 84 students. In subsequent academic sessions, evaluation of the module and its refinements has continued, but in a manner that is consistent with standard institutional practice at Southampton. Thus, in 2005/2006, when the module was taken by 33 students, and 2006/2007 (51 students) evaluation has been based solely on student responses to the online questionnaire.

Observations in the 2004/2005 Academic Session

During the introductory lecture, the lecturer spoke about the potential excitement for, and engagement of students, as a rationale for developing a new course. They would be producing a conceptual model of interacting processes and should begin to question the received wisdom in this field. The suggestion that, in the context of being exactly halfway through their "training," they should start to be more "critical" was met with a murmur of approval. Although the announcement that there would not be a field trip seemed to disappoint some students, the lecturer promoted the use of digital library resources and interrogation of online models to progress both science and student learning, and described the online reflective learning diary as a "field notebook for a campus-based course." The few student questions at the end of the lecture were all to do with clarification of the assessment regime.

Monitoring Student Use of the Virtual Learning Environment and Reflective Diaries

Student contributions to the discussion boards and reflective diaries were monitored for seven weeks. The discussion boards were well used to clarify terminology and to report/resolve a number of technical problems with the online learning activities (OLAs). Lecturer responses were timely and students also helped each other. Entries in the diaries were a mix of reflection on elements of the substantive topic and the novel approach to teaching and learning:

Having missed the lecture on the previous Tuesday due to a sporting commitment I feel as though I have caught up sufficiently and that the OLA helped me understand the content more fully. It is clear that some basins have a higher drainage density than others, as they have a greater length of streams per unit area. Runoff does not influence drainage density, it is more dependent on the permeability and the gradient or the land. The runoff is dependent highly on the shape of the basin as the more elongated the basin is the higher the peak flow. The digital library was an interesting feature to use. (Student 1)

The course so far has been a lot different from what I had expected. The OLA3 was well annotated and so I knew what was expected of me, this was a worry at the start of the course. The lectures have complimented the practicals well, which has helped. This unit is quite different from the set up of others, but is good. The continuous practicals are better than having more towards the end of a course. I would say I had learnt differently, so it is difficult to compare. Because I find computers slightly hard, I have learnt a lot, but for those who find them easy, it might not be the case so much. (Student 2)

The total number of entries in their reflective learning diaries per student during those months ranged from 0 ($n = 5$) to 6 ($n = 1$) with an average of 2. The evaluator noted and discussed with the teaching team that there were no lecturer comments/entries. It seemed that this discouraged students from making further entries. As there was no credit for using the diary, and no feedback on it, it is understandable that it was not valued by the students (see student comments and *Analysis of questionnaires* below).

Interviews with Individual Students

During a workshop towards the end of the teaching period, informal interviews were conducted with individual or pairs of students who were, at that time, employing a drainage basin model (WILSIM; Luo et al., 2004, 2005, 2006) as part of their assessed research project. The evaluator started by asking if the students had enjoyed the module, whether it was different from others they had taken, and progressed to questions about specific aspects of the module (i.e., the OLAs, discussion boards, reflective diary, and the final project). All interviewees appeared to have enjoyed the OLAs, although some had not used the formative ones effectively. They were positive about the discussion boards, but negative about the reflective learning diary. Although the drainage basin modelling was interesting and challenging, they would have preferred it to come earlier (often a difficulty in a one semester module). Several students commented that the unit had fulfilled the expectations set by the lecturer.

The unit was quite easy until the project. I was happy with the lectures and the early OLAs. I didn't do the last OLA properly. With the formative OLAs I just keep choosing answers until I get 100% but I realise now that if I'd done the last one properly it would be useful for the project. The project is challenging and too close to the exam. (Student 3)

The unit has been good, interesting. It is good to learn and then apply the learning. The OLAs build up in difficulty. Earlier use of models would help. The diary is a "waste of time." The project is difficult. Discussion boards are good. (Student 4)

The unit is very different. It's better because everything is all in one place on Blackboard, but I feel slightly disadvantaged because I'm not good with computers especially if there are technical problems. The formative OLAs were useful. The diary was not a productive use of my time and there was no feedback so I stopped doing it. The project and model is tricky. I'm not sure what the model is showing. I used the discussion boards occasionally. I would recommend this unit to other students. It would be improved if the (WILSIM) model was available earlier. (Student 5)

I don't really like computers, especially for reading big chunks of text. I like to have materials on paper so that I can highlight bits and write notes. I didn't use the diary at all after the first OLA. It's not useful. I did post questions and got "pretty quick" and "useful" replies. It is very useful to see other students' questions and the answers. It is a good unit, interesting. I enjoyed the lectures and the background reading. The project is too late, too near exams. (Student 6)

Analysis of Completed Questionnaires

The generic questionnaire (see Chapter XIII) was mounted in the course Blackboard site (Table 5) and supplemented with a further 21 questions (Table 6). Sixty-five (77%) of the eighty-four students who took the unit in its first year (2004/2005) completed the questionnaire. Responses were particularly positive about the content, embedded tools, support mechanisms, and other students' and lecturer's contributions to the discussion boards. Responses were particularly negative about the reflective learning diary. Although there were mixed views about enjoyment of, and learning from, the OLAs, 75% of respondents would not prefer to be assessed by essays. The small number of year 3 students who responded were generally less positive about the innovations than the year 2 students. There were no gender differences in the responses.

Students contributed many more comments via the online questionnaire than was typical on paper-based ones used in previous DialogPLUS evaluations. As with other DialogPLUS cohorts, there was less comment on the content of the learning activities than on the delivery mode, navigation, assessments, and support. Many students liked the variety on offer, but others commented on technical problems, and/or that the OLAs were very time consuming, particularly for students trying to access them off-campus or for those who found it difficult to work online. Some students found the formative quizzes trivial and guessed at answers, but others did find them a useful way of testing their understanding of theory. Comments about the blended approach were overwhelmingly positive.

All student comments, and suggestions for improvements, from the questionnaire and observation sessions were made available to the teaching team. These led to a number of refinements in the content of the module and its structure. These focused on correcting the teething problems encountered in making the content available across a wide range of platforms, abandoning the reflective learning diary (and instead encouraging reflective discussion through use of the message boards), and addressing institutional problems with timetabling in 2004/2005 (the OLAs were timetabled prior to the lectures that were designed to introduce them!). Subsequent evaluations (for the 2005/2006 cohort only: at time of writing the 2006/2007 version of the module was being delivered) were undertaken using the same online questionnaire.

CONCLUSION

In this chapter we have presented the basis for, and content of, a Level 2 module in Drainage Basin Geomorphology. The key innovative feature of the module is its reliance on the use of a range of digital library and modeling resources, accessed online, as a means of inculcating the skills and

Table 5. Student responses to standard online evaluation questionnaire about the OLAs for the launch (2004/2005 academic session cohort). The statistics produced in the course VLE (Blackboard) are shown. Note that sixty-five (77%) of the eighty-four students who took the unit completed the online evaluation.

Generic quality questions	Number and percentage of the students giving each score				Mean response
	0 – No	1 – Somewhat	2 – Yes	N/A	
1. Did you find that there was a full description of the online learning activities, including the learning objectives?	1 1.5%	14 21.5%	48 74%	2 3%	1.7
2. Did you find the OLA interface (this refers to the usability of the activities, not of Blackboard) easy to use?	0	22 34%	42 65%	1 1.5%	1.6
3. In your view, were all the tools required to complete the OLAs included (e.g., database, spreadsheet, note making, discussion board)?	1 1.5%	6 9%	57 88%	1 1.5%	1.9
4. In your view, did the content of the OLAs match your preferred learning style?	10 15%	22 34%	32 49%	1 1.5%	1.3
5. Did you find that the content of the OLAs was relevant, appropriate and clear?	1 1.5%	15 23%	48 74%	1 1.5%	1.7
6. Did you find that all the embedded materials (e.g., hyperlinks, spreadsheets, and references) were easily accessible?	3 5%	23 35%	37 57%	2 3%	1.5
7. Did you find that mechanisms were provided for information and support in the event that you needed them?	0	14 21.5%	49 75%	1 1.5%	1.8
8. Did the instructions and/or the OLAs define the maximum response time to learner queries?	6 9%	25 38.5%	20 30.5%	10 15%	1.1
9. Were the tutors' responses to your queries helpful?	0	11 17%	48 74%	6 9%	1.7
10. Did you find the online activities were helpful in improving your drainage basin geomorphology knowledge and skills?	4 6%	22 34%	38 58.5%	1 1.5%	1.5

Table 6. Student responses to supplementary online evaluation questionnaire about the OLAs for the launch (2004/5 academic session cohort). The statistics produced in the course VLE (Blackboard) are shown. Note that sixty-five (77%) of the eighty-four students who took the unit completed the online evaluation.

Supplementary quality questions 0 – No 1 – Somewhat 2 – Yes	Mean Responses				
	All (65)	Yr2 (59)	Yr3 (6)	Female (30)	Male (31)
1 Did you find that the unit's Blackboard site (as opposed to the OLAs) was easy to use?	1.82	1.81	1.83	1.90	1.74
2 Did you find that all queries about the administration of the course (e.g., hand-in dates, etc.) were resolved in good time?	1.74	1.73	1.83	1.70	1.77
3 Did you find that technical (e.g., computer) queries and problems were resolved in good time?	1.42	1.41	1.50	1.53	1.26
4 Did you find that the technical problems had a negative effect on your learning?	0.46	0.42	0.83	0.43	0.45
5 Did you enjoy doing the OLAs?	1.22	1.27	0.67	1.20	1.29
6 Do you feel that you learnt a lot from doing the OLAs?	1.43	1.47	1.00	1.30	1.65
7 Do you find that the time suggested for completing the OLAs was accurate?	1.45	1.44	1.50	1.37	1.52
8 Did you think that the weighting of 25% each for the two summatively assessed OLAs (OLA3 and OLA6) was fair?	1.57	1.59	1.33	1.50	1.71
9 Did you find that the assessed elements of the summative OLAs (OLA3 and OLA6) were appropriate for the learning objectives?	1.74	1.76	1.50	1.73	1.77
10 Would you prefer to be assessed by essays rather than by the summative OLAs?	0.34	0.27	1.00	0.23	0.42
11 Did you find that the assessed elements (the MCQs) of the formative OLAs (OLA1, OLA2, OLA4, OLA7, and OLA8) were appropriate for the learning objectives?	1.57	1.63	1.00	1.53	1.68
12 Did you find that the MCQs for each of the formative OLAs provided you with useful feedback that helped you learn?	1.47	1.53	0.80	1.40	1.60
13 Did you find that the MCQs for each of the formative OLAs helped you to identify what parts of the course you did and did not understand?	1.25	1.31	0.67	1.23	1.32
14 Did you find that overall the formative OLAs helped you learn about drainage basin geomorphology?	1.54	1.56	1.33	1.43	1.65
15 Did you enjoy contributing to the discussion boards?	1.11	1.14	0.83	1.14	1.10
16 Did you find that the contributions that you made online helped your learning?	1.18	1.19	1.17	1.17	1.16
17 Did you find that the contributions that other students made online helped your learning?	1.78	1.78	1.83	1.87	1.71
18 Did you find that the tutor's contributions to the discussion boards were helpful to your learning?	1.86	1.86	1.83	1.97	1.77

continued on following page

Table 6. continued

18	Did you enjoy writing in your reflective diary?	0.23	0.25	0.00	0.13	0.35
19	Did you find that the reflective diary made an important contribution to your learning?	0.23	0.25	0.00	0.20	0.29
20	Would you use your reflective learning diary on other courses?	0.18	0.20	0.00	0.07	0.32
21	Did you find that the blend of online activities with traditional lectures and a project was a good way to learn?	1.66	1.69	1.33	1.67	1.68

knowledge necessary to understand how and why drainage basin geomorphology varies through time and space in predictable manners. We have shown that this approach is unlike more traditional courses within the physical geography curriculum. Detailed and ongoing evaluation of the module suggests that many of the students who have taken the module enjoy and value it, though inevitably it is not possible to say whether the positive response is actually induced by these differences.

In the past, practitioners have sometimes been keen to embed e-learning within the curriculum due to the perceived advantages that it offers in terms of aspects such as efficiencies in working with large class sizes, in marking, and in the timely provision of detailed feedback. While these are valuable elements of the day-to-day experience in delivering the module, we have attempted to provide a realistic assessment of the work that is involved in authoring and maintaining the online learning resources. Even if there is no net increase in staff workload (when averaged over a period of a few years after the creation of the course), equally it is clear that modules of this type should not be attempted by practitioners who are seeking to reduce their workload. Instead, this type of module offers the opportunity to invest staff time in different ways. For example, and perhaps ironically, the use of e-learning devices such as discussion forums and the automated marking of a significant component of the coursework has enabled us to invest more time into face-to-face contact with students and in providing significant amounts of feedback. However, not all the learning outcomes

of the module can legitimately be assessed in this way, and it is clear that the "blended" learning model employed herein retains much pedagogy that is common with more traditional modules in physical geography. Instead, where our approach fundamentally differs is in its emphasis on using the potential for digital libraries to enable students to expand their geographical horizons and skills virtually.

REFERENCES

Arnell, N. W. (2003). Effects of IPCC SRES emissions scenarios on river runoff: A global perspective. *Hydrology and Earth System Sciences, 7*, 619-641.

Biggs, J. (2003). *Teaching for quality learning at university* (2nd ed.). The Society for Research into Higher Education and the Open University Press.

Brown, S., Race, P., & Bull, J. (1999). *Computer-assisted assessment in higher education.* London: SEDA and Kogan Page.

Church, M. (2005). Continental drift. *Earth Surface Processes and Landforms, 30*(1), 129-130.

Fox S., & MacKeoch K. (2003). Can eLearning promote higher-order learning without tutor overload? *Open Learning, 18*, 121-134.

Gold, J. R., Jenkins, A., Lee, R., Monk, J., Riley, J., Shepherd, I., & Unwin, D. (1991). *Teaching geography in higher education.* Oxford, UK: Blackwell.

Honey, P., & Mumford, A. (1992). *The manual of learning styles.* Maidenhead, UK: Peter Honey.

Intergovernmental Panel on Climate Change (IPCC). (2000). *Special report on emissions scenarios.* Cambridge, UK: Cambridge University Press.

Jorum. (2007). *Jorum user terms of use.* Retrieved May 31, 2007, from http://www.jorum.ac.uk/user/termsofuse/index.html

Kerres, M., & De Witt, C. (2003). A didactical framework for the design of blended learning arrangements. *Journal of Education Media, 28,* 101-113.

Laurillard, D. (2002). *Rethinking university teaching: A framework for the effective use of educational technology* (2nd ed.). London: RoutledgeFalmer.

Luo, W., Duffin, K. L., Peronja, E., Stravers, J. A., & Henry, G. M. (2004). A Web-based interactive landform simulation model (WILSIM). *Computers and Geosciences, 30,* 215-220.

Luo, W., Stravers, J. A., & Duffin, K. L. (2005). Lessons learned from using a Web-based interactive landform simulation model (WILSIM) in a general education physical geography course. *Journal of Geoscience Education, 53*(5), 489-493.

Luo, W., Peronja, E., Duffin, K., & Stravers, J. A. (2006). Incorporating non-linear rules in a Web-based interactive landform simulation model (WILSIM). *Computers and Geosciences, 32,* 1512-1518.

Mutch, A. (2003). Exploring the practice of feedback to students. *Active Learning in Higher Education, 4,* 24-38.

New, M., Hulme, M., & Jones, P. D. (1999). Representing twentieth century space-time climatic variability. Part 1: Development of a 1961-1990 mean monthly terrestrial climatology. *Journal of Climate, 12,* 829-856.

Pattinson, S. (2004). *The use of CAA for formative and summative assessment – student views and outcomes.* Paper presented at the CAA Conference. Retrieved March 22, 2005, from http://www.caaconference.com/

Peltier, L. C. (1950). The geographical cycle in the periglacial region as it is related to climatic geomorphology. *Annals of the Association of American Geographers, 40*(3), 214-236.

Perkin, M. (1999). Validating formative and summative assessment. In S. Brown, J. Bull, & P. Race (Eds.), *Computer-assisted assessment in higher education* (pp. 29-37). London: Kogan Page.

Rust, C. (2001). *A briefing on assessment of large groups.* Assessment Series No. 12. York, UK: LTSN Generic Centre.

Schon, D. (1987). *Educating the reflective practitioner: Toward a new design for teaching and learning in the professions.* San Francisco: Jossey Bass.

Seale, J. K., & Cann, A. J. (2000). Reflection on-line or off-line: The role of learning technologies in encouraging students to reflect. *Computers and Education, 34,* 309-320.

Stefani, L. A., Clarke, J., & Littlejohn, A. H. (2000). Developing a student-centred approach to reflective learning. *Innovations in Education and Teaching International, 37,* 163-171.

Tooth, S. (2006). Virtual globes: A catalyst for the re-enchantment of geomorphology? *Earth Surface Processes and Landforms, 31,* 1192-1194.

Chapter VI
Engaging with Environmental Management:
The Use of E-Learning for Motivation and Skills Enhancement

Jim Wright
University of Southampton, UK

Michael J. Clark
University of Southampton, UK

Sally J. Priest
Middlesex University, UK

Rizwan Nawaz
University of Leeds, UK

ABSTRACT

There is an inherent antithesis between environmental management as professional practice and as concept or philosophy. Not only does this antithesis pose a problem in teaching environmental management, but also learners often have difficulty with the broad-based, multi-disciplinary nature of the subject and the value-laden nature of many environmental management decisions. Furthermore, field experience is an inherent part of environmental management and fieldwork is thus a necessary component of most environmental management modules. E-learning offers a mechanism through which to address these potential problems, through virtual practical experience and by serving as a virtual management laboratory. In this chapter, the undergraduate focus of a module on Upland Catchment Management and on environmental management is compared with e-learning for postgraduate delivery (a module on GIS for Environmental Management). The differing styles of delivery highlight the flexibility of e-learning as a vehicle for acquiring skills and knowledge, and underpin the claim that the result is an enhanced engagement with the practice of informed management.

THE CHALLENGE OF TEACHING ENVIRONMENTAL MANAGEMENT AS A DISCIPLINE

Environmental management is a discipline that seeks to co-ordinate development so as "to improve human well-being and mitigate or prevent further damage to the Earth and its organisms" (Barrow, 1999). Environmental management emerged as an academic subject during the 1970s and 1980s in response to a growing awareness of environmental degradation. The scope of the discipline is contested, with some arguing that many environmental management courses have a rather technocentric and state-led focus that should be broadened to consider human interactions with attitudes to the environment more generally (Bryant & Wilson, 1998). Others (Diduck, 1999) have argued that the discipline should facilitate public engagement with the environment, enabling both empowerment of local communities and social action in response to emerging environmental problems. In this chapter, we follow Barrow's definition of environmental management, but paying particular attention to the different perspectives on the environment of the public, environmental managers, business, and other stakeholders.

Within geography, environmental management is being increasingly recognized within the discipline as an area of importance and is considered by some to be the third category of geography, incorporating the interactions between the physical and human elements (QAA, 2000). As a subject within geographical education, environmental management is seen as a potential means of enhancing the employability of geography graduates, as well as meeting a growing student interest in issues of sustainability. Clark (1998), for example, describes the use of student work placements involving environmental auditing of different companies as one means of boosting employability.

However, teaching and learning in environmental management presents several discipline-specific challenges, namely:

- The difficulty of teaching an inherently practical, applied subject such as environmental management;
- Providing effective fieldwork as a necessary part of an environmental course;
- The breadth of understanding required by students, because of its multi-disciplinary nature;
- Enabling students to appreciate the value-laden nature of many environmental management decisions.

We will consider each of these challenges in turn, before exploring some of the e-learning solutions developed in three case studies.

Environmental management in the three case study modules described here is a practical subject, since it focuses on solutions to specific environmental problems such as identifying an appropriate location for siting wind turbines. It should be noted that some have contested the practical, pragmatic element of environmental management, for its over-emphasis on compromise and inter-organizational collaboration (Prasad & Elmes, 2005). Nonetheless, in this context, all three case studies entailed students working towards the solution of environmental problems and there is an inherent contradiction in trying to teach this within the classroom. How can the "real world" problems and decisions be brought into classroom activities? A constructionist approach has generally been adopted within the development of learning materials for environmental management where the main focus of unit development is to ensure that environmental management learning is embedded within "realistic and relevant contexts" (Honebein, 1996, p. 11). However, ensuring workplace relevance in environmental management education remains a significant challenge.

There is a long tradition of field-based learning in both geography and environmental management, which is often regarded as an essential part of the undergraduate experience (Boyle at al., 2003; Munowenyu, 2002) due to a belief that learners are only able to understand and therefore engage with complex environmental management decisions when faced with the physical site characteristics themselves. It is also recognized that the fieldwork experience adds unique value to classroom-based learning, not matched by entirely virtual fieldwork (Spicer & Stratford, 2001). A challenge in both environmental management and geography is therefore developing linkages between concepts taught in the classroom and experience gained in the field.

Environmental management is a broad discipline, requiring not only an understanding of environmental science but wider issues of policy, resource economics, and decision-making too (Grumbine, 1994). A USA-based study (Hogan, 2002) found that after beginning by considering a wide range of issues, students quickly became very narrowly focused in trying to resolve an environmental management problem, relative to an expert group of environmental decision-makers. This suggests that many students have difficulty in developing the broad, multi-disciplinary perspective necessary for environmental decision-making.

Often, students bring a specific set of values to the discipline and these values can shape their approach to problem solving. A Spanish study (Jimenez-Aleixandre, 2002), for example, found that whilst a group of 16 to 17 year old students were able to develop quite detailed understanding of concepts in environmental management, their decision-making differed markedly from an "expert" manager. This was because the "expert" based decisions on a value set that took more account of the economic impact of management decisions, whereas the students took greater account of the ecological consequences of a decision. This evidence suggests that exposing students to other value sets may also be an important component of teaching and learning in environmental management.

This chapter aims to explore the e-learning innovations developed within the Schools of Geography at both the University of Leeds and Southampton. In both instances, e-learning resources have been developed to teach different aspects of environmental management within the undergraduate curriculum. These learning experiences will be compared with Southampton's development and delivery of an environmental management unit at the postgraduate level. In exploring these case studies, we will assess the extent to which each module was able to address the four challenges in environmental management education outlined above.

DEVELOPING E-MODULES IN ENVIRONMENTAL MANAGEMENT

In the following sections, we describe the use of e-learning in three different environmental management modules developed through the DialogPLUS project. This project brought together geographers from four different universities in the UK and USA (University of Southampton, University of Leeds, Pennsylvania State University, and the University of California at Santa Barbara) to develop digital learning resources collaboratively (http://www.dialogplus.soton.ac.uk/). As shown in Table 1, one postgraduate module (*GIS for Environmental Management*) was written explicitly from the outset for students studying wholly online, part-time, and at distance. Materials for the two remaining modules (*Upland Catchment Management* and *Environmental Management*) were modified to incorporate an e-learning element, enabling a "blended" module delivery to take place.

Table 1. Characteristics of three environmental management units delivered through the DialogPLUS project (http://www.dialogplus.soton.ac.uk/)

Unit Title	Upland Catchment Management	GIS for Environmental Management	Environmental Management
Level	Undergraduate	Postgraduate	Undergraduate
Materials development	Revision of an existing unit	New unit	Revision of an existing unit
Original mode of delivery	Face-to-face tutorials, practicals, and fieldwork	N/A	Face-to-face lectures and practicals
Revised mode of delivery	Blended delivery, with original face-to-face teaching methods supported by e-learning	Wholly online delivery to distance learners	Blended delivery, with original face-to-face teaching methods supported by e-learning
Institution	University of Leeds	University of Southampton	University of Southampton
Duration of teaching	10 weeks	10 weeks	1 semester
Number of credits	15	15	15
Number of unit deliveries studied	1	3	3

Introducing E-Learning into an Undergraduate Module: *Upland Catchment Management*

The environmental management module *Upland Catchment Management* (UCM) is an optional BSc Level 3 geography undergraduate module. It is an online campus-based module running over 10 weeks. The reason for setting up an online module that is campus-based was so that the responsiveness of undergraduates (who do not enroll on distance learning courses) to online learning could be better assessed. Students on this module are expected to have attained four main objectives upon completion. These are summarized below:

1. Demonstrate an appreciation of the context, issues and management pertaining to upland catchments in general;
2. Understand the management tools that generate sound policy and practice for upland catchment management;
3. Understand the scientific basis of the key environmental processes operating in upland catchments through a wide-ranging grasp

of the literature (academic, practitioner, and policy-related) and appropriate experience gained from fieldwork;
4. Use this scientific understanding, both general and catchment specific, in combination with both policy understanding and consideration of stakeholder views, to develop a realistic catchment management plan for a case-study catchment.

The delivery and assessment format included:

*   Face-to-face sessions during week 1 and the final two weeks
*   Seven online lectures
*   One-day field trip mid-way through the module
*   Formative assessment: three online multiple choice question exercises during weeks 2-4
*   Summative assessment: five practical exercises, a group presentation, and a project report

It was decided that fieldwork should form an important component of the module aided by the setting up of an online discussion room to help students engage in post fieldwork dialogue. The students were taken to Upper Wharfedale, in the Yorkshire Dales for a day. The whole group was assigned a series of tasks during the morning. In the afternoon, the students were split into two groups and assigned an almost identical task. The aim was to measure river flow on a stretch of river. The groups were required to conduct river flow measurements at a short distance apart – the idea being that, in an ideal world, both groups should come up with similar values of river flow rate. In reality, the measured rates would differ owing to differences in techniques and calculations.

Students were asked to document their findings in a project report in fulfillment of the summative assessment component (10% weighting of total module marks). It was essential that the report contained results obtained by both groups and reasons for differences. Since the students had little opportunity for information exchange during the fieldwork, it was explained that an online discussion room would be set up to facilitate this.

The student cohort was relatively small, comprising ten undergraduates of which six were male. Consent was sought from the students to investigate and report on their online discussion.

It was decided to adopt the five-stage model of Salmon (2000) to support a student-centered approach to online learning. A summary of the model is provided below:

- **Stage 1: Access and motivation.** Students at this stage need to easily be able to pinpoint the discussion room and have the right motivation for its use.
- **Stage 2: Online socialization.** Students should begin to gain confidence in using the discussion room. Students who know each other should have no trouble in passing this stage.
- **Stage 3: Information gathering and dissemination.** Students should start to acknowledge the strength of the discussion room as a resource for information exchange, peer support, and feedback.
- **Stage 4: Knowledge construction.** A key stage at which both student and tutor work together to develop new ideas through dialogue
- **Stage 5: Development.** Student has sufficient confidence to be independent online with minimal instructions. Students should start to think about the discussion room as their own space rather than the tutor's.

The first stage in getting the students to use the discussion room was to e-mail all students access instructions. It was explained to the students that if required, assistance was available from the help-desk of the virtual learning environment (VLE) facility at the university (known as Bodington Common). The merits of using the discussion room were then emphasized. It was made clear that the project report needed to contain details of the other group's methodology and results. The more fully the differences in the two methodologies were explained, the higher the marks to be gained in the project report.

As noted by Salmon (2000), during stage 2, students do appear to gain confidence in using the discussion room. This was found to be the case here, and as shown in Table 2, almost everyone had contributed to the discussion within a week. This was despite there being some pressure but no compulsion to do so. Given this apparent success, it was felt that the tutor needed minimal input. Students were communicating with each other quite freely, although e-mail was also used to exchange information between those who knew each other well.

During stage 3, those students who had not yet contributed were identified. Student I was sent an e-mail and did later post a query on the discussion room. The other student (Student J) who had not posted to the room was a special case, as he was also the tutor's dissertation student. He therefore preferred to see the tutor directly rather than post

Table 2. Project report-related discussion room postings

Subject	Days since discussion went online	Days before project deadline	Name
Field Trip	0	13	Tutor
	0	13	Student A
	2	11	Student B
	3	10	Student C
Group B data	4	9	Student D
			Student E
Formula	6	7	Student F
	10	3	Student G
	10	3	Student D
	10	3	Tutor
Group A and B methodologies	10	3	Student D
	10	3	Student G
	10	3	Student D
	10	3	Student H
	10	3	Student I
Error with formula	11	2	Student F
	12	1	Student B
Uncertainty Framework	12	1	Student G
	12	1	Student F
	12	1	Student B
	12	1	Student F
	12	1	Tutor
Transects	13	Few hours before deadline	Student E
	13	Few hours before deadline	Student H

queries on the discussion room. Despite some encouragement, Student J did not use the discussion room on this occasion, although he would use it to support his learning later.

Some attempt was made by the tutor at the key stage 4 to work with the students. A comment had been posted by Student F (a very bright and confident character) that a formula was incorrect.

The formula was in fact misunderstood by Student F, and rather than responding immediately, the debate was opened. Fellow students were asked to suggest possible reasons as to why the formula might be deemed incorrect, and to make suggestions on its validity. With online teaching, it can sometimes be all too easy to neglect students and this is what probably happened during stage 4. It is felt the students could have been engaged more but time constraints did not allow this.

It is believed that the final stage of Salmon's five-stage model was reached. Towards the end, most of the students did start to function as independent learners online.

To gain students' view on the benefits and drawbacks of the discussion room, feedback was collected. Results (see Table 3) indicate the group discussion clearly proved to be a valuable exercise for almost all students on the UCM course. The main drawback, as already noted, was the lack of tutor contribution.

Students still continued to use the discussion room for information/ideas exchange related to other aspects of the course, once the fieldwork exercise had been completed. As noted by Salmon (2000), extra effort is needed during the early stages of implementing a discussion room to provide lasting benefits.

Resource Development within an Existing Unit: *Environmental Management*

As with the University of Leeds module, the revisions to an *Environmental Management* unit delivered by the University of Southampton were made to an existing level two unit. This is an optional unit, but is usually taken by the majority

Table 3. Student feedback on discussion room use

Student Response		Number and percentage of the students present giving each score				Mean Response
		0 – No	1 – Somewhat	2 – Yes	N/A	
1	I enjoyed contributing to the discussion forum.	0	4 40%	6 60%	0	1.6
2	I found the contributions that other students made online helped my learning.	0	3 30%	7 70%	0	1.7
3	The tutor's contributions to the discussion forum were helpful to my learning.	2 20%	3 30%	5 50%	0	1.3
4	Separate group discussions were useful for the poster assignment.	1 10%	1 10%	8 80%	0	1.7
5	I used e-mail regularly with colleagues to exchange relevant information/ideas.	0	7 70%	3 30%	0	1.3
6	E-mail allowed good contact with tutor.	1 10%	4 40%	5 50%	0	1.4

of students taking BSc Geography, and had an uptake of around 120 students (both geography and external students). The course runs in the third semester (i.e., the first semester of the second year) of a student's university career and for most is their first main unit exploring the management of the environment. The learning objectives for this unit are presented in Table 4.

The delivery and assessment on this unit comprised:

- 18 face-to-face lectures held throughout the semester
- E-learning practical clinic sessions (one associated with each practical) held in weeks 3 and 6
- Summative assessment: 2 online practicals (25%) (due in weeks 4 and 7); coursework essay (due in week 10) (25%), and two-hour unseen written examination (50%) (end of semester)
- Online forums available for the students to participate in and ask questions.

The original module included mainly paper-based practical elements based on some electronic data, but it was accepted that a more interactive approach would allow different data to be used and provide a more integrated and enhanced student learning experience. A "blended" approach to e-learning was adopted (Kerres & De Witt, 2003). Two new e-learning practicals were developed; the first to explore how to manage and query environmental data and the second focused on environmental decision-making and the use of environmental indicators and indices. These datasets and activities were developed to enhance students' appreciation of some of the issues involved within environmental decision-making and to put them in a similar role to a professional environmental manager. Therefore, the practicals were primarily built around an existing dataset, the Environment Agency's River Habitat Survey (RHS), which holds consistent and reliable data for over 15,000 river sites across the UK. The Environment Agency kindly permitted the developers of this practical to reuse and reconfigure these datasets for educational purposes.

Table 4. Learning objectives for physical geography in environmental management

LEARNING OBJECTIVES

Group A: At the end of this unit, you will be able to display *knowledge* and *understanding* of:
- The relationships between physical and human processes in linking people to their physical environment.
- The application of geographical knowledge to the understanding of environmental management problems, and to the sustainable management of biophysical environments.
- The value and need for multi-disciplinary approaches in advancing knowledge.

Group B: At the end of this unit, students will be able to display a range of *subject specific intellectual skills*:
- Analyse critically the literature at the intersection between human and physical geography.
- Assess the merits of contrasting theories, explanations and policies in the area of environmental management.

Group C: At the end of this unit, students will be able to display a range of *subject specific practical skills*:
- Assemble, analyse, and understand data at the intersection between human and physical geography using appropriate computational approaches.
- Understand the ways in which data of various types can be combined and interpreted in support of environmental management.

Group D: At the end of this unit, students will have developed a range of *transferable/general (key) skills*:
- Pursue knowledge in an in-depth, ordered, and motivated way.
- Provide fluent and comprehensive written reports on complex topics.
- Confidently use a range of relevant forms of IT software.
- Marshall and retrieve data from library and Internet resources.

The practical topics therefore are restricted to river management examples. This is not seen as a negative aspect however as a broad range of issues are explored including the science of river environments highlighting the importance of river forms and processes, policy-driven habitat and species conservation and preservation, resource-based decision-making for environmental improvement, and the selection of river sites for leisure use.

A main advantage of the e-learning practicals was that they used actual environmental data, which provided students with the opportunity to develop more practical and problem-based skills within an independent-learning situation (Stefani et al., 2000; Laurillard, 2002; Biggs, 2003). Although the practical elements (and specifically the RHS dataset) were introduced within the lecture component of the unit, students were asked to work through the practicals at their own pace and independently. However, it was recognized that students might require a higher degree of support and therefore clinic sessions were held for both of the practicals to assist them with problems and to answer queries.

The e-learning sessions had various assessment exercises. There were straight multiple choice tests, tasks where the students had to use newly acquired skills to query data in order to answer questions, as well as text-based questions, where the students needed to use interpretation and explanation skills. These different types of assessment and different types of tasks (assimilative, analytic, communicative, and reflective [Laurillard, 2002]) were selected not only to develop and assess different skills, but also to address and accommodate different learning preferences and facilitate the engagement of all students.

In addition to introducing students to real world data, the practicals were designed to introduce a broad range of issues including asking students to think through the scientific, policy and economic dimensions of a scenario and how to resolve the potential conflicts of interest that may arise.

This was achieved by initially introducing them, through the first e-learning practical, to different skills for evaluating and assessing environmental information, such as the use of spreadsheets and databases. For instance, students were asked to undertake advanced searches for information and then use different tools to investigate and analyse this data. This allowed the students to come away with a range of practical skills. Following the acquisition of these skills, more in-depth examples were introduced through which students assessed the science of the problem, the associated policy issues, and also how they would provide solutions using limited resources. For example, students were asked to decide whether they would concentrate on the redevelopment of a small number of river reaches to improve the worst cases and get them to a high standard, or spread the financial resources wider and bring a larger number of reaches up to a lower standard.

The value-laden nature of making environmental decisions is a well-known and accepted difficulty. The second e-learning practical concentrated on providing the students with different skills and techniques that aim to make environmental decision-making more objective, transparent, and fair. The practical introduced the use of indicators and indices to assess environmental data with a specific problem in mind. Students were asked to not only apply an environmental index provided to them, but also to design and implement their own index for a specific scenario, namely identifying those river sections best suited for recreational activity. They were asked to use all of their scientific knowledge to answer a specific policy question, before using the index to select specific sites for the allocation of resources.

On the whole, feedback from student questionnaires (a sample of 43% of the student cohort), informal feedback, and staff perspectives suggested that the student experience had been enhanced. Table 5 illustrates the output from the student questionnaire for the second e-learning practical and provides examples of the questions asked and responses given. Although not always an appropriate measure, the high attainment of students (the average grade being 72% and the range being between 50 and 94%) indicates that in most cases the learning objectives were met. Many students commented that they had gained additional skills through the completion of the practical; for instance with practical 2, 73% of

questionnaire respondents stated that their skills and knowledge had increased.

From a more practical perspective, the student questionnaires suggested that students appreciated the formative feedback that the practicals provided. Marks and comments from the first practical were returned to the students after two weeks, prior to the second practical and, similarly,

Table 5. Student questionnaire responses for practical 2

For the e-learning activity, ***Environmental Indicators***, which you have just completed, please score each statement.	Number and percentage of the students present giving each score				Mean response
	0 – No	1 – Somewhat	2 – Yes	N/A	
1 There was a full description of the learning activity, including learning objectives.	1 2%	5 9%	49 88%	1 2%	1.9
2 The interface was easy to use.	2 4%	17 30%	36 64%	1 2%	1.6
3 Required tools were included (e.g., database, spreadsheet, note making, bulletin board).	1 2%	4 7%	50 89%	1 2%	1.9
4 The content met the needs of my preferred learning style.	2 4%	25 45%	28 50%	1 2%	1.5
5 The content was relevant, appropriate and clear.	0	29 52%	26 46%	1 2%	1.5
6 All embedded materials were easily accessible.	2 4%	24 43%	29 52%	1 2%	1.5
7 Mechanisms were provided for information and support.	1 2%	7 13%	47 84%	1 2%	1.8
8 Maximum response times to learner queries were defined.	4 7%	16 29%	31 55%	5 9%	1.5
9 The assessed elements of the activity were appropriate for the learning objectives.	0	14 25%	40 71%	2 4%	1.7
10 The activity improved my environmental indicator knowledge and skills.	0	14 25%	41 73%	1 2%	1.7

Data collected and analysed by Karen Fill.

the second practical was returned to students before the end of semester examinations. This permitted both students and staff to gauge progress and to some degree assess their understanding of the issues.

There was initially some resistance and nervousness about the delivery of activities online, as in many cases it was the first time that students had been taught in this manner. This was addressed through reassuring the students that they would be supported through this process and the introduction of clinic sessions. Following this initial hesitancy, the majority of the students coped well with the new medium. Although there were a handful of students who did struggle with the mode of delivery, the majority did appear to appreciate the benefits that the approach allowed—the use of real world data, the practical skills developed and the opportunity to provide formative feedback. Perhaps more importantly, there was also anecdotal evidence (and also evidence from the subsequent examinations) to suggest that some students had benefited from the process of interpretation and making decisions where values were important, as these issues were much more commonly expressed and discussed in a more comprehensive way.

From a wider perspective, this example highlights that it is possible to overcome many of the difficulties in teaching such a broad discipline as environmental management. Although initially labor intensive at the outset, the establishment of online learning elements has proved to enhance the learning experience of students through the introduction of various skills, and the use of real-world data and current problems. However, it is necessary to ensure that students appreciate the relevance of what they are doing and are reminded of the difficulty of environmental management. Rather than appreciating the parallel to real environmental decision-making, it would be quite easy for students to become disheartened and frustrated when trying to answer a problem that does not have an easy and neat solution. Although this

was experienced to some degree, students were reassured and reminded that the purpose of the exercise was for them to "step into the shoes" of professional environmental managers and experience the difficulties involved. Ensuring sufficient student support is essential to the successful teaching of environmental management.

A New Postgraduate E-Module: *GIS for Environmental Management*

GIS for Environmental Management was designed as a new unit for postgraduate students studying part-time by online distance learning. The unit thus attempted to meet the challenge of combining "real world" environmental management problems and decision-making with relevant concepts and reflection through part-time study, alongside workplace experiences. Alongside developing generic, transferable skills, the learning outcomes for the unit were to enable students to:

- Undertake spatial data analysis, including network tracing, terrain modeling, and spatial dynamic modeling of the environment;
- Critically analyze the literature relating to environmental remote sensing and spatial analysis;
- Abstract and synthesize information from a range of different sources in relation to environmental impact assessment;
- Analyze and critically interpret primary and secondary environmental data.

Materials for the course were team-written and made use of learning objects. We have described the process by which these materials were created elsewhere (Wright et al., under review). In brief, however, in writing our courses we have used learning objects, using a definition in line with IEEE: "Any entity, digital or non-digital, which can be used, reused and referenced during technology-supported learning" (IEEE Learning Technology Standards Committee, 2006). As originally designed, the module in question was intended to be worth 15 credit points at the Universities of Southampton and Leeds (a system in which a full Masters degree is worth 180 credits) or 150 hours of student learning. The module was divided into broad units, each of which contained multiple learning objects: discrete, self-contained blocks of learning material. Cross-referencing from one learning object to another was explicitly prohibited, so that learning objects could subsequently be recombined with maximum flexibility. The exceptions to this were the "sacrificial learning objects" designed to introduce each unit. Following an approach taken elsewhere (Fernandez-Young et al., 2006), these did refer to the content of the subsequent objects and were thus designed from the outset to be non-reusable.

Each unit was designed with an accompanying discussion forum and each object contained at least one activity for students to undertake. Some activities required students to post to the associated discussion forum. There was also a general discussion forum and a forum through which students were invited to introduce themselves to one another.

The course was delivered over 10 weeks, once every 6 months. Between October 2005 and December 2007, 19 online students took the module, of whom 5 students were working in environmental management and a further 3 students were employed in areas relevant to the discipline. We assessed the extent to which these students,

already working in environmental management, were able to reflect on their practice and experience through an analysis of discussion board postings. This also provides some indication of the extent to which those students not working in the environmental management sector were exposed to the practices and experiences of their peers who did work in this sector. In assessing discussion board postings, we focused on two deliveries of the course with at least two students from an environmental management background. Table 6 summarizes our findings. Of 148 postings in one delivery of the course, 13 postings involved students reflecting on their own environmental management-related experiences in the workplace and the tutor or other students responding to such experiences. In a second delivery, 8 out of 98 postings involved such reflection. Only a subset of the students working in environmental management referred to their workplace experiences in the discussion boards.

There were 3 mechanisms that triggered this type of reflection:

- As students introduced themselves at the start of the module;
- In response to an *ad hoc* query from the tutor, tangentially related to the formal module syllabus;
- When specific online activities were relevant to activities or experience in a particular student's workplace

Table 6. Summary of discussion board postings during two deliveries of GIS for Environmental Management

Delivery	No. students	Total discussion board postings	Postings related to workplace experience
Oct – Dec 2006	6	148	13
Oct – Dec 2007	6	98	8

The "general discussion" forum, which was made available for student discussions outside of the module syllabus, triggered no such reflection. Nonetheless, there was therefore evidence that overall, the online materials were triggering some student-student learning and reflection on workplace practice.

PERSPECTIVES ON E-TEACHING ENVIRONMENTAL MANAGEMENT

Having described the three environmental management modules developed or enhanced under the DialogPLUS programme, we now review the extent to which the e-learning activities introduced met the challenges in environmental management education outlined in the introduction.

E-Learning to Facilitate Reflection on "Real World" Problems and Practice

The postgraduate module *GIS for Environmental Management* (discussed above) was designed as a wholly online, distance learning resource, which would enable those working in environmental management to reflect on their workplace experience and other students to learn from this experience. Evaluation of discussion board interaction suggests that at least some reflection on workplace activity was taking place by students working in environmental management. Previous work has found that collaborative tools such as e-mail and discussion boards were effective in promoting interaction between student groups located in different states of the USA (Hurley et al., 1999). Here, discussion boards enabled communication between those working in environmental management and those outside the profession, rather than between groups in different locations. In both situations, however, there appear to be elements of constructivist learning, with students "actively involved in constructing meaning from

their experiences and prior knowledge" (Stein et al., 1994).

E-Learning to Facilitate a Broad Multi-Disciplinary Perspective

The blended approach adopted for teaching environmental management permitted the theories introduced within a traditional lecture environment to be reinforced and demonstrated through a practical approach. One of the main successes and achievements of the delivery of environmental management in this way was to challenge the students to make the difficult value-laden decisions for real environmental problems using real environmental data. Allowing students to experience being an "environmental manager" or "decision-maker" forced them to think about how they would justify the decisions that they were making and examine how different stakeholders would view the same situation. In addition, the blended approach to learning allowed different elements of environmental management to be introduced within the lecture series including physical science, economic, conservationist, policy, and social science perspectives, which were then reinforced within the practical elements. Ideally, these ideas would be further revisited within the wider study of the students either within other courses or through dissertation-level study. However, this was outside of the remit of the unit presented. It would also be beneficial for the students to follow through some of the concepts and ideas explored in relation to environmental management within other real-life decision or policymaking; perhaps through attending or participating in stakeholder engagement meetings for decisions being taken. This however is very impractical with high student numbers on the course (often around 120) and the timeframes involved for its delivery. The practicals therefore work well as a substitute for this kind of involvement.

E-Learning in Support of Fieldwork

The UCM course described above used e-learning to support fieldwork, meeting one of the disciplinary challenges in environmental management outlined in the introduction. Three approaches have been taken to using e-learning in support of fieldwork (Fletcher et al., 2007): as a means of preparing students for fieldwork before a trip, during a trip to provide additional context to experience gained in the field, and after fieldwork to provide greater opportunity for reflection, outside of busy fieldwork schedules. Internet-based technologies have generally been used either pre- or post-fieldwork (Ford, 1998) and the discussion forums used following the UCM course fieldwork fit in with this overall pattern. In the UCM course, the latter approach was adopted and student evaluations suggest that it was broadly successful. This contrasts with previous work suggesting that student enthusiasm and impetus can be lost if post-fieldwork activities are delayed too long (Gardiner & Unwin, 1986).

CONCLUSION

These findings have several practical implications for instructors of environmental management. The MSc module case study suggests that online learning can be an effective means of linking those already working in the discipline with those who do not yet have such practical experience. However, where students did draw on their workplace experiences in the discussion boards, this was largely triggered by specific activities within the module. No reflection on workplace experience took place outside of the module framework. This implies that if such reflection is to take place, it needs to be explicitly considered during the course design stage when designing modules for environmental professionals.

The delivery of environmental management works best when students are able to complete as close to real-world examples as possible, including the use of real environmental data on real environmental problems. This not only reinforces the relevance of the exercises but will also introduce the difficulties involved in making environmental decisions. But because students are therefore directly exposed to these difficulties and the inherent uncertainties and frustrations in this process, there is sufficient support to reassure them that they are able to successfully complete the exercises and succeed from a summative assessment perspective.

The three case studies described here are based either on small groups of students (in the case of the MSc module) or on single module deliveries. To gain a fuller understanding of the role of e-learning within teaching and learning, it would be useful to extend the period studied to several academic years. Similarly, in all three case studies, the assessment of the effect of the teaching method changes on learning was largely through indirect means, such as analysis of student discussion board postings. It would be useful to supplement these indirect analyses of student interactions with direct interviews with the students to gain a more rounded perspective on the impacts of introducing e-learning.

E-learning is thus one tool that can be used to help overcome some of the potential difficulties in teaching environmental management. Although previous studies have reported student fears and resistance to the introduction of e-learning materials in natural resource management (Åkerlind & Trevitt, 1999), there was little such student resistance in the three case studies presented here. In the context of part-time, distance education, it can provide a framework for those already employed in environmental management to share their practical experiences with those outside of the profession. In a blended delivery, it can provide a mechanism for reflecting on fieldwork and for exploring the different disciplinary and stakeholder perspectives on a given environmental management problem. Although there are other tools apart from e-learning for overcoming these inherent difficulties in environmental management education, e-learning provides a further, flexible suite of solutions for teaching and learning in the discipline.

REFERENCES

Åkerlind, G. S., & Trevitt, A. (1999). Enhancing self-directed learning through educational technology: When students resist the change. *Innovations in Education and Teaching International, 36*(2), 96-105.

Barrow, C. J. (1999). *Environmental management: Principles and practice.* London: Routledge.

Biggs, J. (2003). *Teaching for quality learning at university.* Buckingham, UK: The Society for Research into Higher Education and the Open University Press.

Boyle, A., Conchie, S., Maguire, S., Martin, A., Milsom, C., Nash, R., Rawlinson, S., Turner, A., & Wurthmann, S. (2003). Fieldwork is good? The student experience of field courses. *Planet Special Edition 5: Linking Teaching and Research in Geography, Earth and Environmental Sciences,* (11), 48-51.

Bryant, R.L., & Wilson, A. (1998). Rethinking environmental management. *Progress in Human Geography, 22*(3), 321-343.

Clark, G. (1998). Maximising the benefits from work-based learning: The effectiveness of environmental audits. *Journal of Geography in Higher Education, 22*(3), 325-334.

Diduck, A. (1999). Critical education in resource and environmental management: Learning and empowerment for a sustainable future. *Journal of Environmental Management, 57*(2), 85-97.

Fernandez-Young, A., Ennew, C., Owen, N., DeHaan, C., & Schoefer, K. (2006). Developing material for online management education - a UK eUniversity experience. *International Journal of Management Education, 5*(1), 45-55.

Fletcher, S., France, D., Moore, K., & Robinson, G. (2007). Practitioner perspectives on the use of technology in fieldwork teaching. *Journal of Geography in Higher Education, 31*(2), 319-330.

Ford, C. E. (1998). Supporting fieldwork using the Internet. *Computers and Geosciences, 24*(7), 649-651.

Gardiner, V., & Unwin, D. (1986). Computers and the field class. *Journal of Geography in Higher Education, 10,* 169-179.

Grumbine, R. E. (1994). What is ecosystem management? *Conservation Biology, 8*(1), 27-38.

Hogan, K. (2002). Small groups' ecological reasoning while making an environmental decision. *Journal of Research in Science Teaching, 39*(4), 341-368.

Honebein, P. (1996). Seven goals for the design of constructivist learning environments. In B. Wilson (Ed.), *Constructivist learning environments* (pp. 17-24). New Jersey: Educational Technology Publications.

Hurley, J. M., Proctor, J. D., & Ford, R. E. (1999). Collaborative inquiry at a distance: Using the Internet in geography education. *Journal of Geography, 98*(3), 128-140.

IEEE Learning Technology Standards Committee. (2006). *Standard for information technology, education and training systems, learning objects and meta-data.* Retrieved November 10, 2006, from http://ltsc.ieee.org/wg12/

Jimenez-Aleixandre, M. P. (2002). Knowledge producers or knowledge consumers? Argumentation and decision-making about environmental management. *International Journal of Science Education, 24*(11), 1171-1190.

Kerres, M., & De Witt, C. (2003). A didactical framework for the design of blended learning arrangements. *Journal of Education Media, 28*(2-3), 101-113.

Laurillard, D. (2002). *Rethinking university teaching: A framework for the effective use of educational technology* (2nd ed.). London: RoutledgeFalmer.

Munowenyu, E. M. (2002). Fieldwork in geography: A review and critique of the relevant literature on the use of objectives. *Educate, 2*, 16-31.

Prasad, P., & Elmes M. (2005). In the name of the practical: Unearthing the hegemony of pragmatics in the discourse of environmental management. *Journal of Management Studies, 42*(4), 845-867.

Quality Assurance Agency for Higher Education (QAA). (2000). *Geography: Subject benchmark statements.* Cheltenham, UK: Quality Assurance Agency for Higher Education. Retrieved January 31, 2005, from http://www.chelt.ac.uk/gdn/qaa/geography.pdf

Salmon, G. (2000). *E-moderating: The key to teaching and learning online.* London: Routledge.

Spicer, J. J., & Stratford, J. (2001). Student perceptions of a virtual field trip to replace a real field trip. *Journal of Computer Assisted Learning, 17,* 345-354.

Stein, M., Edwards, T., Norman, J., Roberts, S., Sales, J., & Alec, R. (1994). *A constructivist vision for teaching, learning, and staff development.* Detroit, MI: Detroit Public Schools Urban Systemic Initiative. Retrieved March, 18, 2008, from http://www.eric.ed.gov/ERICWebPortal/content-delivery/servlet/ERICServlet?accno=ED383557

Stefani, L. A., Clarke, J., & Littlejohn, A. H. (2000). Developing a student-centred approach to reflective learning. *Innovations in Education and Teaching International, 37*(2), 163-171.

Wright, J. A., Treves, R. W., & Martin, D. (under review). Challenges in the reuse of learning materials: Technical lessons from the delivery of an online GIS MSc module. *Journal of Geography in Higher Education.*

Chapter VII
Earth Observation:
Conveying the Principles to Physical Geography Students

Louise Mackay
University of Leeds, UK

Samuel Leung
University of Southampton, UK

E. J. Milton
University of Southampton, UK

ABSTRACT

In our experience of earth observation (EO) online learning we highlight the usefulness of the World Wide Web in terms of its software, functionality, and user accessibility for developing and delivering a range of activities and delivery modes to both undergraduate and advanced learners. Through the mechanism of developing teaching materials and adapting them for the online classroom, EO learning can become highly interactive and well-illustrated by linking to online image processing software and relevant image data, make use of the Web's graphical interface to reinvigorate DOS-based remote sensing programs to be more student-friendly, and with the advent of collaborative Web software, such as Wiki, provide a networked community for EO learners. In this chapter we showcase a variety of delivery modes for our EO materials—online lectures delivered within a blended learning module for the undergraduate to individual online activities (remote sensing practical exercises and an electronic learning diary) for the advanced EO learner. Examples of our learning materials are discussed in this chapter to show how adapting to online delivery and making use of Web technology has supported our teaching of EO.

INTRODUCTION

Earth observation (EO), the science of remotely investigating the earth's surface from airborne or satellite imagery, is a topic ripe for learning online. It is a visual, contemporary topic in geography, which relies heavily on computer-assisted digital processing and analysis. EO e-learning can encompass a variety of learning activities from comprehensive content-oriented modules to smaller topic-specific e-activities. In this chapter we illustrate the wealth of online materials that we have created: materials aimed for the novice undergraduate and for the advanced EO learner. We discuss the development and delivery of an undergraduate module delivered predominantly online and illustrate how this module is greatly enhanced by the media-rich nature of the Web. We also show how the World Wide Web is of benefit as a highly adaptable medium where legacy and DOS-based EO programs and courseware can be adapted to provide a user-friendlier interface. We illustrate how, with the advent of Wiki software, it is now possible to make the Web a learner community in which participants in field courses or other group-based activities can contribute equally and work collaboratively. From our experience of developing and delivering a variety of EO e-learning materials we provide knowledge for the student and guidance to the aspiring e-learning tutor.

The chapter will address the need for good online EO materials, explain the structure and delivery of the blended delivery module *Earth Observation of the Physical Environment* with an online lecture example and illustrate our smaller e-activities: *Atmospheric Correction, Visualizing Directional Reflectance,* and use of an *Electronic Learning Diary.*

BACKGROUND

What About EO Tutorials Already on the Web?

We want to introduce EO to the undergraduate student who wishes to acquire good quality EO image processing skills, the student with more than a passing interest in image data. For the person asking, "How did they do that?" or "What does NASA's image of the day mean?," the content of existing Web tutorials on EO are of interest to the novice user. In terms of conveying sound principles of EO to the undergraduate student, learning materials based in the foundation of EO as a science that also provide skills and techniques for the advanced EO learner are necessary. The examples we show in this chapter provide that level of learning: both in the basis of the science and in the higher-level detail.

Why Translate Traditional Courses and Create EO Online Materials?

Is EO as a topic ripe for online learning? We would argue, absolutely! EO more than most topics benefits from visual examples, which capture both the spatial and spectral richness of the subject. Add to this the availability of video images to represent dynamic phenomena and the case becomes compelling. Learning in an online mode takes advantage of the Web – we can access image databases or spectral data within the learning materials. That means for the student the subject matter is immediately relevant, well illustrated, and accessible.

What are the EO E-Activities?

EO is one of the core geography sub-areas of the DialogPLUS project. Within our e-learning project online learning materials were created from existing courses consisting of face-to-face lectures and lab-based practical exercises. Having created EO learning materials for the project we were then given the opportunity to usefully embed them in the curricula and support our teaching. Nuggets (our term for a reusable learning object and explained in further detail in the Preface) developed from original EO modules were then delivered in a different format: online and distance-taught.

Online EO in Higher Education

More and more EO scholars are improving teaching through multimedia (Weippert & Fritsch, 2002; Höhle, 2004; König et al., 2005). Many institutions are now introducing the application of online technologies and delivery methods for EO teaching. From the development of e-learning modules in spatial data management (Kruger & Brinkhoff, 2005) to the embedding of Web-based e-learning environments in photogrammetry and remote sensing (Pateraki & Baltsavias, 2005), a trend does seem to be appearing that the move to an online mode of learning for earth observation is being eagerly embraced by teaching faculty. The use of computer-based materials in EO is not new, but has been transformed by two developments in particular: enhanced low-cost (or free to user) access to high-quality EO data and high-speed Internet access from the desktop. Previously, the use of computer-based materials in EO was largely limited to teaching the basics of image processing, often using very small subsets of images of areas of the world unfamiliar to the learners.

Since the late 1990s EO has become a topic for distance learning—with the intention of disseminating EO technology and enhancing undergraduate education (Tempfli, 2003; Höhle, 2004; Ferreira et al., 2005). It could be argued that courses in EO have then quickly adapted to an online delivery more than most as hands-on exercises have been made available through tutorials that make use of online free image processing software and the use of selected datasets

(images, digital maps, topographic data) from readily available public sources. Allied to this the collaborative nature of international students bringing local knowledge (and with that EO data) to the group learning environment (Ferreira et al., 2005), the move to online learning for EO as a topic does seem to be proving very successful as these elements provide very flexible and interactive training (Weippert & Fritsch, 2002; Kruger & Brinkhoff, 2005).

Interestingly, the use of free image processing software by EO scholars for research and teaching purposes and by online learners has created an additional advantage to the EO field, that being the improvement of open-source image processing software, in some instances surpassing the basic functionality of commercial remote sensing software systems like Erdas IMAGINE or ENVI; for instance, the development of the freely available RAT (Radar Tools) to simplify the processing of complex SAR (Synthetic Aperture Radar) data (König et al., 2005). This has fed back to the commercial world in some cases, as mainstream EO software has benefited directly by incorporating advanced tools created by users, for example the IDL routines contributed by users to the ENVI package (http://www.ittvis.com/codebank/index.asp).

Historically, we tend to see the discussion of EO online learning examples in terms of size of learning object: from full online modules that tend to evolve from distance courses (Weippert & Fritsch, 2002; Tempfli, 2003; Ferreira et al., 2005; Kruger & Brinkhoff, 2005) to individual activities/tutorials that adapt Web-based technologies to create interactive processes and virtual experiments (Höhle, 2004; König et al., 2005). In many cases, and irrespective of granularity of the learning object, the development of EO online learning has developed in the wake of e-learning initiatives at the project or institutional level (e.g., the e-learning project Eye Learn at the Swiss Federal Institute of Technology [Pateraki & Baltsavias, 2005]) or from larger international bodies such as the European Organization for Spatial Data Research (EuroSDR, formerly OEEPE) (Höhle, 2004).

The push to online delivery in EO has sparked the imagination of many teachers to develop courses and activities that take advantage of the highly visual nature of the EO topic, its ease of adaptation with multimedia technology, and the freely available nature of useful example data, processing software, and other online teaching resources. Freely available Web tutorials on EO such as the Remote Sensing Core Curriculum (http://www.r-s-c-c.org/) and the Remote Sensing Tutorial (http://rst.gsfc.nasa.gov/) are very useful resources (see the online tutorial links at the end of this chapter). However, as successful as established online remote sensing teaching materials are, their rigid structure and ordering, for example, the volumes and chapters of the Remote Sensing Core Curriculum, can be difficult to repurpose in other geography courses of which EO and remote sensing may constitute only some parts of rather than the whole curriculum.

Where a move to online delivery (which may include adaptation of existing online resources) for EO teaching has evolved, the success of embedding within the curriculum in terms of successful teaching of principles and techniques to undergraduate students has not been comprehensively evaluated. In our case, we have evaluated the student experience of learning EO in an online mode (see Chapter XIII: Evaluating the geography e-learning materials and activities: Student and staff perspectives) and show the effect on student learning when delivery of a full undergraduate module is translated from traditional to blended delivery – providing us with useful indicators on how to improve our delivery and advice for the novice e-tutor (see later). In this chapter we showcase a full range of EO nuggets (an online lecture from a blended undergraduate EO module and, by contrast, advanced EO Web-based activities), developed to provide an adaptable knowledge base for the aspiring EO tutor to draw upon.

TEACHING EARTH OBSERVATION ONLINE

In this section we introduce some of our online activities: an online lecture from a blended learning module introducing the physical principles of EO and online activities in the advanced topics of Atmospheric Correction, Visualizing Directional Reflectance, and use of an Electronic Learning Diary. In addition to the learning resources developed from within the project, we shall also share our experience in embedding other emerging Web technologies such as Java Applet and Wiki in the teaching and learning process.

An Online Module for the Physical Geography Student

The blended EO module example is a Level 2 undergraduate module entitled *Earth Observation of the Physical Environment* delivered to students at the School of Geography, University of Leeds. The module is designed as an introductory course for geography undergraduates in the physical principles of EO; the aim of the module being to provide the basis for understanding the role of EO in physical geography applications. In this blended learning module the online lectures are supplemented by campus-based practical exercises designed to introduce the image processing techniques behind the physical principles of EO (requiring the use of lab-based image processing software). The practical exercises for this module are still presented on campus as we would argue that certain aspects of EO benefit from learning in a traditional mode – for instance when acquiring technical skills in image processing. In our experience the novice student learner greatly benefits from on campus assistance during image processing practical exercises. The module is supported through e-mail contact by the module tutor.

What are the EO Online Lectures Like?

The online lectures adapt existing EO lecture material, but with a slightly more comprehensive narrative so that students can follow the lectures clearly in their own time. The online versions are designed to be visually appealing and accessible; as the look and feel of an online course can be as important as the content and skills it is expected to communicate. As Boshier et al. (1997) would say, appearance can "make or break" an online course.

The materials adhered to a consistent and systematic use of fonts, white space, logos, metaphors, and figures that break up large tracts of text. The online materials were created and stored in the University of Leeds virtual learning environment (VLE) Bodington where lectures and associated resources are "rooms," which branch off through virtual doorways from a parent "room" within that housing the online module (see Figure 1).

The front page of the module (Figure 1) introduces the materials, is easy to load, and provides clear entry/navigation to the learning activities.

It is easy to find your way around from this page. All guiding buttons are visible and options are available for easily exiting, plus saving and downloading of materials for off-line work.

More specifically, the module lectures have a different "tone" to that of face-to-face lectures in order for students to work unaided. The narrative is very descriptive (not overwhelming), warm, and accessible (see Example 1 later in this chapter). When designing and writing these materials, the attention given to the lecture's narrative was paramount as the online student would be engaging with the lecture independently.

The online versions of the lectures are enhanced with added graphical detail and links to other online resources (e.g., image databases) to add to the learning experience and to provide very clear and visually stimulating EO examples. So in effect using the functionality of being online to enhance teaching and interest in the new EO student.

Each lecture (independent from campus practical exercises) is designed to take the student one to two hours to complete and is structured with

Figure 1. The front page of the module room in the University of Leeds VLE where all class resources originate from

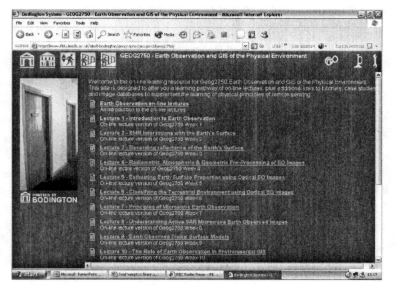

an introduction, learning outcomes, signposted sectioned content with specific query points, links to relevant reading/online resources, lecture summary, and linkage to the lecture that follows. Through the addition of open queries, which in some instances require additional research and postings to the class discussion board, and suggested readings, this consolidates the introduced topics and directs additional self-study. Learners can also test and review their understanding of the material using the self-assessment exercise provided with each lecture.

The entire online component of the module (series of 10 lectures) also comes with a separate introduction, which provides the student with the lecture contents, a suggested method with which to address the lectures, affiliations, and points of contact/correspondence.

During delivery of the module, the online lectures are supplemented by on-campus practical exercises. They are designed to reflect weekly lecture content and can be used as a form of assessment. Weekly practical exercises motivate the online student by providing a schedule for accessing and learning the material. Most importantly, the weekly practical exercises provide face-to-face contact with fellow students and on-campus technical support. However, the online lectures are designed as learning units that can be used independently from the practical exercises.

Resources that Aid the Learning of Lectures Online

In any online course it is vitally important to provide a variety of ways for students to access and interact with information (Schweizer, 1999). In this EO module we provide an assortment of activities and resources to provide a dynamic online classroom delivered via a VLE.

Within the module "room" (Figure 1) resources were added to supplement the material content and to manage the module. As well as the main teaching resources for the module, additional learning resources were added; that is, links to online examples, tutorials, and image resources, all so that the students could supplement their learning of the lectures and fuel their interest in the topic.

So now that we have introduced the module content, let's take a walk through one of our lectures (Example 1).

Example 1: Reflectance Characteristics of Earth Surface Types

Opening Lecture 2 from the main contents page (Figure 2) we can get a feel for the look and design of an online lecture. When we first open Lecture 2, we can see the title, aim and list of contents of the lecture. We are then introduced to the content in Lecture 2, signposted for ease of navigation (sections are hyperlinked to the first list of contents). Working through the lecture content (Figure 3), we are introduced to the lecture concept that earth surface land covers exhibit spectral behavior (reflectance of electromagnetic radiation at different ranges and in different intensities) based on their physical and chemical composition. For example, the chlorophyll content of green vegetative matter produces a characteristic increase in reflectance from the red to green visible light transition in the spectrum. Alongside these new facts, examples of image spectral data of land surface cover types are provided.

Once the concept has been introduced, the student is then asked to reflect and apply their newfound knowledge (but no assessment at this point). Query points (Figure 4) in the lecture ask the student to look at reflectance plots of surface types and asks why they exhibit the characteristics they show. A suggestion, but not the full explanation is provided; links to further resources where the student can research the material (e.g., a hyperlink to relevant pages in an online spectral library) provide an answer to the query. The understanding of land cover reflectance and spectral properties is a recurring theme in the entire module

Figure 2. The "look and feel" of the first page of a lecture with a clear breakdown of the content allowing for ease of navigation

Figure 3. Introducing basic principles of EO in online lecture 2 of the module – spectral reflectance of land surface cover types

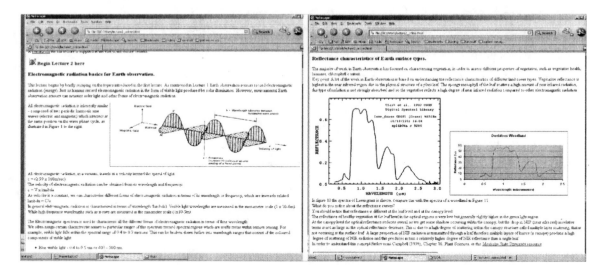

and a very important element in understanding the physical principles of EO. (Similarly in other lectures [Figure 5] action points prompt students to access image database material or further case study material and post their findings to a class discussion board. Random student groups can be set different query tasks and asked to post findings to the discussion room.)

Scrolling to the end of Lecture 2, the introduced concepts are summarized, the lecture is linked to the next in the series and a hyperlinked (to the home library) reading list is provided. Fol-

Figure 4. Introducing query points in online lectures

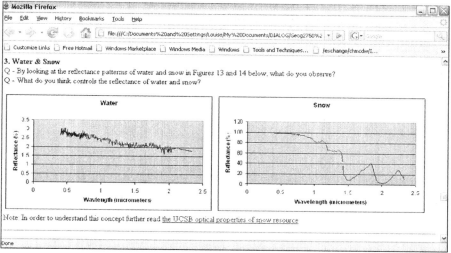

Figure 5. Linking to image data sources: Providing a strong visual example to the introduced topics

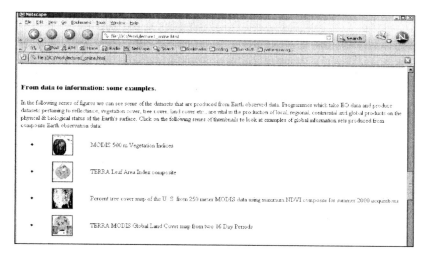

lowing the lecture students can then assess their understanding and revise the material using the lecture self-assessment, for this lecture an online multiple choice question (MCQ) with feedback is provided (Figure 6).

So in the format introduced in Example 1 lectures are concise units in a style to engage the student with the material, but with the additional embedded resources and a lecture self-assessment test they consolidate and test the student's understanding. We have provided you with an overview of an online lecture from a blended learning undergraduate module in earth observation. In our next example we showcase online EO materials of a different nature, which cover smaller and more advanced individual EO topics.

Figure 6. Testing understanding of the introduced content – the lecture self-assessment test

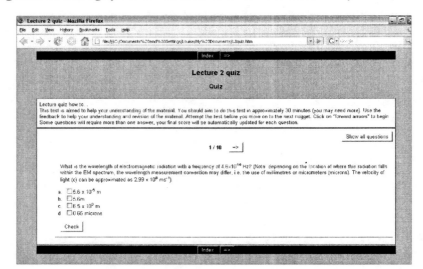

E-Activities in Advanced EO Topics

The EO e-learning development work with our DialogPLUS partners at the University of Southampton is comparatively fine-grained in terms of granularity (Hernández-Leo et al., 2006) and characteristically focuses on a specific activity or task. The work supports the teaching of a level 3 module entitled *Calibration and Validation of EO data* aimed at final year undergraduates and taught master of science (MSc) students.

This module aims to inform students how to acquire and assess accurate and reliable data from the terrestrial environment using modern optical remote sensing techniques. In particular, the quality and integrity assessment of remotely-sensed imagery and the validation of products subsequently derived from those data are examined in detail. To vividly explain and demonstrate the theories and techniques underlying quantitative remote sensing and image processing, a series of computing software and tutorials are embedded into the course curriculum. Some of them are well-established DOS-based resources, but their primitive command-prompt interface makes

them less accessible to students who are used to the conveniences of the graphical user interface (GUI) of Apple Macintosh or Microsoft Windows. The arrival of the Web in the early 1990s has not only revolutionized network computing but also popularized the uses of hypertext and scripting languages for communicating and sharing information. At the same time, more and more institutions opt for a managed desktop environment (MDE) to combat the threats posted by viruses and worms introduced by the Web. MDE increases the network security by forcing a unified system configuration in which non-standard software packages cannot be installed locally. It was against this backdrop of resource accessibility and network security, together with the university's desire to formalize the use of the VLE for teaching, we started to adapt existing resources including those in atmospheric correction and the visualization of directional reflectance; and explored how electronic learning diary technologies can be used to support students.

Atmospheric Correction

Remote sensing from satellite platforms is only possible because the energy from the ground surface is relatively unaffected by its passage through the atmosphere. This is achieved by choosing regions of the electromagnetic spectrum that are relatively transparent, such as the microwave region, or the "atmospheric windows" in the optical region. However, this does not mean that the signal measured from space is identical to that leaving the ground surface, and correction for the effect of the atmosphere is an important topic in quantitative remote sensing.

There are several different approaches to atmospheric correction in the optical region, ranging from simple empirically-based methods to those based on highly-complex radiative transfer models. In developing the atmospheric correction nugget described here, we have sought to raise students' awareness of the more advanced radiative-transfer (R-T)-based approach, and in particular, to open their eyes to the limitations and assumptions of this approach. This is important, because increasingly R-T models are being incorporated into commercial GUI-based image processing packages and the danger is that students will adopt them in an uncritical fashion.

The stimulus for this nugget came from two learning tools developed by Milton (1994a; 1994b) to teach the physical principles of remote sensing. The first was a folded card slide-rule designed to teach the basics of energy-matter interactions with the surface and the atmosphere, and the second was a spreadsheet to introduce students to the "Chavez-modified Dark Object Subtraction" method of atmospheric correction. These have been used very successfully in the undergraduate course at Southampton for over ten years. The nugget described here is an attempt to extend the learning outcomes achieved with the atmospheric correction spreadsheet, through exposure to a widely used radiative transfer model, the "Second Simulation of the Satellite Signal in the Solar Spectrum (6S)" (Vermote et al., 1997).

The Fortran computer code for the 6S model is available in the public domain at ftp://kratmos. gsfc.nasa.gov/pub/6S/, but it is a very large program, comprising of over 100 subroutines and over thirty-four thousand lines of code. A GUI-based front-end is available (http://www-loa.univ-lille1. fr/SOFTWARE/Msixs/msixs_gb.html), but this requires knowledge of Linux, which most undergraduates do not have. However, a simplified MS-DOS version of the 6S model has been created by Mauro Antunes (http://www.ltid.inpe. br/dsr/mauro/6s/), formerly of the Brazilian Space Centre (INPE). This is much more accessible, although the command-line nature of MS-DOS presents a barrier to contemporary undergraduates (Mitchell, 2004).

To help students to focus on understanding and applying the model rather than battling in the unfamiliar MS-DOS environment, a more friendly and informative Web form (http://www. dialogplus.soton.ac.uk/eors/6s/inputform.html) was devised and made available through the institutional Blackboard VLE. The form aims to simplify the parameters input stage for the execution of the correction program.

To run the 6S program in the DOS command line environment, a parameter file and a control file must first be prepared and stored in the same directory as the executable file (i.e., 6S_atms_Corr. exe). The parameter file contains the input choices related to the modeling and satellite conditions. The control file only has four lines, which are the names of the two input and two output files needed for the running of the 6S program. If one of the files is absent, the program will not work and will return at least two file not found errors such as that below.

```
forrtl: The system cannot find the file
specified
forttl: severe (29): file not found, unit
9, file...
```

Moreover, if there is a syntax error, as minor as a typographical error or an extra whitespace, the program will not work either and result in an error like the one below.

```
forrtl: severe (59): listed-directed I/O
syntax error, unit 5, file...
```

For students who have limited experience with DOS, these error messages are not particularly useful for them to diagnose the problem and to correct their mistakes. There is also a lack of interactive guidance and checks in the DOS environment for them to verify the parameters file prior to the program execution. The introduction of the interactive online form at the input stage has overcome these two major shortcomings. After completing the form, the two correctly formatted text files required for the program are generated. All students need to do is to type the command in the DOS command prompt to complete the process.

```
C:\Work\6S>6S _ atms _ Corr
```

The online form does not replace but complements the existing 6S program in DOS. First, the software program and data files are easily distributed to students via the Web page where the online form is hosted. Sending files via e-mail attachment is no longer feasible as increasingly e-mail programs are automatically treating any executable file as a potential threat. Large file transferral via e-mail can impinge on network bandwidth and is therefore not recommended by the central computing services of the institution. While placing the files on the network drive could be a solution, a localized drive is less accessible compared to the Web, which is particularly important for off-campus and distance students. The second advantage is the amount of additional information that can be provided on or linked from the online form. It is the very nature of hypertext that a Web page can be easily linked to other re-

sources. As literature and library catalogues are increasingly being digitized, the online version can offer the attraction of direct linkage and the possibility of extensible instructional steps that may be required by some but not all students. Such accessibility is not easily achieved from printed handouts, which are often didactic and linear in nature.

Besides these two improvements that are generally applicable for most online learning resources against their paper-based counterpart, the greatest asset of the online input form for the 6S program is the ease by which students can generate the necessary parameter and control files. The program requires students to create two text files at specific stages, one of them containing 15 parameters that are used to define the atmospheric model (see Figure 7). Some of these parameters are self-evident, for example, the date the image was acquired. Others require the student to access other sources of information on the Web (e.g., utilities to calculate the solar zenith angle at the time of data acquisition), and some require the student to make value judgments or inferences based on their prior knowledge. For example, one of the parameters is the type of aerosol distribution, which depends to some extent upon the latitude of the site and the season. Given the available choices and the accompanying explanation, students should be able to judge whether a generic aerosol model (e.g., mid-latitude rural maritime) is the most appropriate, and if not, how they would determine the parameters most applicable to their particular image.

In summary, the greatest strength of the online form is the ability to check and verify inputs before the program is executed. The form is designed to provide students with intuitive feedback throughout the process. From student reaction during the practical exercise and subsequent postings to the discussion board, it is clear that the online form is easier to work with compared to the DOS command line prompt. More importantly, students are able to focus on the assumptions and results of the

Figure 7. *A screenshot of the online 6S input form where students can enter the parameter values of their chosen atmospheric correction model. The remarks on the rightmost column of the form help students to determine the suitable range of input values (source: http://www.dialogplus.soton.ac.uk/eors/6s/inputform.html).*

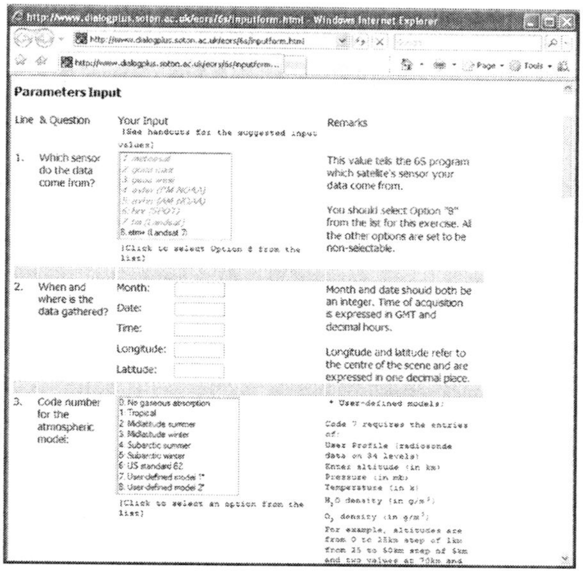

model, rather than be side-tracked by mending any syntax errors they may create.

The atmospheric correction nugget is generally used as part of a computer-based practical exercise in which students are provided with an extract from a Landsat ETM+ image which they have to correct for the effect of the atmosphere without any contemporaneous ground data (see Figure 8). This replicates the situation that many of them encounter when doing their own research projects, so is of immediate practical benefit to them.

A detailed example of the nugget applied to a Landsat ETM+ image is available online at http://www.ncaveo.ac.uk/special_topics/atmospheric_correction/example1/.

Visualizing Directional Reflectance

Casual observation of the stripes on a newly mown lawn reveals something fundamental about the way in which the reflectance proper-

ties of earth surface materials vary according to the geometry of viewing and illumination. The grass leaves are identical, but the reflectance of the grass canopy varies, primarily because the amount of shadow presented to an observer varies. In the same way, a patch of broadleaf woodland does not have a single "spectral signature," but has an ensemble of reflectance properties that vary with view/illumination angle, and which may be represented by what is known as the bidirectional reflectance distribution function (BRDF). It is not possible to measure the BRDF directly, as it is a mathematical function defined over infinitesimally small angular increments, but it can be modeled mathematically, and it can be approximated by the bidirectional reflectance factor (BRF), which can be measured in the field with a spectroradiometer.

This nugget was created to help students visualize the way in which the reflectance of a bare soil surface varies according to the angles from which it is illuminated and viewed. A sec-

Figure 8. Results from the atmospheric correction nugget applied to a Landsat ETM+ image. The left-hand graph shows a spectrum from an area of grass in the raw data (data points are squares) compared with the corrected image (data points are triangles). The right-hand graph shows the same from an area of water: raw data (data points are squares), corrected (data points are triangles).

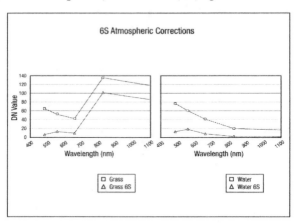

ond objective was to introduce the idea that this variation can be represented in a mathematical model, and that the model could then be used to estimate the BRDF, which is a fundamental physical property of the surface.

The learning tool comprises two parts. First, there is a visualization module that generates assemblages of objects based on geometric primitives (cones, spheres, cubes), which are then rendered as realistic three-dimensional scenes using a public domain ray-tracing program called Persistence of Vision (www.povray.org). Second, there is an analytical module, which uses an established BRDF model (Cierniewski, 1987) to calculate the relative reflectance of the scene at various view angles in the solar principal plane. The outputs of these modules are then combined on screen so that students can see both a visual representation of the surface and its predicted BRDF. They can then vary the conditions of the simulation, for example by testing how the BRDF would vary if the properties of the surface were to vary. A simple bare soil surface is simulated, and the properties that can be changed by the student are (1) the elevation of the sun, (2) the roughness of the soil surface (i.e., average size of

modeled spheres) and (3) whether the surface is shiny or matte. Students first visualize a surface configuration, and then use their understanding of physical processes to predict what the BRDF would be like, before finally running the model and viewing the results.

The first version of this learning tool was created in 1992 by Milton and Cierniewski and used an early hypertext program called Guide. It was later re-coded for the Asymetrix Toolbook package. Although this allowed the distribution of the package via a run-time version, it became increasingly difficult to maintain and impossible to update, consequently thought was given to creating a major new version, optimized for the Web.

POV-Ray was retained as the means to generate the soil surface visualization, and JavaScript was used to provide the user interaction and to link the visualizations with the outputs from the Cierniewski model (see Figure 9). This provided cross-platform operability and accessibility to most Web browsers.

As with the Atmospheric Correction nugget, the Web version has made the BRDF resources much more accessible. Students no longer need

Figure 9. The screen below illustrates a typical investigation of the effect of surface roughness on the BRDF of a soil surface. The solar elevation angle remains fixed at 60 degrees and the aggregate type is set to "matt" for all runs. The top ray-traced image shows the simulated "smooth" surface, the BRDF of which is shown in the black line (flat, therefore the surface is almost Lambertian). The middle ray-traced image shows the simulated "moderate" surface, the BRDF of which is shown in the green line. The bottom ray-traced image shows the simulated "rough" surface, the BRDF of which is shown in the red line.

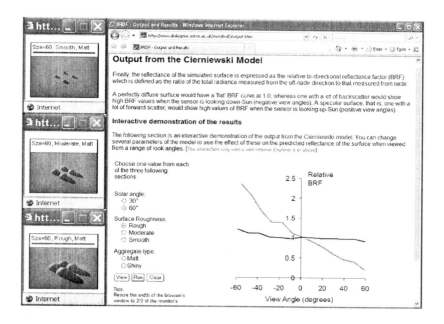

to install any software or viewer plug-in to access the materials. A compatible browser capable of supporting animated content is all that is required. From the tutor's point of view, the content is no longer locked in a proprietary software nor editable only by a specialist. Anyone who has a Web editing tool and basic knowledge of hypertext mark-up language (HTML) can revise and update the resources. More importantly, the full resource set can easily be packaged, repurposed, and uploaded to any mainstream course management system (CMS) or VLE.

BRDF is a difficult concept for many students to understand; yet its effects are evident in the world all around us. This nugget takes as its starting point a commonplace, readily observable phenomenon and then reveals how this can be understood by modeling. Although the main remote sensing outcome is a better appreciation of the role of geometric-optical models, and especially that devised by Jerzy Cierniewski, an important secondary outcome is an appreciation of the wider role of computer-based models in representing the real world and in simulating imagined worlds.

Electronic Learning Diary

The third nugget developed for this project is an Electronic Learning Diary (ELD) to support MSc students learning about remote sensing. The ELD is based on a program called Tiddly-

Wiki, described by its author Jeremy Ruston as a reusable non-linear personal Web notebook. TiddlyWiki is written in HTML, cascading style sheet (CSS), and JavaScript to run on most Web browsers without any additional programs. In the present example, the TiddlyWiki has been used to replace many of the functions of a traditional paper-based notebook that fulfilled two roles, that of laboratory notebook and learning diary. For many years, remote sensing students at the University of Southampton have recorded their progress in a learning diary, which is assessed at the end of the course. The diary records their reflections on the subject matter as well as the results of practical classes and other set activities. The purpose of the ELD was to provide a more convenient format, allowing students to record their thoughts electronically and to link more easily to Web-based and other digital resources, without sacrificing the portability and freedom of expression of a traditional notebook. The TiddlyWiki offers several advantages in this application:

- **Highly portable:** The entire ELD is stored on a USB stick.
- **Searchable:** All the content entered by the student can be searched electronically.
- **Audit trail:** Material entered by the student is automatically date-stamped, making it ideal for assessing whether the learning has been progressive or whether it is the result of pre-deadline panic.
- **Hyperlinks:** Students can link the sections of material they create, and also link to external resources.
- **Customizable:** The appearance as well as the content of the ELD is highly customizable, which appeals to students' desire to express their individuality.

The ELD was trialed in the 2006/2007 session of an international master's level course in Geo-information Science for Environmental Modeling and Management. This course was taken by 34 overseas students who took two compulsory modules in Remote Sensing. Most of these students were early career scientists, al-

Figure 10. The basic Electronic Learning Diary provided to the students in the first week of the remote sensing module

ready familiar with IT, who had come to the UK as part of their professional development training in remote sensing and geographical information systems (GIS).

Figure 10 shows the basic ELD provided to the students in week one of the module. This had been configured by the module tutor to provide a basic three-column layout with a clickable calendar and main menu in the left-hand column, a menu of "housekeeping operations" in the right hand column, and the main content in the central column. Students were given some basic training in adding and linking content during an early lecture on the module, but were then left to develop the ELD according to their own ideas.

Towards the end of the module, the students submitted their ELDs electronically for formative assessment. A feedback session was held with the group to share how different students were using the resource and to discuss any problems that students had encountered. This was very useful for both the students and the teaching staff. For the former it provided inspiration and encouragement as well as solving a few technical issues. For the latter, it revealed technical problems with running the TiddlyWiki software on different versions of Internet Explorer and on Apple Macintosh computers. Most of these were overcome eventually, but this was an unexpected problem that detracted from the overall experience. However, despite these problems, most students were enthusiastic about the ELD and very proud of their final products, two examples of which are shown in Figure 11.

After completing their studies at Southampton, the students moved to the University of Lund to continue their master's course, and as part of their coursework for a module on *Distributed Modeling* the group decided to use the Tiddly-Wiki program again. Each member of the group agreed to research and summarize five topics, which were then combined by eight students and one chief editor to produce the TiddlyWiki shown

in Figure 12 (available online at http://www. natgeo.lu.se/Personal/Jonas.Ardo/ER_2007/student_stuff/gem.html).

How Can Others Access the Resources?

All EO materials are available to download for registered academic users from JISC's digital repository Jorum (http://www.jorum.ac.uk). The full EO online lecture series consists of the following 10 topics:

1. Introduction to earth observation
2. Measuring the spectral response of the earth's surface
3. Image acquisition and image structure
4. Radiometric, atmospheric, and geometric processing of multispectral images
5. Estimating earth surface properties from multispectral images
6. Classifying the terrestrial environment using earth observed images
7. Principles of microwave earth observation of the natural environment
8. Understanding earth observed microwave images
9. Earth observed digital surface models
10. The role of earth observation in environmental GIS applications

The advanced EO e-activities available are:

1. Atmospheric correction
2. Visualization of directional reflectance
3. Electronic learning diary

ADAPTATION OF EO TEACHING RESOURCES

In this section we summarize the usefulness of our resources for EO teaching, provide advice on

Figure 11. Example screens from two of the Electronic Learning Diaries produced by students on the 2006/07 remote sensing module

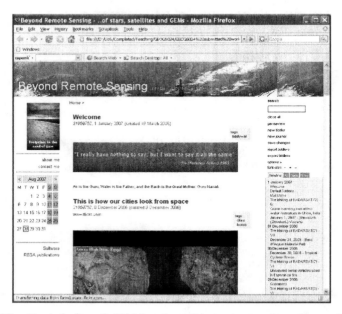

Example A shows a highly visual ELD written in the form of a "blog", and containing many perceptive comments about life as a mature student in Britain, and about the relationship between remote sensing and international politics (Bajinder Pal Singh, 2006).

Example B contained much more textual material, mainly organised into short blocks of information about topics of particular interest to the student, in this case modelling and land cover classification (Xinping Ye, 2006).

All the ELDs made effective use of links to internet sites, but the best were much more than simply lists of links. They provided added value through comments and reflections on the material and about its relationship to material studied in lectures. Many ELDs included information of particular use to the student, e.g., translations of remote sensing terms into their own native language.

Figure 12. The front page of a TiddlyWiki-based resource produced by the group of master's students introduced to this method of creating an Electronic Learning Diary

implementing their delivery and provide lessons learnt from our own experience of delivering the activities.

Why use the Blended Delivery Module in EO?

The blended learning module in EO described above is for a basic/intermediary level—the undergraduate new to the topic of earth observation. The online lectures provide a framework for the understanding of the physical principles of EO and subsequently how such data/processing techniques can be applied in a wider context (e.g., land cover/land use mapping). They are also useful as supplementary material to other geography modules. For instance, many physical geography modules include application examples that incorporate some EO or image-processing element, but do not require an in-depth knowledge of the field. For example, the use of digital elevation models in flood modeling is a popular case study within GIS and catchment dynamics courses; or the use of microwave imaging and analysis in glaciological studies. The provision of these lectures acts as an additional tutorial for those students not enrolled on an EO module.

Due to the design of the lectures they are highly adaptable for different needs. As the series of lectures are clearly signposted and each individual lecture is clearly mapped for navigation through the material, the learner can pick and choose their relevant sections. They can be easily "dipped" into for a specific need (although cumulative knowledge from early introductory lectures is required for the later application-centered lectures). Additionally the entire series of lectures would also make a useful learning tool for dissertation students, who wish to incorporate an EO element into their dissertation topics. In summary the lecture series can be used in its entirety as a complete introduction to EO tutorial or as individual topic-specific lectures.

Why use the Advanced E-Activities in EO?

The three advanced e-activities supplement the online EO module in different ways. The *Electronic Learning Diary* is a general tool adaptable to different levels and for a range of subjects. Creation of the ELD is an open-ended activity and encourages students to take responsibility for their own learning; it therefore fits well with the ethos of online and distance learning. The use of the learning diary has been well received and pointed out by students in various course evaluations to be highly beneficial to their learning. For example, one student found the learning diary "very constructive and helpful" and another was motivated by it to "read additional articles" in order to understand the subject better. As increasingly EO resources in the academic and commercial sectors are available online, the introduction of the electronic diary helped students keep pace with the developments within the field and allowed them to integrate those resources found on the Web more readily. Instead of viewing the requirement of filling the learning diary as a laborious and inflexible task, students valued the process – "really helpful when coming to revise as no real extra work needed to be done" – and a helpful step "to understand lecture and reading content" as the module was being taught.

The two analytical modules deal with topics that many students, especially those from a non-mathematical background, find difficult. The approach taken is interactive, engaging the student in interpreting the results from the models and questioning their assumptions. However, these materials work best when students work in small groups in a conventional practical situation,

bouncing ideas off each other and asking questions. They are not intended for individual study in their present form.

Implementation Issues

The EO materials developed at the Universities of Leeds and Southampton are available for academic users from Jorum (http://www.jorum.ac.uk). However, you may need to adapt them for your learning objectives. Although our activities provide substance, content, and material with which an online student can interact, a developer and tutor must take care to provide relevant resources within the context of meaningful learning.

Learner Support and Engagement (via Assessment and Evaluation)

As with any context of learning (whether face-to-face, wholly online, or blended), it is vital to gauge if students have successfully attained the intended knowledge and to evaluate if the student experience, especially in terms of learning, can be improved.

In both our blended module and with the advanced EO activities, students are assessed in a formal manner through assignments and examination and also in an informal manner through ongoing summative assessment (e.g., MCQ tests), thus providing feedback on student learning.

In a more directed approach our material delivery has undergone rigorous evaluation. For example, in the first run of our blended learning module in academic year 2004/2005 the module was evaluated and reviewed in depth with the aid of pedagogic support from our colleagues at the University of Southampton. A fuller description of the evaluation of this module is discussed in detail in Chapter XIII: Evaluating the geography e-learning materials and activities: Student and staff perspectives. In summary (and without giving too much away) student feedback was particularly positive with respect to design and content of the online lectures. Although the delivery of the full module was challenging in its first run it did provide the students with their first experience of distance learning and student marks were not adversely affected when compared to the previous year's fully face-to-face version of the module. Evaluation such as this helps the process of revision and update of online activities.

Technical Considerations

Many of our learning resources are written in Microsoft Word™ or HTML formats. The textual elements in both formats can be easily transferred to other word processing or Web authoring applications. The graphics are in either GIF or JPG, which again are highly transferable. For tutors who intend to adapt and use the materials in their institution's VLE, it is recommended that they should edit and then reassemble the materials into a standard-compliant learning package. The approach will minimize the editing at the VLE level and thus ensure the interoperability and transferability of the resources for any future use.

The quiz element is slightly different and the adaptation will largely depend on how the quizzes are used. They are currently written in HTML and can be modified using any Web editor. The quizzes at present are all standalone and purely formative so there is no result reporting from them to a server. Should tutors decide to use any of the quizzes for summative purpose, it will be necessary to transfer the questions to the specific assessment engine that is available in their institution's VLE.

Our materials are easy to download as Microsoft Word documents, HTML files, PDF files or JavaScript snippets. The materials can then be easily uploaded within any virtual learning environment, courseware package, or simply placed on a class/module Web site that the would-be tutor designs and manages. Individual files can be read off-line, although some files may be better viewed within a Web browser in order to access

any embedded resources that link to external Web sites. Like any other evolving technologies, newer and more refined Web standards and tools are being made available continuously. It is therefore important that tutors are aware of the issues of browser compatibility and accessibility requirements.

Lessons Learnt from Delivering our EO E-Activities

In our experience the content of e-activities has to be of a high standard in order to inspire the independent learner and to successfully meet the activity's learning objective. However, the delivery method may be one that is unfamiliar and daunting to the undergraduate student. Where the learner is new to distance learning, it may be important to manage their expectations with regards to their ability to learn in a distance-based format. This requires positive advertisement (e.g., student flexibility, one-on-one support with an expert via e-mail) from the home department. In our undergraduate module example some form of face-to-face contact would aid further delivery of the module, for example, by increasing the number of tutorials. This would allay student fears with regards to distance learning and assure students that they were indeed engaging with the material in the correct manner. This is also true in blended learning where face-to-face teaching is the mainstream delivery mode that is supported by individual self-paced learning components.

Where the tutor is also new to distance learning or online delivery, support needs to be in place to provide training and advice from faculty or the parent institution on best practice techniques for design, delivery, and management of online learning.

CONCLUSION

An online classroom needs to be fun, empowering, full of choices for students and a place where students feel that they 'belong.' When you create an online environment that meets these four basic needs, you will create an environment where real learning can take place (Schweizer, 1999, p. 12).

Our EO module fulfills these criteria for a successful online classroom and provides an important element of the curricula for current geography undergraduates. The module provides students with the foundations of EO understanding and image processing skills: a basis for further EO study and dissertation work. The use of Web-based technology increases the flexibility of delivering EO e-activities. The learning of advanced EO concepts is supported by digital technologies through improving the overall quality, broadening access, and widening student engagement via the introduction of an ELD. While it is still too early to assess the educational value of the emerging collaboration tools such as Wikis, the voluntary adoption of the TiddlyWiki approach by some international master's students formerly at Southampton in their work at Lund University in Sweden is a good example to show that many students are willing to take more control of their learning.

By placing our resources online we have been able to experiment with a mixture of delivery styles in teaching and preserve learning materials (vital components of the curriculum) from year to year. Of course, linking to image examples and making use of the distributive nature of the Web are inarguably two important aspects of the teaching and learning of EO. While we may not yet be in the position to substantiate the claim that the Web has vastly improved students' experience, the integration of online learning has led to an improved teaching interface for students to access satellite and aerial photographic data, compare real world examples and, above all, the necessary confidence to work with the emerging technologies in the field of earth observation and remote sensing.

The design and method of delivery of our activities provide successful learning of EO and by depositing our materials within a digital library and allowing open access to them, we invite would-be learners and tutors to use them too. We hope that our introduction to the resources sparks your interest in earth observation as a subject and inspires you to learn online.

REFERENCES

Boshier, R., Mohapi, M., Moulton, G., Qayyum, A., Sadownik, L., & Wilson, M. (1997). Best and worst dressed Web courses: Strutting into the 21st Century in comfort and style. *Distance Learning Education: An International Journal, 18*(2), 327-348.

Cierniewski, J. (1987). A model for soil surface roughness influence on the spectral response of bare soils in the visible and near infrared range. *Remote Sensing of Environment, 23*, 97-115.

Ferreira, H. S., Florenzano, T. G., Dias, N. W., Mello, E. M. K., Moreira, J. C., & Moraes, E. C. (2005). Distance learning courses for disseminating remote sensing technology and enhancing undergraduate education. In G. König, H. Lehmann, & R. Köhring (Eds.), *ISPRS Workshop on Tools and Techniques for E-Learning* (pp. 110-113). Potsdam, Germany: Institute of Geodesy and Geoinformation Science.

Hernández-Leo, D., Harrer, A., Dodero, J. M., Asensio-Pérez, J. I., & Burgos, D. (2006). Creating by reusing learning design solutions. In *Proceedings of 8th Simposo Internacional de Informática Educativa, León, Spain: IEEE Technical Committee on Learning Technology.*

Retrieved March 11, 2008, from http://dspace.ou.nl/handle/1820/788

Höhle, J., (2004). Designing of course material for e-learning in photogrammetry. *International Archives of Photogrammetry, Remote Sensing and Spatial Information Sciences, 35*(B6), *Commission VI, WG VI/2,* 89-94.

König, G., Jaeger, M., Reigber, A., & Weser, T. (2005). An e-learning tutorial for RADAR remote sensing with RAT. In G. König, H. Lehmann, & R. Köhring (Eds.), *ISPRS Workshop on Tools and Techniques for E-Learning* (pp. 28-32). Potsdam, Germany: Institute of Geodesy and Geoinformation Science.

Krüger, A., & Brinkhoff, T. (2005). Development of e-learning modules in spatial data management. In G. König, H. Lehmann, & R. Köhring (Eds.), *ISPRS Workshop on Tools and Techniques for E-Learning* (pp. 18-22). Potsdam, Germany: Institute of Geodesy and Geoinformation Science.

Milton, E. J. (1994a). A new aid for teaching the physical basis of remote sensing. *International Journal of Remote Sensing, 15,* 1141-1147.

Milton, E. J. (1994b). Teaching atmospheric correction using a spreadsheet. *Photogrammetric Engineering and Remote Sensing, 60,* 751-754.

Mitchell, T. (2004). *How do you bridge the CLI vs. GUI gap in app. design?* O'Reilly Digital Media. July 17, 2004. Retrieved August 21, 2007, from http://www.oreillynet.com/digitalmedia/blog/2004/07/how_do_you_bridge_the_cli_vs_g.html

Pateraki, M., & Baltsavias, E. (2005). Eye Learn – An interactive WEB based e-learning environment in photogrammetry and remote sensing. In G. König, H. Lehmann, & R. Köhring (Eds.), *ISPRS Workshop on Tools and Techniques for E-Learning* (pp. 23-27). Potsdam, Germany: Institute of Geodesy and Geoinformation Science.

Schweizer, H. (1999). *Designing and teaching an on-line course, spinning your Web classroom.* Boston: Allyn and Bacon.

Tempfli, K. (2003). LIDAR and INSAR embedded in the OEEPE's distance learning initiative. In A. Grün & H. Kahmen (Eds.), *Optical 3-D measurement techniques VI* (Vol. II, pp.16-22). Institute for Geodesy and Photogrammetry, ETH Zürich.

Vermote, E. F., Tanre, D., Deuzé, J. L., Herman, M., & Morcrette, J. J. (1997). Second simulation of the satellite signal in the solar spectrum, 6S: An overview. *IEEE Transactions on Geoscience and Remote Sensing, 35*(3), 675-686.

Weippert, H., & Fritsch, D. (2002). Development of a GIS supported interactive "Remote Sensing" learning module. *GIS, 14*(9), 38-42.

LINKS TO OTHER EO ONLINE TUTORIALS

http://rst.gsfc.nasa.gov/

(The online version of a classic remote sensing course devised by Nicholas Short, formerly of NASA GSFC)

http://www.gisdevelopment.net/tutorials/tuman008pf.htm

http://www.crisp.nus.edu.sg/~research/tutorial/rsmain.htm

http://ccrs.nrcan.gc.ca/resource/tutor/fundam/index_e.php

http://www.r-s-c-c.org/ (The Remote Sensing Core Curriculum)

http://levis.sggw.waw.pl/Rsbasics/index.html

(An online RS course devised by Jan Clevers, University of Wageningen)

http://www.cas.sc.edu/geog/rslab/751/

http://www.cas.sc.edu/geog/rslab/551/

(A complete RS course devised by John R Jensen, University of South Carolina. The first URL is a set of image processing exercises based on Erdas Imagine, the second is the associated lectures.)

Chapter VIII
Generic Learning Materials:
Developing Academic Integrity in Your Students

Helen Durham
University of Leeds, UK

Samuel Leung
University of Southampton, UK

David DiBiase
The Pennsylvania State University, USA

ABSTRACT

Academic integrity (AI) is of relevance across all academic disciplines, both from the perspective of the educator and the student. From the former perspective there is the need to increase the awareness of AI amongst the student population whilst monitoring and enforcing the rules and regulation regarding plagiarism within their institution. On the other hand, students need a full appreciation of the importance of AI and a clear recognition of the penalties for flouting the regulations in order to steer a successful passage through higher education and on into their professional career. By repurposing learning materials originally developed by the Pennsylvania State University (USA), the Universities of Southampton and Leeds (UK) have developed academic integrity guidelines to support students in their studies and provide an assessment of their understanding of AI concepts. This chapter describes the development of these learning activities and examines the technical and content issues of repurposing materials for three different institutions. It also reflects on the success of embedding the guidelines and assessment in geography programmes at two UK universities, examines the effect of using the online plagiarism detection service, Turnitin, to police plagiarism cases and summaries the lessons learnt in helping geography students to enhance their study skills.

INTRODUCTION

What do we mean by academic integrity? Academic integrity refers to a standard of behavior expected from all members of the academic community in which all work produced should be one's own and if ideas and material are used from other sources, then these should be attributed appropriately. It is not just plagiarism (the representation of someone else's ideas, words or intellectual property as one's own) that is under discussion; hence the broader terminology "academic integrity" is used in this chapter. The term academic integrity is a more positive view of the concept emphasizing the need for correct citation, full referencing, relevant and well-punctuated quotations, and accurately represented paraphrasing. These, in addition to avoiding plagiarism, all constitute academic integrity. Hinman (2002) identifies academic integrity as consisting of five core values required for academic life to flourish: honesty, fairness, trust, respect, and responsibility. Anyone employed or studying in the academic sector is expected to show integrity in their work whether, for example, they are students submitting assignments or academics producing papers. Submitting work that is not wholly your own, or using ideas or quotes from other sources without citing and referencing correctly, is cheating.

Understanding and agreeing to the terms of academic integrity is an important step to help students avoid inadvertent plagiarism resulting from ignorance, lack of understanding, causal attitude, and cultural differences. All of these have become sensitive issues facing the academic community in the light of widening access and an internationalized education market. When cheating has been detected, is the transgression a result of poor scholarship or rather an intention to gain unfair advantage? Larkham and Manns (2002) discuss the difficulties of distinguishing between these two situations and highlight the complexity and practical problems of identifying plagiarism cases and penalizing appropriately.

McCabe (2000) suggests that there is a growing perception among students that many of their peers are plagiarizing to gain unfair advantages without being caught; while most students would prefer a truly honest and level studying environment, many are "unwilling to become moral heroes" and thus reluctantly follow suit. This view of gaining unfair advantage by plagiarizing is supported by studies carried out among staff and students at UK universities by Dordoy (2002) and Barrett and Cox (2005). Student perceptions of cheating and plagiarism are also reported in Ashworth et al. (1997), concluding with the message that it is important to stress to students the positive reasons for correctly attributing work and that students should consider themselves part of a scholarly community. From these studies it can be seen that it is essential for students to realize that academic integrity goes beyond a good grade and that the adherence to it is an important quality in their career and personal development.

This chapter sets out to emphasize the importance of academic integrity across the board of learning, teaching, and research, and describes the repurposing of existing learning resources to guide the academic community in their understanding and application of "good practice". This is achieved through focusing on a case study from the Joint Information System Committee (JISC) and National Science Foundation (NSF) funded DialogPLUS project, part of the Digital Libraries in the Classroom program of work. It draws on experiences of academic staff from one United States of America (USA) and two United Kingdom (UK) universities and explores the development and embedding phase of repurposed learning material from the initial identification of a generic learning resource through to the examination of usage in current study programmes.

BACKGROUND

Academic Integrity Transgressions

The World Wide Web has introduced new and potent temptations for students to pass off the work of others as their own. While it is still inconclusive that Internet technology should be solely blamed for the increasing cases of academic dishonesty offences (Ercegovac & Richardson, 2004; Hansen, 2003), Park (2003) is concerned that the proliferation of paper mills and the cut-and-paste online culture have clearly facilitated digital plagiarism and transformed the Web into a more accessible environment for those who are prepared to cheat. The availability of online resources, such as articles and learning materials, may be too alluring for some and, whether premeditated or accidental, the ease with which whole sentences, paragraphs and pages can be cut and pasted into one's own document without proper references may prove too tempting. Another challenge to consider is the ease at which term papers or assignments can be bought via the Web on essay banks, cheat sites, and paper mills as warned by Austin and Brown (1999), Groark et al. (2001), and Underwood and Szabo (2003). Plagiarism detection software exist that can help identify cases of this kind and will be discussed in greater detail in the next section.

Jocoy and DiBiase (2006) found that 12.8% of 429 assignments submitted by adult students in an introductory course on geographic information between July 2003 and June 2004 contained one or more copied sentences or two or more sentences improperly paraphrased from another online source. This investigation was driven by concerns that active Internet users may be more likely to copy other author's work verbatim without correctly citing the author. In the UK, Szabo and Underwood (2004) concluded that the Web is both a significant personal, as well as situational, factor in favoring academic dishonesty. In their investigation, over 90% of the 291 students who took part in a study to examine their beliefs and attitudes to Internet plagiarism were familiar with using the Internet to search resources. Alarmingly, only about half of these students vowed that they would not use the copy/paste function from Web-based resources to prevent themselves failing a module. Approximately 50% of the students in this study indicated that they might be prepared to cheat online in order to avoid academic failure. This uncomfortable finding suggests that there is a positive correlation between active Web exposure and plagiarism risk as suggested by Jocoy and DiBiase (2006).

It is not just in higher education that plagiarism is an issue. There is growing awareness at secondary education level that robust policies are required to reduce plagiarism (BBC News, 2008). A survey carried out in 2007 among teachers of sixth-form students in schools and further education institutions in England, Wales, and Northern Ireland highlighted that plagiarism was a concern. Over half the teachers participating in the survey, which was carried out on members of the Association of Teachers and Lecturers (ATL), believed that 25% of work submitted by students contained sections copied from the Internet, even to the extent of Web site advertisements being included in one instance. Many of these teachers believed it was lack of understanding and ignorance on what constitutes plagiarism that was the root of the problem, and that many students were unable to discriminate between plagiarism and legitimate research. Hart and Friesner (2004) cite practices at secondary school as a possible cause for the rise in plagiarism in higher education; teachers and parents may encourage students to use the Internet without emphasizing the importance of correct attribution.

Whatever the cause of transgressions at any level of study, combating plagiarism by ensuring that students understand the principles of academic integrity, along with robust methods of detecting cheating, should provide the foundation for reducing and removing the problem.

Detection Methods

Plagiarism cases may be brought to the attention of tutors by fellow students, or material may be recognized by assessors who, as knowledgeable in their field, may well recognize phrases or sentences and be alerted to possible misconduct. Such discoveries are often made when suspicious work is searched against educational CD-ROMs, online bookstores, or the "hits" of popular search engines (Austin & Brown, 1999). In light of growing popularity of blogging and collaborative Web-authorship, tutors have now extended their search list to Wikipedia and its multi-language sites.

Identifying cases when students are copying from each other is problematic especially when student work is marked by many members of a teaching team. This is where electronic support in detecting plagiarism is most urgently needed. In the UK, the JISC Plagiarism Advisory Service (JISCPAS), http://www.jiscpas.ac.uk/, was launched in September 2002 to help raise awareness of plagiarism in the academic community and to provide electronic resources for detecting plagiarism in student work. The online detection service provided by JISPAS is TurnitinUK, a simple, Web-based system of plagiarism detection used by many academic communities. After discovering that manual evaluation of student assignments overlooked four out of five academic integrity violations (Jocoy & DiBiase, 2006), Penn State faculty adopted Turnitin as a routine step for all student project assignment evaluations. Students or tutors submit assignments to the service and they are checked against a huge database of material from Internet sources, books, journals, newspapers, and other student assignments. After comparing content with current and archived material, an "originality report" is produced that itemizes sections that match against database material and may thus require further investigation. Inspection of this report may show highlighted passages to be correctly quoted or referenced, so a level of interpretation is required by the recipient of the report before any action is taken.

In addition to TurnitinUK, there are now many plagiarism detection software packages available to help the academic community in the fight against plagiarism. Two useful summaries of existing tools are given by Culwin and Lancaster (2001) and Devlin (2002) in their work in the UK and Australia respectively.

Penalties and Consequences

The findings of a U.S. Department of Education report that "students will not internalize ethical values if they believe faculty are apathetic or uninformed about the process of detecting and sanctioning offenders" (as cited in Schulman, 1998, Effective Enforcement section, para. 3). These findings support the need for a transparent and highly visible enforcement system to be put in place to deal with any incident of integrity breach. Transgressors of the academic integrity rules should be left in no doubt that they will be caught and penalized, fairly and swiftly, in accordance with their institution's regulations. The penalty could range from a written warning through reduced marks, zero marks, or even exclusion from the University. For that reason it is a serious offence and as educators we should support and provide guidance to students and researchers in maintaining academic integrity in their work.

However, many students, particularly those new to further and higher education, may not be fully aware of the severity of not maintaining academic integrity. Many do not fully understand the concepts or know how to apply them. For many, the easy accessibility to material via the World Wide Web is too alluring and the temptation to cut and paste, without proper referencing, may be too great.

THE DEVELOPMENT OF A RE-USABLE AND GENERIC E-LEARNING RESOURCE

Selection of the Resource

This section describes the reasoning behind the adoption of academic integrity concepts as a generic learning resource by the three universities.

Although this chapter focuses on a case study taken from a geography-based project and, as such, the content of the quiz in particular reflects this bias, academic integrity covers all academic disciplines. Due to its broad applicability across all disciplines, project members could see the potential of using non-discipline specific content in the case study. The issues in falling academic integrity and increasing cases of plagiarism may have been exacerbated by the misuse of the Web. There is a genuine demand for all partner institutions to consider and enforce an academic integrity guideline that can work effectively in a virtual learning environment (VLE). It is therefore logical that the tested and established academic integrity nugget at Penn State was seen as an ideal candidate to answer the call and hence the adaptation began. Not only can the academic integrity nugget be integrated easily in the existing curriculum, its existence will also greatly improve the overall quality of the existing materials, which is cited by Boyle and Cook (2001) as the most important factor when considering the worthiness of learning object reuse.

The excitement of finding a reusable learning resource was then tested by a reality check on the issues of content compatibility and technical interoperability when adapting the Penn State nugget in the other institutions. Currier and Campbell (2005) identified in another JISC project that the three basic factors in defining reusability are the technical format, contextual dependency, and technical dependency. Their study measured the degree of learning object reusability horizontally and vertically. Vertical reusability concerns whether a resource can be adapted and reused within a specific subject at different levels whilst horizontal reusability looks at how broad a resource can be used across different subject fields and disciplines. It is clear that the Penn State nugget had a high degree of horizontal reusability and the context dependency could be readily overcome as the institutional settings and target audience were similar among all the institutions. The atomic nature of the original nugget, whereby the background information, policy, and procedures as well as the test and reporting are all segregated, paved the way for content to be readily reassembled and reused in any new hosting environment.

The Process of Repurposing

This process by which the original Penn State material was repurposed to the versions now available at Southampton and Leeds will be described in this section. As outlined previously, the academic integrity guidelines and quiz originally developed at Penn State were intended for use in a graduate online geography course. The material was subsequently repurposed by the Universities of Southampton and Leeds in the UK and extended to campus-based and distance learning students at undergraduate and graduate levels. A detailed description of this process is given in Fill et al. (2006).

The framework of all three versions of the resource is in two parts. The first part consists of guidelines in the form of a narrative, which describes the academic integrity concepts. The configuration of the resource was tailored to individual institutions and reflected the local institution's rules and regulations related to academic integrity and penalties for infractions. Customizations were also made to address the specific needs of the target audience and to utilize local expertise and available resources (Leung et al., 2005). The second part is a multiple-choice quiz (MCQ), which allows self-assessment. For students who prefer a more problem-solving learning style, they are encouraged to take the quiz without first consulting the narrative resources. The initial result can help them to identify areas of concern and seek appropriate guidance in a more structured way. The inclusion of more comprehensive and answer-specific feedback in the Southampton and Leeds versions also aims to help students learn from their wrong answers.

When the Penn State nugget was introduced, it was developed entirely within a commercial VLE known as ANGEL (Angel Learning, 2008) and pre-dates any established learning object standards such as IMS or Shareable Content Object Reference Model (SCORM). These are the two most common standards adopted by the learning community to develop a consistent and interoperable approach when creating, assembling, and exchanging online learning materials (CETIS, 2005). This packaging standard was also greatly assisted by the adoption of metadata standard borrowed from the library profession. Due to the lack of standard elements in the Penn State's nugget, the repurposing exercise required extracting the original resource from the ANGEL VLE, manual content decomposition, rewriting, and repackaging into an IMS or SCORM standard-compliant format before finally uploading the new version to another VLE in the individual institution. The whole process was democratically divided between the tutors who were responsible for the pedagogy of the content and developers to explore and tackle the technical and implementation issues.

The Technical and Content Issues

Technical issues were bound to arise in the course of repurposing and reusing material that has been developed for one online learning environment, but was then mounted in other VLEs. In addition,

issues related to repurposing content could also hinder the experience. This section will report on some of the issues and reflect on how future experiences could be improved.

To explain clearly the entire repurposing process, it is important to understand the structure of the original nugget and the subsequent developments in terms of addition, removal and modifications made to it and the rationales behind each decision.

Because of the diverse background, different levels of accessibility to the Internet, and restricted provision of campus-based facilities, the original Penn State nugget was designed to be highly standalone and self-explanatory for distance learners. Such conditions conveniently result in a clean and pedagogically sound structure for the design of the academic integrity nugget. The access to the resource is through the course site available to all students enrolled on the unit. The easy-to-read homepage of the nugget is one-page length providing students with a clear and concise definition of academic integrity and the possible consequences when the rule is violated. It further contains links to the full academic integrity policy of the College of Earth and Mineral Sciences of which the Department of Geography is part. Also available is a link to guidelines for citations and references that is particularly useful for students who have received little or no training in the area in the past. The last two sections of the nugget homepage see the link to the academic integrity quiz and a table displaying the submission date and time and the grade attained in each attempt. A pass of the quiz will then allow the students to unlock their first written assignment.

A very similar framework is adopted in both the Southampton and Leeds versions. The quiz component remains detached from the narrative content so that the discrepancies of the quiz engines in different VLEs can be dealt with separately. The first real hurdle was to untangle the core content of the original nugget from any formatting and corporate elements that are specific to Penn State. As there is no export facility in ANGEL (and indeed many other well-known VLEs), compressing and unzipping the resource is the quickest way to transfer the entire resources from Penn State to Southampton and Leeds. The repackaging of the newer version of the nugget has since been improved as a result of the adoption of the Reload editing and metadata tool (Reload, 2006).

Working with the stripped-down version, tutors at Southampton started to substitute policy and procedures documents, customize references that are locally available, and review the questions and feedback of the quiz. Careful adjustment was made so that any modification could reflect the needs of one institution but at the same time bring more complexity to future adaptation. Questions utilizing local resources and addressing frequently asked questions were added to the question pool. It is also felt that students could learn from their mistake so answer-specific comments, as opposed to general comments, were provided. The number of questions was increased from nine to thirteen.

On the technical side, all the content pages were authored in the standard hypertext markup language (HTML) with detached template and cascading style sheet (CSS) that were appropriate to the local context and learning environment that are approachable to the intended audience. The approach used in both the Southampton and Leeds cases was to "cut and paste" the textual element of the nugget into a plain text editor before the content was modified using an HTML editor, such as Dreamweaver. While it was quite possible and seemingly quicker for the content to be recreated directly within a VLE such as Blackboard (used by Southampton) or Bodington (used by Leeds), this approach would have made the revised nugget totally proprietary to one particular system and defeated the object of any further reuse.

Much of the development time was spent on the "versioning" process, which involved itera-

tive attempts to adapt and fine-tune any learning resource to its new context (Thorpe et al., 2003). The lack of compliance with the question and test interoperability (QTI) standard of the three VLEs among Penn State, Southampton, and Leeds at the time was the first main hurdle to overcome. Unlike the narrative element where the division of text is largely cosmetic and presentational, tests and quizzes, even in different styles such as multiple-choice, filling in the blank, matching, ordering, essay, and so forth are composed of core elements including question number, question text (technically known as the stem), answers, score, and feedback. Repurposing the quiz in a QTI-compatible format became a clearly necessary way to move forward.

While there are many QTI-editing tools available within the academic community, a commercial package called Respondus was used in Southampton because the tool provides a seamless linkage to Blackboard. The whole quiz creation process involved three main steps: saving the question text and answers in plain text; importing the text file into Respondus; and assigning a score and publishing the quiz to a neutral platform, for example, QTI-quiz server or a specific mainstream VLE, for example, Blackboard. As Bodington was not compatible with Respondus at the time, the quiz developed at Southampton was distributed to Leeds in the QTI format for further repurposing.

Repurposing at Leeds took place in a similar time frame to the work carried out at Southampton with some wider repurposing work having been carried out prior to the QTI-compliant quiz being delivered by Southampton. Aimed at distance learning and campus-based geography students at all levels of study, the Leeds version was expanded to include information on TurnitinUK (the JISC Plagiarism Detection Service) and Endnote referencing software. The content was further adapted to reflect regulations in place both within the School of Geography at Leeds and also at University level and included discipline-specific exemplar material and quiz questions and links to resources that were of relevance to Leeds registered students. During repurposing, the main technical difference between Leeds and Southampton was that the interactive quiz was developed using the composing tool within the Bodington VLE in place of Respondus.

Accessing the Resource

This resource is available to registered students and staff of the three universities involved in its

development via their respective virtual learning environment or content management system. In general, this will require user authentication to access the resource. To enable other students and colleagues to access the resource and use or repurpose it in their own learning and teaching, the resource is also available via several other sources.

One such source is Jorum (Jorum, 2007) a free online repository service, which requires UK Higher or Further Education institution registration and then individual registration to

Figure 1 Example screenshots of the academic integrity resources as developed at the Universities of Penn State, Southampton and Leeds

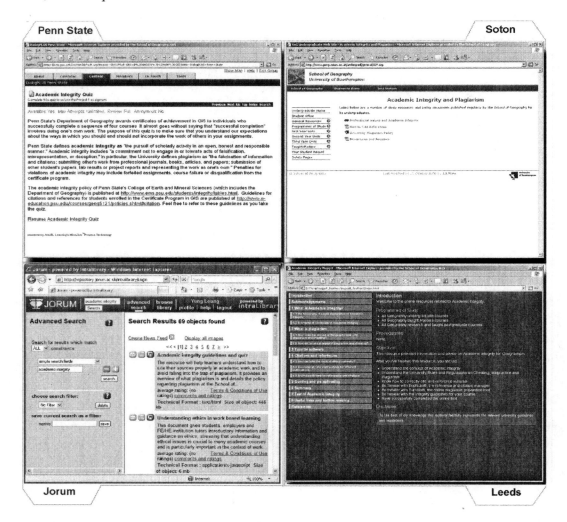

contribute or download material. The Universities of Southampton and Leeds' versions of the resource have been uploaded to this repository and are available for use by registered users. An alternative source is a University of Southampton-maintained Web site versions of the quiz from all three Universities are available at http://www.dialogplus.soton.ac.uk/aig/ai-home.html. This Web site is available to the public and gives a brief description to the background and development of the resources. Screenshots of the resources within the three institutional VLEs and Jorum are shown in Figure 1.

Embedding the Resource in Programmes of Study

The three institutions involved in the development and repurposing of this nugget approached the embedding in programmes of study in different ways, from compulsory to optional participation and from summative to formative assessment. These variations in adoption are summarized in this section, highlighting the flexibility with which the nuggets can be embedded.

The Penn State academic integrity resource was prepared for students enrolled in Geography 482: The Nature of Geographic Information course. Geography 482 is the compulsory first course in Penn State's professional certificate and master's degree programs in GIS, which are offered online through the University's "World Campus." Penn State first established an online certificate program on a non-credit basis in 1999. In 2004, the University's Graduate School approved the proposed conversion of the certificate program to a for-credit basis, and established a professional master's degree in GIS toward which the post-baccalaureate certificate courses could be counted. The decision to award academic credit at a graduate level increased concern about academic integrity among program faculty and administrators. Working with an instructional design specialist, faculty members designed the

academic integrity guidelines and quiz to ensure that students understood the faculty's concern that students perform in compliance with University standards. All incoming students to the current certificate and master's programs in GIS must complete the academic integrity quiz with a score of 100 to "unlock" the project assignment instructions.

In a similar vein, the University of Southampton introduced the guidelines in a compulsory unit to all first year undergraduate students in the School of Geography. There has also been a plan to extend the requirement to students of three other schools including Electronics and Computer Science, Engineering Sciences as well as Oceanography and Earth Science (Leung et al., 2008). Geography students were required to study the guidelines and the related study skills resources. To fulfill the course requirement, it was obligatory for all students to pass the academic integrity quiz and agree with the plagiarism statement, both of which were packaged as a SCORM learning object embedded in the Blackboard VLE. From the beginning of the 2006/2007 academic year, the guidelines were also included in the orientation program for the taught master's students who increasingly have different understanding and training in the areas of academic integrity and plagiarism. The inclusion of the guidelines serves well to remind and, possibly, re-educate, students in the importance of scholarly ethics in the rich media online learning environment.

At the University of Leeds a non-compulsory approach has been taken. The guidelines and quiz are available to all members of the School of Geography via the University VLE, Bodington. Its purpose is to provide support and advice, in the form of a tutorial, to undergraduate and post-graduate students, and provide an easily accessible reference facility to academic and research staff who might need to assess student work or seek guidance to maintain the academic integrity of their own work. The material, available as a self-study guide, can be accessed either as a PDF document or as a Web resource, facilitating use in either a practical session or tutorial. Assessment is not mandatory and students can repeat the exercise as many times as they wish until they are happy with the concepts.

ALTERNATIVE RESOURCES

At the time when the Universities of Southampton and Leeds implemented the adaptation and reuse of the material developed by Penn State, online resources to support the instruction of good practice in any discipline, but more specifically in the geography discipline, was limited. Typing "academic integrity" or "plagiarism" into a search engine now will produce a plethora of Web-based tutorials. Most of these tutorials are produced by higher education institutions and many of them created by North American universities. We speculate that the rise of for-profit higher education institutions in the USA, coupled with widespread concerns within and beyond the academy about the commodification of learning (e.g., Noble, 1998), motivated established non-profit institutions in the USA to protect their brands by stressing the importance of academic integrity to their constituencies.

Subsequent to, but not necessarily as a result of, the development and adaptation of these academic integrity nuggets within this DialogPLUS project, some of the consortium universities have created generic resources providing non-discipline based advice and assessment for use by the community. One example of this is the University of Leeds, which hosts an online plagiarism awareness Web site, which includes examples of good and bad practice, the formal procedures for dealing with transgressions, and an online quiz to help the student test their knowledge and understanding (Learning Development Unit, 2007). Another example is the Academic Skills Web site that is maintained by the University of Southampton (University of Southampton, 2007). This Web site incorporates interactive guides to help students avoid plagiarism, provides advice on the fine-line between helping and cheating, and allows the Web user to guide a "virtual" first year student through decisions affecting the success of his education.

Other UK online resources, available for reuse, are available but tend to be focused on one aspect of academic integrity such as referencing. Jorum (2007) includes several study skill tutorials on the Harvard referencing system, which explain the meaning of plagiarism, describe how to reference work, and create a bibliography and a referencing exercise. Jorum also contains a resource for students on report writing, which includes a brief introduction to plagiarism.

The increasing availability of readily accessible resources that provide advice and allow knowledge to be tested in both the generic and subject-specific arenas should leave students in no doubt that educational institutions take bad academic practice very seriously.

EVALUATING THE BENEFITS OF ACADEMIC INTEGRITY

For an education to be meaningful, it must be one obtained legitimately. Learning to think critically and gaining skills is a continual process and maintaining academic integrity on the journey is the hallmark of a quality education (Virtual Academic Integrity Laboratory, 2005).

This sentiment is reflected in Box 1, which is an excerpt from the University of Southampton's academic integrity quiz. This question highlights the importance of students understanding the implications of academic misconduct in relation to their educational prospects. The consequences of transgressing the regulations should not simply

Box 1.

Why are violations of academic integrity so damaging to the educational prospects of an individual?

A.   Because plagiarisers are likely to be caught and face punishment
B.   My grades will suffer if my reference lists are not presented in the correct format
C.   They're not damaging—plagiarism makes it easier for me to achieve a better degree classification
D.   Because, amongst other things, the skills involved in referencing work correctly will help me more clearly distinguish facts from interpretations

The correct answer is **D**.

Achievement is not measured by the attainment of grades per se. It is the development of professional values that enhances an individual's chance of succeeding, both at University and in later life.

The responses to the incorrect answers were as follows:

A. While it is true that plagiarised material is almost always detected, and that transgressions will be punished, even a large deduction to the final degree classification does not in itself harm your personal development. Instead it is a failure to develop professional values that undermines an individual's chances of succeeding, both at University and in later life.

B. While it is true that incorrectly formatted reference lists result in the loss of marks, even a large deduction does not intrinsically harm your personal development. Instead it is a failure to develop professional values that undermines an individual's chances of succeeding, both at University and in later life.

C. Cheats sometimes (albeit rarely) do prosper in this way, but the degree classification is not the real issue. What is at stake is whether you have developed the skills that society seeks. A first class degree might open the door to employment, but if you can't write or analyse an argument, you are unlikely to succeed in the job.

Source: Academic Integrity Guidelines, School of Geography, University of Southampton

be about the potential for being caught and penalized. Although that in itself should be a deterrent from cheating in any form, more important to the future of any student wishing to develop professional skills is understanding the ethics behind producing work that is appropriately referenced and accurately presented.

The guidelines also work very well alongside other online academic skill resources such as the JISC Plagiarism Detection Service. The service is commercially known as TurnitinUK and has been used in many geography courses at Southampton. Rather than simply deploying TurnitinUK as a policing tool, students are assigned the responsibility to submit their work to the service. Armed with the originality report generated by the TurnitinUK server, students can then satisfy themselves as well as the tutors that the work submitted is their own and all the source materials have been properly referenced. Should the report reflect any inadequacy of referencing and citation students are encouraged to revisit the academic integrity guidelines, allowing for resubmission before the deadline. From the tutors' point of view, the originality reports of the whole class do not only allow them to assess each individual assignment, they also help to highlight any frequently-used online resources or cheat sites and diagnose certain common problems in referencing for that particular cohort. Based on these observations, tutors can reformulate the assignment and revise the academic integrity quiz in order to address the issues identified.

Box 2. Where do you draw the line?

The following statements describe a range of practices from plagiarism through to good citation (i.e. not plagiarism). Identify which of these statements describe practices that would *not* be accused of plagiarism? Select *all* that apply.

A.   Copying a paragraph verbatim (word-for-word) from a source without any acknowledgement.

B.   Copying a paragraph and making small changes (e.g. replacing a few verbs, replacing an adjective with a synonym). The source is given in the references.

C.   Cutting and pasting a paragraph by using sentences of the original but omitting one or two, and putting one or two in a different order, no quotation marks; in-text acknowledgement e.g. (Jones, 1999) plus inclusion in the reference list.

D.   Composing a paragraph by taking short phrases of 10 to 15 words from a number of sources and putting them together, adding words of your own to make a coherent whole; all sources included in the reference list.

E.   Paraphrasing a paragraph with substantial changes in language and organisation; the new version will also have changes in the amount of detail used and the examples cited; in text acknowledgement e.g. (Jones, 1999) and inclusion in the reference list.

F.   Quoting a paragraph by placing it in within quotation marks, with the source cited in the text and the list of references.

The correct answer is **E** and **F**.

In order to avoid plagiarism you must ensure that

1.   The original work is either
- *Paraphrased and entirely re-written in your own words, or*
- *Quoted word-for-word, and placed in quotation marks.*

2.   The source of the ideas, arguments, quotes, etc. is acknowledged with a citation in the text.

3.   The details of the source are included in a reference at the end of the essay or report.

Source: Academic Integrity Guidelines, School of Geography, University of Leeds. Adapted from an exercise in Swales and Feak (1994).

ACADEMIC INTEGRITY AMBIGUITIES

The contents of the resources described in this chapter are geography-focused but the concepts are discipline free and can be used to foster a student culture whereby academic integrity is maintained and upheld. Unfortunately, these concepts are not necessarily consistent across all disciplines resulting in a level of fuzziness; acceptable levels of what does and does not constitute plagiarism vary from faculty to faculty, institution to institution, and country to country. Brutoco and Genereux (1997) (cited in Schulman, 1998, Collaboration, paragraph 1) highlight this potential confusion when they state, "What some consider acts of cheating, which give an unfair advantage, others consider resourcefulness. For example, while some faculty members allow a paper from one course to be the building block for a new paper, others consider that to be cheating." MacDonald Ross (2005) suggests that subject specific needs are responsible for this ambiguity of definition, citing an example that students helping each other in checking grammar and essay clarity may be acceptable practice in some disciplines but would not be considered appropriate for language students.

A poll, based on an exercise by Swales and Feak (1994) and developed at Oxford Brookes University by Jude Carroll, was carried out on students and staff of the University of Leeds to quantify their perceptions of plagiarism, and where they consider the boundary between good practice and academic misconduct should be drawn. This produced interesting results (Learning Development Unit, 2007). Evidence suggests that the student and staff communities identified a different threshold for what constitutes plagiarism with the students (perhaps not surprisingly) taking a more lenient view than the staff. The results also allow reflection on whether views of acceptable practice are more closely consistent with one community of respondents or another.

Box 2 contains an adapted version of the question posed in the University-wide poll that has been incorporated into the School of Geography (University of Leeds) version of the academic integrity quiz. The answer given conforms to the regulations adhered to by this one school in one academic institution, but as readers from a (potentially) different discipline, university, or country, where would you draw the line?

CONCLUSION

This chapter has highlighted the importance of academic integrity in terms of developing and maintaining professional values and also from the students' perspective that understanding and applying academic integrity may increase an individual's chance of success in later life. Academic success goes beyond achieving a good grade—it is supported by acquiring core values, which can be carried forwards into any career and personal development.

Upon submitting any piece of assessed work, students must sign an academic integrity form, stating that the work is all their own. Advancements in technology have made the detection of infringements easier with the availability of plagiarism detection software. Assignments can be compared against a database of current and archived material and a report highlights any potential plagiarism cases. Ignorance, lack of understanding, or different cultural backgrounds are not acceptable excuses for transgression by students. As an accredited educational establishment we should not only use Internet technology to catch those who break the rules but we should use the Internet to take our role as educators responsibly by ensuring online resources are available to make students aware of the importance of good practice.

Defining a threshold of good academic practice is not clear-cut. This chapter has highlighted some of the ambiguities between disciplines within a single institution and has suggested that this is a common difficulty across and between all academic institutions that are responsible for defining and enforcing their own rules and regulations on academic integrity. As a result of these ambiguities it is not always possible to reuse resources developed by one faculty or institution but may need adaptation of content to reflect local thresholds and enforcement policies.

In addition to adapting content there are technical issues relating to distribution of material via different VLEs. In the three adaptations of the nugget discussed in this chapter, all are available online using the VLE in place at each institution, and the use of a VLE to permit access to the resources is pertinent to the subject matter in discussion. The increased availability of online resources has potentially fuelled opportunities for cheating in the form of cut and paste of sections of text and figures or to obtain pre- or custom-written assignments. It is therefore appropriate that guidelines and quizzes that advise and consolidate understanding of the rules of academic integrity are available in a format consistent with the transgression.

REFERENCES

Angel Learning. (2008). *Angel learning: Recognized innovator of enterprise eLearning software.* Retrieved March 20, 2008, from http://www. angellearning.com/

Ashworth, P., Bannister, P., Thorne, P., & Unit Students on the Qualitative Research Methods Course. (1997). Guilty in whose eyes? University students' perceptions of cheating and plagiarism in academic work and assessment. *Studies in Higher Education, 22*(2), 187-203.

Austin, J., & Brown, L. (1999). Internet plagiarism: Developing strategies to curb student academic dishonesty. *The Internet and Higher Education, 2*(1), 21-23.

Barrett, R., & Cox, A. L. (2005). At least they're learning something: The hazy line between collaboration and collusion. *Assessment and Evaluation in Higher Education, 30*(2), 107-122.

BBC News. (2008). *Teachers voice plagiarism fears.* Retrieved February 8, 2008, from http://news.bbc.co.uk/1/hi/education/7194772.stm

Boyle, T., & Cook, J. (2001). Towards a pedagogically sound basis for learning object portability and reuse. In G. Kennedy, M. Keppell, C. McNaught, & T. Petrovic (Eds.), *Proceedings of ASCILITE 2001* (pp. 101-109). Retrieved August 7, 2007, from http://www.ascilite.org.au/conferences/melbourne01/pdf/papers/boylet.pdf

Brutoco, D., & Maurissa, G. (1997). *Making the grade: Cheating at Santa Clara University.* Unpublished report. Santa Clara, CA: Santa Clara University.

CETIS. (2005). CETIS briefing on e-learning standards. Retrieved August 26, 2008, from http://zope.cetis.ac.uk/static/briefings.html

Culwin, F., & Lancaster, T. (2001) *Plagiarism, prevention, deterrence and detection.* Retrieved June 30, 2007, from http://www.heacademy. ac.uk/resources.asp?process=full_record&section=generic&id=426

Currier, S., & Campbell, L. M. (2005). Evaluating 5/99 content for reusability as learning objects. *VINE: The Journal of Information and Knowledge Management Systems, 35*(1/2), 85-96.

Devlin, M. (2002). *Plagiarism detection software: How effective is it?* Retrieved June 30, 2007, from http://www.cshe.unimelb.edu.au/assessinglearning/03/Plag2.html

Dordoy, A. (2002). Cheating and plagiarism: Student and staff perceptions at Northumbria. In *Proceedings of the Northumbria Conference: Educating for the Future, Newcastle, UK.* Retrieved February 18, 2008, from http://www.jiscpas.ac.uk/images/bin/AD.doc

Ercegovac, Z., & Richardson, J. V. (2004). Academic dishonesty, plagiarism included, in the digital age: A literature review. *College and Research Libraries, 65*(4), 301-318.

Fill, K., Leung, S., DiBiase, D., & Nelson, A. (2006). Repurposing a learning activity on academic integrity: The experience of three universities. *Journal of Interactive Media in Education, 2006*(1). Retrieved June 30, 2007, from http://jime.open.ac.uk/2006/01/

Groark, M., Oblinger, D., & Choa, M. (2001). Terms paper mills, anti-plagiarism tools, and academic integrity. *Educause Review, 36*(5), 40-48.

Hansen, B. (2003). Combating plagiarism: Is the Internet causing more students to copy? *CQ Researcher, 13*(32), 773-796.

Hart, M., & Friesner, T. (2004). Plagiarism and poor academic practice — a threat to the extension of e-learning in higher education? *Electronic Journal on e-Learning, 2*(1), 89-96.

Hinman, L. M. (2002). Academic integrity and the World Wide Web. *Computers and Society, 32*, 33-42.

Jocoy, C. L., & DiBiase, D. (2006). Plagiarism by adult learners online: A case study in detection and remediation. *International Review of Research in Open and Distance Learning, 7*(1). Retrieved June 30, 2007, from http://www.irrodl.org/index.php/irrodl/article/view/242/466

Jorum. (2007). *Jorum user terms of use.* Retrieved May 31, 2007, from http://www.jorum.ac.uk/user/termsofuse/index.html

Larkham, P. J., & Manns, S. (2002). Plagiarism and its treatment in higher education. *Journal of Further and Higher Education, 26*(4), 339-349.

Learning Development Unit. (2007). *Plagiarism – University of Leeds Guide.* Retrieved March 20, 2008, from http://www.ldu.leeds.ac.uk/plagiarism/index.php

Leung, S., Fill, K., DiBiase, D., & Nelson, A. (2005). *Sharing academic integrity guidance: Working towards a digital library infrastructure* (LNCS 3652, pp. 533-534). Retrieved June 30, 2007, from http://www.springerlink.com/content/2x2x1wuc6dw3n1ev/

Leung, S., Harding, I., Wang, S., & Moloney, J. (2008). *Encouraging academic integrity to discourage plagiarism.* Paper presented at the 3rd International Plagiarism Conference (June 23-25), Newcastle-upon-Tyne, UK.

MacDonald Ross, G. (2005). Plagiarism the Leeds approach. *Learning & Teaching Bulletin, (8).* Leeds, UK: University of Leeds. Retrieved August 7, 2007, from http://www.ldu.leeds.ac.uk/l&tbulletin/issue8/ross.htm

McCabe, D. (2000). New research on academic integrity: The success of "modified" honor codes. *Synfax Weekly Report, (17),* 975.

Noble, D. F. (1998). Digital diploma mills: The automation of higher education. *First Monday, 3*(1). Retrieved March 20, 2008, from http://www.firstmonday.dk/issues/issue3_1/noble/

Park, C. (2003). In other (people's) words: plagiarism by university students – literature and lessons. *Assessment and Evaluation in Higher Education, 28*(5), 471-488.

Reload. (2006). Reusable eLearning object authoring & delivery. Project Web site. Retrieved March 21, 2008, from http://www.reload.ac.uk/

Schulman, M. (1998). Cheating themselves. *Issues in Ethics, 9*(1). Retrieved June 30, 2007, from http://www.scu.edu/ethics/publications/iie/v9n1/cheating.html

Swales, J., & Feak, C. (1994). *Academic writing for graduate students.* Ann Arbor, MI: University of Michigan Press.

Szabo, A., & Underwood, J. (2004). Cybercheats: Is information and communication technology fuelling academic dishonesty? *Active Learning in Higher Education, 5*(2), 180-199.

Thorpe, M., Kubiak, C., & Thorpe, K. (2003) Designing for reuse and versioning. In A. Littlejohn (Ed.), *Reusing online resources: A sustainable approach to e-learning* (pp. 106-118). London, UK: Kogan Page.

Underwood, J., & Szabo, A. (2003). Academic offences and e-learning: Individual propensities in cheating. *British Journal of Educational Technology, 34*(4), 467-477.

University of Southampton. (2007). *Academic skills.* Retrieved March 20, 2008, from http://www.academic-skills.soton.ac.uk/

Virtual Academic Integrity Laboratory. (2005). *Academic integrity and dishonesty policies: What every student needs to know.* University of Maryland. Retrieved June 30, 2007, from http://www.umuc.edu/distance/odell/cip/vail/students/AI_policies/

USEFUL LINKS

University of Southampton-maintained Web sie of AI quizzes: http://www.dialogplus.soton.ac.uk/aig/ai-home.html

Jorum: http://www.jorum.ac.uk

University of Leeds AI Resources: http://vle.leeds.ac.uk/site/nbodington/geography/geogall/geog_ai/ (access restricted to University of Leeds staff and students)

Penn State Academic Integrity Guide: http://www.e-education.psu.edu/natureofgeoinfo/resources/c2.html

Section III
Software Support for Learning Material Design

Chapter IX
A Toolkit to Guide the Design of Effective Learning Activities

Karen Fill
KataliSys Ltd, UK

Gráinne Conole
The Open University, UK

Chris Bailey
University of Bristol, UK

ABSTRACT

The DialogPLUS Toolkit is a web-based application that guides the design of learning activities. Developed to support the project's geographers, it incorporates well-researched pedagogic taxonomies that are presented as drop-down lists with associated 'help' pages. Toolkit users are encouraged to consider and specify factors including learning and teaching approach, environment, aims and outcomes, assessment methods, learner and tutor roles and requisite skills as they design any number of tasks within a learning activity and select the tools and resources needed to undertake them. The output from the toolkit is a design template that can then be used to guide the instantiation and implementation of online learning activities. The designs are saved within the toolkit, forming a database of designs, which other toolkit users can view. This chapter will present the rationale for the toolkit and the detailed taxonomies. It will describe and illustrate the software design, development and implementation, including the approach to contextual 'help', provide examples of learning activity designs created using the toolkit; and present and discuss feedback from users.

INTRODUCTION

Pedagogic knowledge results from the study of how we learn. Ever since the earliest Greek philosophers started asking questions about the world we live in, people have been interested in understanding the nature of how we absorb, process, and apply information. This research can help improve learning, both maximizing the knowledge transfer for learners and increasing the effectiveness of teachers. In today's digital world, the challenge of applying these ideas becomes even more difficult as, like it or not, technology moves us forward; enhancing the tools and resources we use and evolving the communities that surround us.

To keep up with this challenge, tools are needed to help today's learning designers make best possible use of the resources available to them when designing or developing units of learning. The DialogPLUS Toolkit was developed by a team of educationalists and computer scientists working with subject specialists (geographers) to help understand how practitioners approach the task of designing learning activities.

The aim of this group was to design and deliver an easy-to-use system that would guide teachers and learning technologists as they created learning resources, tasks, and activities so that these would support effective learning. Drawing on previous work and the wide body of literature about approaches to teaching and learning, the toolkit is underpinned by a comprehensive pedagogic taxonomy. Thus, toolkit users are encouraged to consider and specify factors including the learning and teaching approach, aims and outcomes, assessment methods, learner and tutor roles, and requisite skills as they design the tasks that make up a learning activity, and select the tools and resources students will need to undertake them. The output from the toolkit is a design template that can then be used to guide the actual creation and implementation of the online learning activities or can be used by others as an example, which can be repurposed to create a new learning activity.

The toolkit is available on the Web for any practitioner to access (DialogPLUS, 2006). Designs are saved within a database and can be easily viewed by other toolkit users.

The specific objectives of this chapter are to present the background and rationale for the toolkit; explain the underlying taxonomy; describe the software design, development and implementation, including the novel approach to contextual "help;" provide examples of learning activity designs created using the toolkit; and to present and discuss feedback from users. Finally, the future of this and similar tools to support designs for learning is considered.

BACKGROUND

Practitioners are faced with a potentially bewildering array of tools and technologies to support learning and teaching. However, "while it is clear that technologies are having an increasing impact on institutions … it is equally apparent that their potential for enabling new styles of learning is not yet being realised" (Conole et al., 2005, p. 3). Practitioners designing and creating learning activities need to make a complex set of inter-related decisions. Examples of the questions they ask themselves, and others, include:

Which tools should I use to promote dialogue between students – chat, a discussion forum, video conferencing?

How can I set up an activity to encourage students to collaborate on a shared problem?

What is the best way to create a collaborative shared space of resources?

Some practitioners are confused by the ever-expanding range of tools and theories and need support in deciding which might be appropriate for a particular learning activity. As a result, in recent years there has been considerable interest in gathering examples of good practice and providing scaffolds, or design support, to aid practitioners as they create new learning activities. However, defining case studies of good practice and articulating the steps and processes involved in learning design is far from trivial. Our aim was to create a toolkit which would guide teachers through the decision making process of creating learning activities. This built on previous work (Conole & Oliver, 1998; Conole et al., 2001), with the intention that the learning activity toolkit would be designed to present relevant information to the user on a needs basis. The output would be an outline design of a learning activity and the key associated components. This outline would be held in the toolkit database so that it could be searched by other users.

Rationale and Requirements Analysis

In order to ensure that development of the toolkit was grounded in real practice, the design specification was based on requirements gathered from the DialogPLUS geographers in a series of semi-structured interviews. The purpose of the data collection was to:

1. Follow individuals through a series of decision-making processes over a period of months in terms of designing a new course, component of a course or individual learning activity.
2. Elicit information on individual thought processes as part of the decision making, identification of trigger points, support mechanisms, and barriers to design.

3. Understand the process of design and the types of presentations individuals used to facilitate their design process.

The practitioners chosen provided a spectrum of potential users of the system. They included an established teacher/researcher who wanted to create a new course that encapsulated aspects of current research as part of a student portfolio in an undergraduate final year optional unit; an established teacher who was redesigning elements of a course based on evaluation and feedback from previous years; a novice teacher taking over an existing course; and a team updating and adapting course content and activities.

A series of semi-structured interviews was conducted. Each session lasted between thirty minutes to one hour. Sessions were deliberately flexible and focused on the stage of the design process that the individual was currently working on. The session was a mixture of a form of "think aloud" protocol supported by a series of prompting questions. Questions covered issues such as:

*   What were the key aspirations inherent in the proposed design of the course?
*   What did they want the students to be able to achieve?
*   How did they find information to support their design process?
*   Where did they find resources?
*   How were resources incorporated into the design process?
*   Were there any explicit or implicit pedagogical models being used?
*   What difficulties or issues were they encountering at that point?

From this iterative process it was possible to gain an understanding of the way in which individuals (from novice through to expert) thought through the design process. Unsurprisingly, this demonstrated that the design process was "messy," with individuals thinking at a number of levels and oscillating between the different factors involved in their decision making. From these sessions the factors involved in design began to emerge and were used to develop an initial specification for the toolkit, as well as an underpinning taxonomy, which described the components involved in creating a learning activity. An outline of the components of the taxonomy is provided in the next section. Versions of the developing taxonomy were iteratively validated with the geographers as well as with other practitioners in the education community (through conference presentations, papers, and workshops). This iterative feedback helped to refine the taxonomy and ensure that its components were realistic and reasonably robust.

The DialogPLUS Pedagogical Taxonomy

At the heart of this approach was a desire to provide a scaffold for supporting practitioners' decision-making processes when creating learning activities.

Beetham (2004) defines a learning activity as *"an interaction between a learner or learners and an environment (optionally including content resources, tools and instruments, computer systems and services, "real world" events and objects) that is carried out in response to a task with an intended learning outcome"* (p.7). However, analysis of the data collected from practitioners on their thinking processes in creating learning activities, coupled with a review of the literature (aspects of which are reported in Conole et al., 2005), showed that not only was there no clear consistency in the terminology being used (learning objects, learning activities, pedagogical approaches, models, theories, and so on) but that little was actually understood about what constituted a learning activity, and the key

components involved. Therefore, we decided it was necessary to use the data generated from the practitioners to help define the components that constitute a learning activity, working them into the form of a pedagogical taxonomy. Currier et al. (2006) provide a comprehensive review of the historical development and use of different pedagogical vocabularies, which helps to contextualise this work.

There is also a substantial amount of work in the more technical arena of standards to support e-learning. Here the bodies of the IMS Global Learning Consortium and Advanced Distributed Learning are working towards producing specifications to aid interoperability in many areas of e-learning. Their goal is to allow materials to be created, packaged, and delivered to students in a manor that transcends the specific delivery mechanisms. This allows materials to be imported into any number of different learning environments and content management systems, such as Blackboard or Moodle, providing such environments support the interoperability specifications. A standard closely related to learning activity design is IMS Learning Design (IMS-LD), a flexible specification for structuring online material to suit many different pedagogical styles. It builds upon existing standards such as content packaging (IMS 2004) and the earlier educational modelling language (EML) (Koper & Manderveld, 2004). These specifications are highly technical, low-level designs to provide computer systems with enough information so that they can deliver online course material to the student in the correct fashion. IMS-LD contains, amongst other things, information for specifying educational content, learning activities, sequencing information to determine the order to display the materials, and details about which roles are required for a given learning design. Our toolkit development team kept a close watch on the IMS-LD specification as it evolved over the course of our project. However, rather then adopt such a technical specification, it was decided early on that to best support real practitioners we needed a higher abstraction that more closely matched their working practice.

The DialogPLUS taxonomy attempts to consider all aspects and factors involved when practitioners develop a learning activity, from the pedagogical context in which the activity occurs through to the nature and types of tasks undertaken by the learner. Figure 1 shows the higher-level components. The taxonomy builds on Beetham's definition of a learning activity, dividing this into three key elements:

1.   The context within which the activity occurs; this includes the subject, level of difficulty, the intended learning outcomes, and the environment within which the activity takes place;
2.   The learning and teaching approaches adopted, namely pedagogical theories and models;
3.   The tasks undertaken, specifically the type of task, the (teaching) techniques used to support the task, any associated tools and resources, the interaction and roles of those involved, and any associated assessments.

The first two elements, context and pedagogical approach, provide an overarching rationale for the learning activity, whereas the task specification represents the heart of the taxonomy. Therefore, learning activities are designed to achieve a series of intended learning outcomes, arrived at through the completion of a series of tasks. Learning and teaching approaches are grouped into three categories—associative, cognitive, and situative—according to a classification developed by Mayes and De Freitas (2004). Learning outcomes are what the learners should know, or be able to do, after completing a learning activity; for example, be able to understand, demonstrate, design, produce, or appraise. These are mapped to Bloom's taxonomy of learning outcomes (Bloom, 1956; Anderson & Krathwohl, 2001) and grouped into three types: cognitive, affective, and psychomotor.

Task Taxonomies

In designing a learning activity a teacher often has a linear sequence of tasks in mind but, especially in an online learning environment, learners will not necessarily follow that sequence. Indeed an early project experience flagged up the need to enable

Figure 1. Key elements of a learning activity

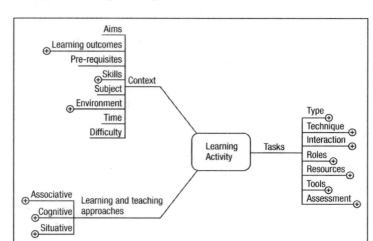

learners to move easily around the resources and tasks. This is not merely a requirement for linear navigation, but illustrates the inherent expectation to jump between pages when accessing online information.

As the tasks are undertaken, roles are required of both learners and tutors. Instances of roles defined within the toolkit include individual learner, pair or group participant, facilitator, presenter, and moderator.

Tasks also involve different resources and, when appropriate, tools to access or manipulate them. Taxonomies for resources and tools were developed based on Laurillard's five principal media forms (2002, p. 90). These, together with the type of learning they support are:

- Narrative →attending, apprehending
- Interactive→investigating, exploring
- Communicative→discussing, debating
- Adaptive→experimenting, practicing
- Productive→articulating, expressing

Narrative media tell or show the learner something (e.g., text, image). Interactive media respond in a limited way to what the learner does (e.g., search engines, multiple choice tests, simple models). Communicative media facilitate exchanges between people (e.g., e-mail, discussion forum). Adaptive media are changed by what the learner does (e.g., some simulations, virtual worlds). Productive media allow the learner to produce something (e.g., word processor, spreadsheet).

It was hoped that two major advantages would accrue from this approach:

1. Based on the classification of media in the five categories, and the desired learning outcomes and strategies, the toolkit would be able to suggest appropriate media types and combinations as a teacher designs a learning activity. The taxonomy itself would be hidden from the end user.

2. The validity of the taxonomy would be tested and, perhaps, expanded or updated.

These aspirations were taken forward in the software development phase.

TOOLKIT DEVELOPMENT AND IMPLEMENTATION

With a model of a learning activity that was using terms familiar to the geographers, but not too tightly bound to any specific practices, materials, or techniques, the next task was how to embed it within a software solution. The four overarching requirements that we wanted to incorporate were community, generality, flexibility, and support. Each of these affected the technical design decisions as the toolkit was developed.

Community

The first requirement, which underpinned all our decisions, was to develop a tool that would be of use not just to individuals, or project team members, but also to the wider teaching and learning community. The DialogPLUS project involved four separate institutions in different locations and so we had to derive a solution that would appeal to all of them. The project was primarily concerned with issues surrounding cross-institutional collaboration and cooperation, necessitating a tool where the design outputs of individuals could be viewed by colleagues both within an individual institution and also be accessible to peers located in other institutions. Our early design concepts focused heavily on a tool where finished designs could be uploaded to a "pool" of community activities such that peers could search for interesting learning activity designs and incorporate whole designs or partial aspects into their own activities. In this vision, designs could be propagated to the community, discussed with peers, refined, and even heralded as best-practice "exemplars."

It would also be useful to track their use within structured lessons and courses and gain feedback on these real world cases. Another mechanism for fostering a strong community element would be to provide colleagues wishing to collaborate with the means to work together via the tool to plan out and design new learning activities.

All these considerations led to the development of a single Web-based learning activity designer, or toolkit, based on a traditional online client-server architecture, with a hypertext markup language (HTML) front end and a database-driven backend to store all the designs. Rather than provide each institution with their own database of designs, it was decided that a single tool accessible to all institutions was more appropriate and would help bring authors together via the use of a single system. It would also make sharing of designs easier and quicker. Finally, an online discussion forum was added to the toolkit to foster a community spirit and provide a means for authors to converse and discuss their works with other toolkit users.

Generality

While the purpose of the toolkit was very much rooted in the need to support the geographers involved in this project, we wanted to create a solution that would appeal to the wider teaching and learning community. We were always conscious of the fact that a toolkit to enable the design of geography learning activities would also be of use to anyone wishing to develop learning activities in a range of subjects. The pedagogical taxonomy and the terminology used within the software toolkit were carefully chosen to be non-subject specific. It quickly became apparent that geography was, in fact, a very good discipline to support in this way due to the diverse nature of existing teaching practices and the practitioners' eagerness to experiment with new ideas and approaches. One of the challenges, both in the design and development of the toolkit, was to

cater for the wide range of existing and planned learning activities that they had in mind. Learning activities used a range of resources from simple, text-only Microsoft Word™ documents, through structured databases, to fully interactive Shockwave Flash™ applications. In addition, the practitioners themselves had a range of different skills in using digital media, software tools, and online resources. Supporting such a wide variety of learning objects and user skills in a terminology-neutral language allows the toolkit to be suitable for the majority of subject disciplines. Another feature of the toolkit is to allow the taxonomy to be extended with new techniques as they are discovered or evolved. Thus, the taxonomy can grow with the community and be adapted to specific disciplines if required.

Flexibility

The approach taken in designing the interface and functionality of the software was to develop a "one-size-fits-all" product that would be flexible enough to be used to describe all the existing learning activities that had been identified for the team to share, in addition to any new ones that would be developed during the project's lifetime. As already mentioned, the range and extent of existing activities was very broad to start with and we did not want to develop multiple tools for specific activity types. This led us to adopt the idea of a software "toolkit." Our team had extensive experience developing similar toolkits in previous projects and understood the advantages here. For our purposes, a toolkit is a configurable software solution that lies on an axis between monolithic software products, which can be used flexibly for a variety of purposes (e.g., Microsoft Word™), and much smaller single-task specific programs (e.g., printer wizards). Our toolkit exists somewhere between these two extremes (see Figure 2) and hence provides some degree of structuring and support but is also flexible to some extent.

The toolkit was built as a series of small

Figure 2. The toolkit lies on this spectrum

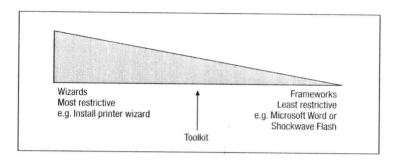

form-based components each designed for editing specific parts of the pedagogic taxonomy. These components can then be used in several different ways, either in isolation to edit an aspect of the design outside the context of the whole activity, or incorporated into larger modules that embed a structured authoring workflow. The resulting toolkit then allows two authoring "paths." The first is a bottom up approach where users can create and edit multiple tasks independently of any learning activity and then, at a later stage, create the surrounding activity framework. The second approach allows users to choose a more traditional "top down" style whereby the high level metadata–such as activity type, learning level, or difficulty–are defined first and then they drill down to define learning outcomes and then tasks for those outcomes. While the first approach might be more useful for people who already have existing tasks they want to incorporate into new activities, the second approach was more interesting as it provides more guidance to users who have less of an idea about the kinds of learning activities they want to design.

Support

The final requirement for the toolkit was to act as a pedagogical guidance tool supporting users in creating new learning activity designs. The toolkit offers several levels of support to its users. Firstly, although users can start at the bottom creating tasks and then building up learning activities from those tasks, there is a stronger emphasis placed on choosing a top down design approach. If one chooses this option then a pre-defined workflow is imposed onto the learning activity design authoring process. This workflow was designed to closely follow the process existing designers chose when developing new activities so that the options the toolkit presented at each stage would seem familiar and understandable to users. Secondly, we incorporated a help and support section in the Web site, which documents each of the stages required to develop a design and what type of information is required for each stage. The most detailed form of support offered by the toolkit is found in the help buttons attached to each of the required data entry fields. These buttons provide dynamic context-sensitive help to the user. When a user clicks on one of these links, the system examines the existing, partially completed, design. Using this information it compares the current design to a set of help and support rules that have been previously set up by pedagogical specialists. The system finds the most specific rule that matches the existing state of the design and then presents the help attached to that rule. In this way the system uses as much of the surrounding design context as possible when presenting help. For example, if the user wants to try a problem-based approach in their learning and teaching method, when they come to create a task and are asked what type of technique to use, then the system can provide suggested tasks that have specific applicability to problem based learning (PBL).

Another form of support that is available to users is the ability to view designs created by the other users of the system. The toolkit adopts a visible-to-all policy with the designs so that at any given time a user's designs are visible to (but not editable by) all the other toolkit users. There is also a free-text search mechanism to search and view these designs. Novice users or practitioners wanting to try new approaches in their teaching can quickly search the repository of existing designs to see how others have taught the same subject or used a specific learning and teaching approach in their designs. The final support for toolkit users is in the form of the community-level discussion forums mentioned previously. Via this method users can quickly get in contact with a variety of other practitioners, ask questions to other users, and read existing messages.

Pilot Learning Activities

In the course of deriving and refining the definitions and taxonomy, two real life examples were modeled as learning activities and discussed with teaching colleagues. The representation proved both robust and informative. One activity was for individual learners and entailed reading two online documents, answering questions in a multiple-choice quiz and receiving automatic

feedback. The other involved learners viewing a wide range of online resources, data, text and images, accessing a public database, manipulating extracted data using a spreadsheet, deriving and submitting numeric answers, selecting and ranking choices from tutor supplied drop-down lists, and composing and submitting written critiques. Tutors were involved in responding to learner comments and questions via an asynchronous message board and in marking written submissions. Numeric answers and some multiple-choice selections were automatically marked.

Once these learning activities had been modeled, it was clear that the resulting information would enable other teachers to review them quite swiftly and decide whether to adopt or adapt them. Furthermore, the modeling of resources and tools facilitates repurposing. For example, the first activity was based on two documents internal to one of the USA partners. It would be easy for the other partners to substitute their own documents while keeping the surrounding activity structure intact. In the second activity, all the resources pertained to river habitats in the UK. Colleagues in the USA could substitute indigenous resources without changing the sequence of tasks and outcomes the learners are required to undertake and produce. It would be possible to replace some of the individual analytical tasks with group work, and so on.

These modifications were conceptually possible but remained complex in terms of technical infrastructure. A challenge for the toolkit developers was to resolve interoperability issues. They were aware of, and involved with, research and development in this field. The DialogPLUS project offered an excellent proving ground for potential solutions, as all four partners had different virtual leaning environments. Copyright and permissions processes were also required in the longer term. From a pedagogic point of view, there were also limitations and issues to be resolved. These included measuring tacit knowledge, abstracting out the essence of the learning activity, identifying pedagogical models, and evaluating the quality of a learning activity. Furthermore learning activities produced by working through the toolkit only presented a static view of the learning activity, giving no indication of the dynamic real-time running of the activity.

Design solutions to these challenges proved intractable during the development phase of the DialogPLUS project. Work continues on them in other quarters. However, Version 1 of the toolkit has been freely available on the World Wide Web since 2005 and trialled extensively by a range of users.

Using the Toolkit

The toolkit was introduced to DialogPLUS team members in a series of workshops and one-to-one sessions. They were encouraged to use it to design and collaborate on learning activities. It was unfortunate that much of the early learning activity and course design work for the project had been completed by the geographers before Version 1 of the toolkit became available for use. In the

event, a few of their designs were modeled after development in the toolkit. The most interesting trial by the team is described by Durham and Arrell (2006) who used it, together with partners at Penn State University, as part of the collaborative design of learning activities about global positioning systems (see Chapter III).

The toolkit has also been presented in various conferences and workshops to the wider learning and teaching community. Over two hundred people have visited the Web site and tested the toolkit (see Figure 3). Feedback from the majority indicates that the concept is of interest. It has been suggested that the primary use for Version 1 is as a resource for introducing teachers to pedagogical theories and a proven approach to designing effective learning activities. This is being explored in a separate Joint Information Systems Committee (JISC) funded project, Evaluation of Design and Implementation Tools for Learning (EDIT4L, 2007).

Some users would like functionality in the toolkit to enable them to instantiate a design. As a result of following the development of IMS-LD,

Figure 3. Screen shot of a design created in the toolkit

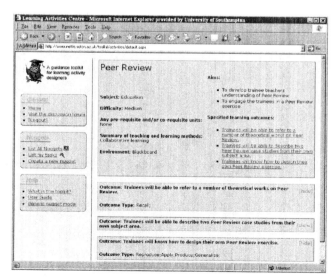

the toolkit team was able to produce an IMS-LD compliant manifest from the toolkit by extending the Reload Editor tool (see Reload, 2006; Bailey et al., 2006). However, it is our view that, while this may be acceptable to a small number of highly technical experts, more user-friendly tools are needed if standards such as IMS-LD are to be adopted by the wider education community. The EDIT4L project is exploring the use of the Learning Activity Management System (LAMS International, 2007) to create activities from DialogPLUS toolkit designs.

The development team is also conscious that Version 1 of the toolkit does not fully meet the aspirations of adaptive guidance at all stages of a design, and for all selections that a user might make. If it ever proves possible to develop Version 2, there is a long "wish list" of required features–many of which reflect progress in the pedagogical and technical fields of design for learning.

FUTURE TRENDS

Pedagogical

A key challenge in today's technology-enhanced educational environment is providing course designers with appropriate support and guidance on creating learning activities that are pedagogically informed and which make effective use of technologies. "Learning design," where the use of the term is in its broadest sense, is seen by many as a key means of trying to address this issue. Beetham and Sharpe (2007) provide a valuable overview of current research in learning design, identifying some of the main areas of activity, along with a set of tools and guidance to support the learning design process.

Two overarching issues emerge in relation to designing for learning. Firstly, how should learning activities be represented which best describe the essence of the learning activity and provide enough information so that this can be used as a basis for designing a new learning activity? Secondly, what support and guidance is appropriate to guide course designers through the process of creating a learning activity?

In relation to the first issue, capturing practice, Conole (2008) argues that there is a range of mediating artifacts that course designers use to guide practice. These range from contextually rich textual mediating artifacts, such as case studies, through to more abstract examples, such as iconic representations of pedagogical models. However, Falconer and Littlejohn (2006) also contend that no one representation will capture all aspects of a learning activity and that course designers tend to use a mixture of approaches depending on individual preferences and needs. With respect to the second issue, the DialogPLUS toolkit provides an example of a form of support and guidance. Indeed as this chapter illustrates this was a fundamental aspiration inherent in the design of the toolkit. However, it is questionable to what degree pedagogically driven support actually works in practice. A recent series of case studies with practitioners from across a range of subject disciplines (Conole et al., 2007), suggests that course designers are driven more by pragmatic needs than pedagogical theories, such as "how can I use a particular set of resources?," "how can I design an activity which will engage this particular group of students?," "how can I get these key concepts across?". The design process was informed by the subject expertise, peer support, and a lot of trial and error. Furthermore, in related work on the creation of a series of use cases for learning design, most of the learning activities created fell within a narrow range of pedagogical approaches (Falconer & Conole, 2006). It appears that course designers in both instances seemed unable to "think outside of the box."

What then does this mean for tools, such as the DialogPLUS Toolkit, which attempt to drive the design process pedagogically? How might this be achieved and will such a pedagogically led approach ultimately lead to a change in practice and better, more innovative learning activities? Finally is there a means by which we can actually map pedagogically approaches to technological tools? The academic and learning technology communities will undoubtedly seek and disseminate answers to these questions over the next few years.

Technological

The toolkit was built several years ago using traditional HTML forms with a database driven back end. However, this presented several difficulties, the biggest of which was the static, state-driven nature of http client-server protocol. Specifically, there are issues around the need to request that the user click on a "Save" button to store data once it has been entered. Issues such as these are not so much a problem with current Web technologies, as solutions such as Ajax (Garrett, 2005) offer alternatives to requiring a user click on a save button to store changes and make large data entry forms easier to manage. We also faced problems concerning the desire for cross-browser compatibility and adherence to Web accessibility guidelines. Our toolkit relies on popup windows to offer context-sensitive help and selections from large taxonomy lists, so that the user is not redirected away from the current edit page. This breaks several accessibility guidelines and makes compatibility more of an issue. Web browsers are increasingly moving towards greater standards compliance and so many of the difficulties we faced would be less important today. Instead new issues arise–such as integrating with existing repositories such as the UK's Jorum (Jorum, 2007) or harnessing the increasingly social nature of recent Web 2.0 techniques, applications, and technologies.

Further Use

At the time of writing (March 2007), use of the toolkit as part of teacher development in higher and further education in the UK is being investigated by the EDIT4L project team as part of the Design for Learning research programme funded by JISC (JISC, 2008).

CONCLUSION

The DialogPLUS Toolkit was conceived and developed by educationalists and computer scientists to support academics as they strive to design effective learning activities. Grounded in the real practice of the project's geography teachers, it incorporates generic features and taxonomies that extend its usefulness to the wider community. Feedback from many members of this community who have had free access to Version 1 suggests that the design aims and approaches are interesting and welcome. It is to be hoped that further work might be undertaken to produce a refined Version 2 addressing user requirements and suggestions; enhancing usability; introducing more features such as the personalised, adaptive "help" originally intended for Version 1; and taking account of current and future technical and pedagogical developments.

REFERENCES

Anderson, L.W., & Krathwohl, D. (Eds.). (2001). *A taxonomy for learning, teaching, and assessing: A revision of Bloom's Taxonomy of educational objectives.* New York: Longman.

C., Zalfan M., Davis H., Fill K. & Conole G. (2006). Panning for gold: Designing pedagogically-inspired learning nuggets. *Educational Technology & Society, 9*(1), 113-122.

Beetham, H. (2004). *Review: Developing e-learning models for the JISC practitioner communities.* Report for the Pedagogies for e-Learning Programme, Version 2.1. Retrieved February 22, 2007, from http://www.jisc.ac.uk/uploaded_documents /Review 20models.doc

Beetham, H., & Sharpe, R. (Eds.) (2007). *Rethinking pedagogy for a digital age. Oxford.* UK: RoutledgeFalmer.

Bloom, B. S. (Ed.) (1956). *Taxonomy of educational objectives, the classification of educational goals—Handbook I: Cognitive domain.* New York: McKay.

Conole, G. (2008). Capturing practice: The role of mediating artefacts in learning design. In L. Lockyer, S. Bennett, S. Agostinho, & B. Harper (Eds.), *Handbook of research on learning design and learning objects: Issues, applications and technologies* (forthcoming). Hershey, PA: Information Science Publishing.

Conole, G., Crewe, E., Oliver, M., & Harvey, J. (2001). A toolkit for supporting evaluation. *Association of Learning Technology Journal, 9*(1), 38-49.

Conole, G., Littlejohn, A., Falconer, I., & Jeffrey, A. (2005). *Pedagogical review of learning activities and use cases.* LADIE Project report. Retrieved December 22, 2006, from http://www.elframework.org/refmodels/ladie/ouputs/LADIE lit review v15.doc

Conole, G., & Oliver, M. (1998). A pedagogical framework for embedding C and IT into the curriculum. *Association of Learning Technology Journal, 6*(2), 4-16.

Conole, G., Thorpe, M., Weller, M., Wilson, P., Nixon, S., & Grace, P. (2007). *Capturing practice and scaffolding learning design.* Paper presented at European Distance and E-Learning Network Conference, Naples, Italy.

Currier, S., Campbell, L., & Beetham, H. (2006). *Pedagogical vocabularies review.* JISC Pedagogical vocabularies project report 1. Retrieved February 28, 2007, from http://www.jisc.ac.uk/uploaded_documents/PedVocab_VocabsReport_v0p11.doc

DialogPLUS. (2006). *Nugget developer guidance toolkit.* Retrieved March 20, 2008, from http://www.nettle.soton.ac.uk/toolkit/

Durham, H., & Arrell, K. (2006). *Introducing new cultural and technological approaches into institutional practice: An experience from geography.* Paper presented at European Conference on Digital Libraries, Alicante, Spain. Retrieved March 21, 2008, from http://www.csfic.ecs.soton.ac.uk/Durham.doc

EDIT4L. (2007). *Evaluation of design and implementation tools for learning.* Project Web site. Retrieved March 21, 2008, from http://www.edit4l.soton.ac.uk:8081/

Falconer, I., & Littlejohn, A. (2006). *Mod4L Report: Case Studies, Exemplars and Learning Designs.* Glasgow, UK: Glasgow Caledonian University. Retrieved February 22, 2007, from http://mod4l.com/tiki-download_file.php?fileId=2

Falconer, I., & Conole, G. (2006). *LADIE gap analysis. Report for the JISC-funded LADIE project.* Retrieved February 22, 2007, from http://www.elframework.org/refmodels/ladie/guides/LADiE 20Gap 20Analysis.doc

Garrett, J. (2005). *Ajax: A new approach to Web applications.* Retrieved, March 01, 2007, from http://www.adaptivepath.com/publications/essays/archives/000385.php

IMS Global Learning Consortium. (2003). *Learning design specification version 1.* Retrieved May 25, 2007, from http://www.imsglobal.org/learningdesign/

IMS. (2004). *Content packaging specification.* Retrieved March 21, 2008, from http://www. imsglobal.org/content/packaging/

JISC. (2008). *Design for learning.* Retrieved March 25, 2008, from http://www.jisc.ac.uk/ whatwedo/programmes/elearning_pedagogy/ elp_designlearn.aspx

Jorum. (2007). *Jorum user terms of use.* Retrieved May 31, 2007, from http://www.jorum.ac.uk/user/ termsofuse/index.html

Koper, R., & Manderveld, J. (2004). Educational modelling language: Modelling reusable, interoperable, rich and personalised units of learning. *British Journal of Educational Technology, 35*(5) 537-551.

LAMS International. (2007). *Learning activity management system.* Retrieved March 25, 2008, from http://www.lamsinternational.com/

Laurillard, D. (2002). *Rethinking university teaching: A framework for the effective use of educational technology* (2nd ed.). London: RoutledgeFalmer.

Mayes, T., & de Freitas, S. (2004). *Review of e-learning frameworks, models and theories.* JISC e-learning models desk study. Retrieved, December 22, 2006, from http://www.jisc.ac.uk/ uploaded_documents/Stage 2 Learning Models (Version 1).pdf

Reload. (2006). *Reusable eLearning object authoring & delivery.* Project Web site. Retrieved March 21, 2008, from http://www.reload.ac.uk/

Chapter X
Concept Mapping to Design, Organize, and Explore Digital Learning Objects

David DiBiase
The Pennsylvania State University, USA

Mark Gahegan
The University of Auckland, New Zealand

ABSTRACT

This chapter investigates the problem of connecting advanced domain knowledge (from geography educators in this instance) with the strong pedagogic descriptions provided by colleagues from the University of Southampton, as described in Chapter IX, and then adding to this the learning materials that together comprise a learning object. Specifically, the chapter describes our efforts to enhance our open-source concept mapping tool (ConceptVista) with a variety of tools and methods that support the visualization, integration, packaging, and publishing of learning objects. We give examples of learning objects created from existing course materials, but enhanced with formal descriptions of both domain content and pedagogy. We then show how such descriptions can offer significant advantages in terms of making domain and pedagogic knowledge explicit, browsing such knowledge to better communicate educational aims and processes, tracking the development of ideas amongst the learning community, providing richer indices into learning material, and packaging these learning materials together with their descriptive knowledge. We explain how the resulting learning objects might be deployed within next-generation digital libraries that provide rich search languages to help educators locate useful learning objects from vast collections of learning materials.

INTRODUCTION

This chapter describes two applications of concept maps. In the first, educators used concept maps to guide the revision of a particularly challenging lesson in an introductory postgraduate class in geographic information science and technology. Student ratings and comments are analyzed to evaluate the effectiveness of the concept mapping approach and the learning object model adopted by contributors to the DialogPLUS project. Using the same lesson example, of map projections and coordinate systems, the second part of the chapter demonstrates how concept maps may be used as a framework for organizing learning objects, and for facilitating knowledge discovery by students.

CONCEPT MAPPING TO DESIGN REUSABLE LEARNING OBJECTS

Institutional Context

After a year and a half of planning and course development, the Department of Geography at the Pennsylvania State University (Penn State) began offering an online Certificate Program in geographic information systems (GIS) through the University's virtual "World Campus" in January 1999 (DiBiase, 2000). At first the program consisted of a year-long sequence of four non-credit, instructor-led classes, each ten weeks in length. Each class required 8-12 hours of weekly student activity on average. Although they were expected to complete weekly assignments, students were never expected to log into the class at any particular time or place. Students completed assignments using educational licenses of desktop GIS software (originally Intergraph's GeoMedia, later ESRI's ArcView). Students showcased their achievements in personal e-portfolios. Penn State instructors directed discussions and read and

responded to student questions daily. All class content delivery and communications were mediated through a Web-based learning management system (originally WebCT, later ANGEL, more recently ANGEL and Drupal). From January 1999 through December 2004, 519 off-campus students earned Penn State's Certificate of Achievement in GIS. The program earned ESRI's Special Achievement in GIS Award in 2004 for innovation in GIS education.

In 2004 Penn State's Graduate School and Board of Trustees approved the Department of Geography's proposal to create a new professional degree: the Master of Geographic Information Systems (MGIS). At the same time, the former non-credit Certificate Program was approved as a for-credit offering for postbaccalaureate students (those who already possess bachelor's degrees). Both offerings were approved for online delivery through the World Campus. An expanded curriculum was designed in consultation with an advisory board composed of industry leaders and scholars from four different academic programs and research centers. Students accepted to the MGIS program complete individual study projects supervised by academic advisors that culminate in public presentations at professional conferences with advisors in attendance. Two years after the new programs were approved, the number of distant students pursuing the MGIS degree and the Postbaccalaureate Certificate of Achievement (121 and 533, respectively, in academic year 2006-2007) now exceeds by a large margin the combined number of undergraduate and graduate students who seek the Department of Geography's on-campus academic degrees (BS, MS, and PhD).

Feedback from Students in an Introductory Class

The compulsory introductory class in both the Certificate and MGIS curricula is Geography 482: Nature of Geographic Information (DiBiase,

2008). This class provides an orientation to the fundamental properties of geographic data and to the practice of online learning. It traverses the wide range of technologies, professions, and institutions involved in producing geospatial data, including land surveying, the global positioning system, aerial photography and photogrammetry, social surveys such as those conducted by the U.S. Census Bureau, satellite remote sensing, cartography, and geographic information systems. In 2006-2007, course enrollment averaged 94 students per quarterly offering.

Instructors who supervise GEOG 482 conduct surveys to assess student expectations, concerns, and satisfaction at the outset, middle, and conclusion of every course offering. Summative surveys conducted at the end of each term use a standardized instrument called the "Student Ratings of Teaching Effectiveness" (SRTE). The SRTE consists of four or more seven-step Likert scales by which students are asked to rate their satisfaction with the quality and effectiveness of classes and instructors. Analyses of student satisfaction data appear in DiBiase (2004) and DiBiase and Rademacher (2005). Overall, students expressed considerable satisfaction with GEOG 482: median ratings on all criteria during all three study periods were only one step below the highest possible rating. Some elements of the class were consistently rated less satisfactory than others, however.

The SRTEs include two open-ended questions that invite students' comments: "What aspect of the course did you like best?" (hereafter referred to as "likes") and "What aspect of the course needs the most improvement?" ("dislikes"). DiBiase and Rademacher (2005) analyzed 55 "likes" and 36 "dislikes" from 50 unique respondents during the 1999-2000 study period (response rate 70 percent), and 71 "likes" and 51 "dislikes" from 55 unique respondents during the 2001 study period (56 percent response rate). Among other findings was a disproportionate number of critical comments about one of the nine course lessons,

namely Lesson 2: Scales and Projections. Among the eighteen lesson-related "dislikes" submitted by students in the 1999-2000 cohort, nearly half (eight comments) felt that the treatment of map projections and coordinate systems was inadequate. The DialogPLUS project (described in the Preface) provided an opportunity to revise the problematic lesson, as well as to rethink the methodology by which effective and reusable educational resources might be designed in the first place.

Reusable Learning Objects

A key objective of the DialogPLUS project was to share digital educational resources among the four participating institutions, and to disseminate resources widely. However, the project team soon discovered that there was little demand for shared resources, in part because participants lacked a shared conception of what a "learning nugget" was or should be. Lack of consensus about this issue was not unique to the DialogPLUS project, of course. The concept of a "learning object" remains "confusing and arbitrary" (Polsani, 2003) despite the efforts of standards organizations like the Learning Technology Standards Committee (LTSC) of the Institute of Electrical and Electronics Engineers (IEEE), and the Instructional Management Systems (IMS) project. It was to avoid the intellectual baggage associated with the term, in fact, that the DialogPLUS project purposely substituted the idiosyncratic jargon "learning nugget" in its successful proposal.

Among the many attempts to define learning objects, Wiley's (2002, p. 6) is among the most widely cited, and the broadest: "any digital resource that can be reused to support learning." To make sense of this extremely inclusive definition, Wiley offers a taxonomy of reusable learning object types, from "fundamental" objects (a single JPEG image file, for instance) to "generative-instructional" objects, which include "logic and structure for combining [lower level] learning objects...and evaluating student interactions with those combinations" (Wiley, 2002, p. 19). Such high-level learning objects, Wiley argues, should be more widely and effectively reusable than lower-level objects.

In clearer and more operational terms, L'Allier (1997) defines a "Learning Object structural component" as "the smallest independent instructional experience that contains an objective, a learning activity, and an assessment." While acknowledging L'Allier's definition as the most clearly articulated of those he reviewed, Polsani (2003, section1.2, paragraph 5) argues that:

Any definition that stipulates the intended use, method and measuring mechanism of a LO beforehand restricts the LO's reusability because the methodology, the intention and the assessment are determined by the instructional situation and not the LO itself.

Polsani therefore proposes that a learning object be understood as "an independent and self-standing unit of learning content that is predisposed to reuse in multiple instructional contexts" (Polsani, 2003).

After considerable discussion and experiment, many participants in the DialogPLUS project adopted a model most similar to L'Allier's, as illustrated in Figure 1. The DialogPLUS learning object conception includes three elements (DiBiase, 2005): a learning activity (e.g., guided exploration of a Web site or other widely-accessible resource, using a Web-based or desktop software application, a paper-and-pencil exercise downloaded as a PDF file, etc.), supporting material (e.g., text and graphics and/or digital video) needed to situate the activity within a knowledge domain and a set of educational objectives, and some form of self-assessment (e.g., an automated quiz that provides immediate feedback) by which students can gauge the extent to which they have achieved the objectives.

As Wiley (2002, p. 9) points out, "the most difficult problem facing the designers of learning objects is that of 'granularity.' How big should a learning object be?" Polsani (2003, section 3.1 paragraph 1) advises that:

A LO, ideally, should include only one or a few related ideas. The rule to be applied is: how many ideas about a topic can stand on their own and can be reused in different contexts?

Figure 1. Elements of reusable learning objects as adopted by the DialogPLUS project and implemented in the revision of the Geography 482 lesson

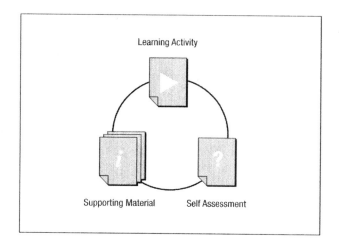

We turned to concept mapping to help parse lessons (whose granularity is determined by the amount of time a student can be expected to devote to his or her studies in a given week) into learning objects (whose granularity should reflect the discreteness of particular concepts).

Concept Mapping

Concept maps employ nodes (representing conceptual elements), connecting lines (representing relations between elements), connecting words (which categorize relations), and patterns (as in multidimensional scaling) to depict the content and structure of a subject area. Concept maps are similar to "mind maps" (Buzan & Buzan, 1996), though the latter tend to be simpler in form, with words and ideas branching radially from a single central key word or idea. Novak (1990) points

out that concept maps can be used in four ways: (a) as a learning strategy, (b) as an instructional strategy, (c) as a strategy for curriculum planning, and (d) as a means of assessing students' understanding of science concepts. Concept mapping now has a very active community of developers and practitioners, as evidenced by software such as Cmap Tools (http://cmap.ihmc.us/) and their growing user communities. In comparison with many applications in teaching, learning and assessment, the literature on concept mapping as a curriculum planning strategy is relatively small (but see Allen et al., 1993; Edmondson 1993; Martin, 1994). Navarro et al. (2005) introduced the notion of using concept mapping to design formal digital learning objects at about the same time the project reported here was underway.

Figure 2 below is the final version (after several revisions) of the concept map that represents topics

Figure 2. Concept map depicting elements of Geography 482 Lesson 2, including scale, coordinate systems, datums, and transformations (including coordinate transformations, datum transformations, and map projections)

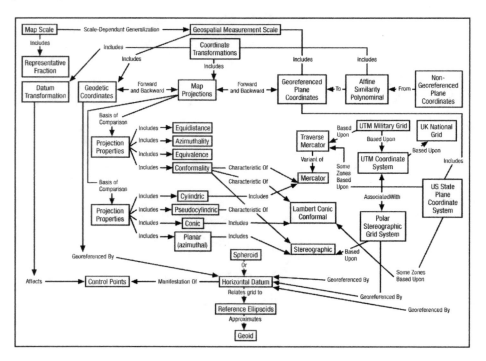

Figure 3. Concept map depicting elements of Geography 482 Lesson 2 overlaid with the boundaries of seven reusable learning objects

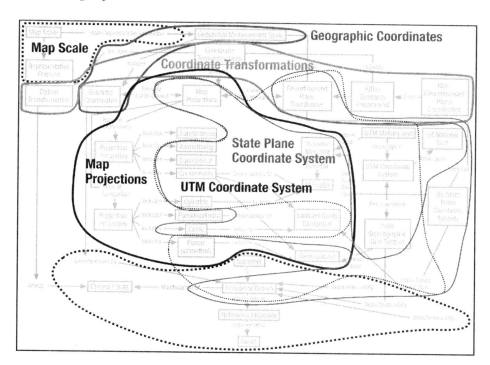

included in Lesson 2 of Geography 482. Figure 3 shows the same concept map, but superimposed with the boundaries of seven learning objects: map scale, geographic coordinate system, datums, coordinate transformations (including plane coordinate transformations, datum transformations, and map projections), the UTM coordinate system, the State Plane Coordinates system, and map projections.

One advantage of using concept maps to define lesson content is that concept maps seem to foster collaboration better than text outlines. The concept map shown in Figure 2, for example, elicited several discussions with colleagues that advanced the object author's understanding of the subject as well as colleagues' appreciation of the author's efforts and insights.

Learning objects are meant to be re-combinable as well as reusable. This implies that the linear sequence of topics embodied in the typical course outline must be broken down and that overlaps

between related topics must be identified and built into freestanding objects. The spatial format of the concept map helps in both ways—linear sequencing is absent and overlaps between topics are revealed. Notice in Figure 3 the close correspondence between the objects Map Projections, UTM Coordinate System, and State Plane Coordinate System, which follows from the fact that the two coordinate systems share a similar basis in conformal map projections.

Example Learning Object

Figures 4, 5, and 6 are excerpts from one of the seven learning objects produced from the concept map and learning object design layout shown in Figure 3. Figure 4 is a learning activity in which students manipulate a Flash application to demonstrate their ability to locate positions specified with randomly generated pairs of geographic

Figure 4. Learning activity element of a reusable learning object on the geographic coordinate system. One of seven learning objects that make up the revised Lesson 2 of Geography 482: The Nature of Geographic Information.

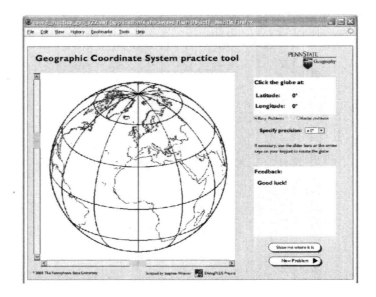

Figure 5. First page of supporting material element for a reusable learning object on the geographic coordinate system. One of seven learning objects that make up the revised Lesson 2 of Geography 482: The Nature of Geographic Information.

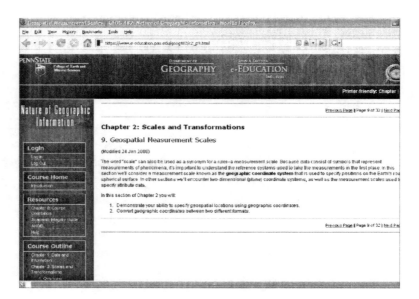

Figure 6. Self-assessment element of a reusable learning object on the geographic coordinate system. One of seven learning objects that make up the revised Lesson 2 of Geography 482: The Nature of Geographic Information.

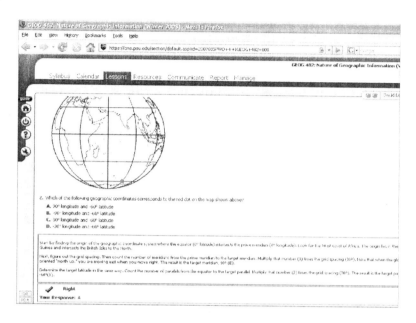

coordinates. Figure 5 is a portion of the supporting material that specifies learning objectives for the activity and situates it within the context of geospatial measurement scales. Figure 6 shows part of an automated quiz that enables students to self-assess the extent to which they have fulfilled the educational objective of the object.

Evaluation of Example Learning Objects

To evaluate the impact of the concept mapping approach to learning object design we analyzed responses to surveys of GEOG 482 students during the first two terms after the revised Lesson 2 was deployed. A voluntary formative assessment conducted midway through the Winter 2005 term (which spans the months of January to March) attracted responses from 35 of 67 students (52%

response rate). Comments such as "...material easily grasped even though I have no background..." and "...consider myself a 'GIS Professional' for 10+ years ... very impressed with amount of 'new' material..." indicated continued satisfaction with the class overall. However, criticisms of Lesson 2 persisted, including:

Lesson 2 is "a little difficult to follow"

Lesson 2 is "difficult to understand" ... additional explanations, examples, supplementary readings would be helpful

"Spreading this [Lesson] out over a couple of weeks" would have made the effort more comparable to other lessons

Maybe lesson two deserves a serious re-write.

One response to the voluntary summative survey conducted at the conclusion of the Winter 2005 term (which attracted a 46% response rate) suggests that students did appreciate the combination of elements that make up the DialogPLUS learning object model: "I appreciate the combination of 'lecture' and 'try it' exercises. The practice quizzes helped greatly in making sure I was mastering the material, and projects took mastery to another level—a good mix."

The midterm formative assessment and final summative assessment were repeated during the Spring 2005 term (April through June). This time, prior to opening the Lesson, the instructor attempted to manage students' expectations by warning them that Lesson 2 would be difficult. Still, criticisms persisted. For example, one respondent to the midterm formative assessment survey (which attracted comments from 20 out of 45 students) observed that "the coordinate transformation section was very informative and gave examples of realistic job-related problems ... [but it] could have been expanded..." The summative evaluation at the end of the term (which attracted seventeen comments overall) elicited two suggests for Lesson 2:

more 'try this' activities might be useful. Especially in Lesson 2...

add a book—and rewrite lesson 2.

We offer several observations about this example. One is that we have no evidence—neither from students' comments nor their performance on assignments—that the concept mapping methodology was effective in rendering difficult concepts more accessible. We do find some evidence, however, that the activity-centered DialogPLUS learning object model did help us to enrich the problematic lesson. Another observation is that the concept mapping methodology was harder work than the more familiar approach of writing a linear narrative. Given the fact that the learning

objects produced for this project were not reused at any of the partner institutions, and that the revised lesson generated no fewer complaints from Penn State students, we cannot in good conscience recommend that colleagues adopt the concept mapping approach. We do, however, promote activity-centered learning object design.

CONCEPT MAPPING TO ORGANIZE AND NAVIGATE LEARNING OBJECTS

The above examples show how learning objects appear when deployed in contemporary learning management and content management systems (in this case, ANGEL and Drupal). In the first part of this chapter, concept mapping is used only in the design stage, since the prevailing technologies enforce linear progressions through prescribed sequences of learning objects. But what if the concept map itself formed a navigable interface onto the learning objects and their descriptive metadata, so that the learner could use it to explore connections between the various aspects of the subject? This section examines the possibilities for learning object technology in the future by giving examples of the same set of learning objects implemented within an interactive concept mapping environment: ConceptVista (GeoVISTA Center, 2005).

To date, most applications of concept mapping tend to be rather "closed" in nature; that is, the concept mapping systems tend to be stand-alone applications that do not leverage burgeoning Web technologies to full advantage. There are, of course, notable recent exceptions. For example the work of McClellan et al. (2004) uses keywords from concept maps to search for learning content in a large collection of learning resources. Newer Web technologies are able to leverage concept maps in many interesting and powerful ways. For example, Henze et al. (2004) show how more formal reasoning, using description logics and ontology (themselves Semantic Web technologies), can provide a useful underpinning to describe the process of learning for individuals and help to connect them with useful resources. Rodríguez-Artacho and Verdejo Maíllo (2004) describe a rich language for annotating educational content as a series of layers that describe aspects from sequencing of learning activities to the structure of the activities themselves. Many other authors have stressed the usefulness of ontology and Semantic Web technologies in describing, and reasoning about, the learning process (e.g., Qin & Finneran, 2002; Buendia-Garcia & Diaz, 2003), so much so that a workshop series is devoted to the subject; the latest proceedings of the Ontologies and the Semantic Web for e-Learning (SWEL '07) are available from http://compsci.wssu.edu/iis/swel/SWEL07/swel07-aied07.html. Concept-Vista is used here to form a bridge between these new technologies and the domain of educational concept mapping.

The ConceptVista project began at Penn State to support collaborative research between geographers studying human-environment interaction,

as a way to capture and share their evolving conceptualizations. During the DialogPLUS project, ConceptVista was heavily modified to support learning object technology. To this end, it supports the following functionalities: (1) interactive exploration of concept maps, directly connected to Web resources such as text, images, video, and assessment and evaluation instruments, (2) an adaptive, automated layout—where the level of detail shown can be controlled so that learners do not become overwhelmed, (3) advanced search capabilities, to help the learner gather additional educational resources from a number of document collections, (4) the ability to connect such discovered resources to the concept map for future use, (5) management of temporal properties for concepts, to show the sequence of topics within a curriculum or syllabus, and (6) close integration of pedagogic descriptions, produced from the Learning Design Toolkit (see Chapter IX), to describe the learning approaches and outcomes in a concise and clear manner. It is an experimental system, as yet to be formally evaluated in a learning setting. Further details of

Figure 7. Concept map of a learning object, shown within ConceptVista; See text for full description

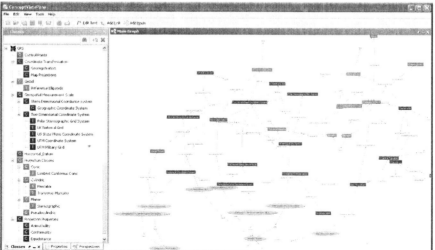

ConceptVista in support of teaching and learning are found in Gahegan et al. (2007).

Some examples follow, captured directly from using the tool described. Figure 7 shows part of the concept map from Figure 3 as it appears in ConceptVista. Learners can interact with the concept map by either clicking on the concepts and relationships shown in the graph to the right, to explore their content and connections, or by selecting from an ordered list of concepts shown on the left. How the concepts are styled and visualized is up to the developers and users. In this diagram, concepts to be learned are shown using dark colors and rectangular boxes, with different colors being used to differentiate various high-level themes (such as "projection classes" and "geospatial measurement scale"). Assessment tools and activities are shown as light gray ellipses, towards the bottom of the diagram. The layout of the concept maps is adaptive; the user can double-click on a concept to expand it and expose any further detail it may contain, or to remove detail where it is no longer required.

Figure 8, below, shows a close-up of some of the Learning Design Toolkit (LDT) metadata tags, themselves visualized as a concept map. The LDT tags can be loaded into ConceptVista as a separate resource, to help describe the kinds of learning that will be engaged when using a specific learning object. The left panel shows the decomposition of learning design concepts: "Assessment," "Interaction," "Learning Approach," "Outcome," and "Task." The educational designer can create links between these learning concepts and the subject matter concepts, for example to show that the outcome from studying a particular topic will be to (say) "Identify_Cause_of" and "Explain" or "Illustrate" it.

Figure 9 shows that the same learning content described above (such as Web content, articles, assessment instruments) can be directly attached to nodes in the concept map. In this case, the learner has selected to browse the Web resources associated with the concept "U.S. State Plane Coordinate System," and this content is shown to the right of the concept map. ConceptVista uses

Figure 8. Concepts from the Learning Design Toolkit are imported into ConceptVista so that pedagogic terms can be associated with subject matter and related activities; See text for full details

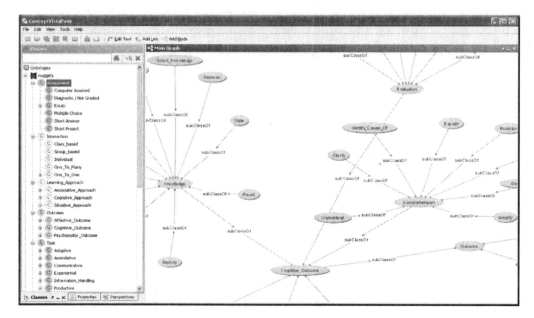

Figure 9. Close-up of the same learning object, with browser at right displaying learning materials connected to a specific concept; See text for full description

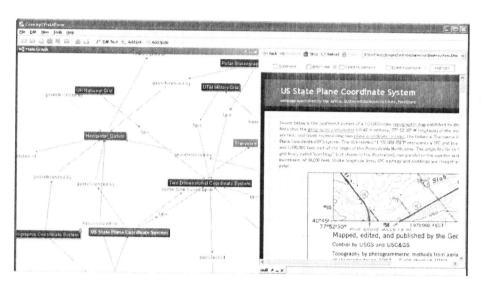

Java Desktop Integration technology to directly embed a Mozilla or Explorer Web Browser within the application. This allows close coupling of the concept map and associated resources, and supports a full range of Web content, including text, images, movies and associated hyperlinks and markup. In addition, the top panel of the browser allows the user to markup (highlight) words in any Web document that are related to any concept of interest; such as synonyms, antonyms, hypernyms (generalizations), or hyponyms (specializations). This can make it easier to understand jargon when it is first encountered. The language corpuses WordNet and OpenCyc are used to provide this markup service, via a remote server.

As a final example, Figure 10 depicts a user searching for additional learning resources that relate to the "Geoid" concept (an idealized model of the Earth's surface). The four boxes to the right of the figure show a search for materials related to "Geoid" in four different remote collections that ConceptVista supports. The bottom right box shows results from Google, via the Google search API. Clicking on any highlighted link

will access the document directly, as one would expect. Bottom left shows CiteSeer (College of Information, Science and Technology, n.d.), a search engine for scientific literature, which has indexed over 700,000 documents. CiteSeer provides links to the text of these documents. Results from a CiteSeer query can also be imported into ConceptVista as a special kind of graph showing authors, keywords, institutions, countries, and other document metadata. Top right is a search of Amazon.com's book catalog, via the Amazon.com search API, thus the returned links are to books indexed by the keyword "Geoid." Top left shows the results returned from the Digital Library for Earth Systems Education (DLESE, n.d.), which contains a vast collection of educational resources for the Earth Sciences, spanning all age ranges from primary school to university. DLESE uses its own distinct metadata geared towards educational needs, and based around national Earth Science curricula, which we have also imported into ConceptVista as a useful resource with which to describe content and activities.

Figure 10. A search for additional learning materials related to the term 'Geoid', showing the four distinct search tools (at right) which ConceptVista supports: Google, Amazon, CiteSeer and DLESE; See text for full description

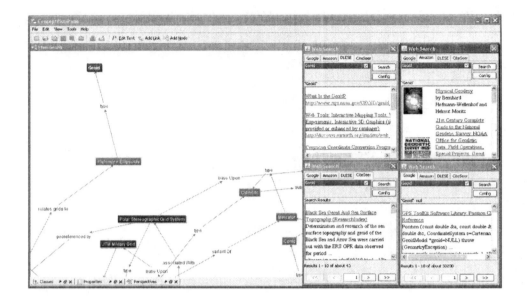

Table 1. Possible structure of a Learning Object developed to address both theoretical and practical considerations involved in sharing learning content between institutions

Learning Object Aspect	Example content
Subject Matter:	learning resources, concept maps, etc.
Assessment Tools:	tests, quizzes, etc.
Pedagogy Descriptions:	learning and assessment styles, pre-requisite concepts and skills, learning outcomes, etc.
Deployment:	mapping to specific Course Management Systems or Virtual Learning Environments, etc.
Experiential:	evaluation tools, user-feedback and ranking, use-cases, etc.
Institutional:	articulation agreements, copyright and fair use, etc.

CONCLUSION

The first part of this chapter describes a methodical approach to the design of course materials for an introductory class in geographic information science and technology. Concept mapping is used to portray the content of a lesson in a way that overcomes the linear sequencing of typical course outlines while revealing explicitly the connections and overlaps among related concept elements. The concept map is then parsed into learning objects by drawing boundaries around the (often overlapping) elements of discrete concepts. The DialogPLUS project team adopted a conception of reusable learning objects that emphasizes student activity, scaffolds activities with supporting text, graphics, and communications, and enables students to self-assess their mastery of the educational objectives that learning objects are designed to achieve. Evidence suggests that the benefits associated with the concept mapping approach to e-learning object design may not justify the extra effort required. However, it is apparent that adult students appreciate the emphasis on student activity and frequent self-assessment opportunities associated with the DialogPLUS learning object design model.

The second part of the chapter demonstrated how concept maps might be used as a framework for organizing learning objects to facilitate learning and additional knowledge discovery and integration by students.

In the longer-term, learning object technology has the potential to embrace next-generation learning environments, to address the need for learner-evaluation and peer-assessment, and even to negotiate institutional and copyright issues. In this expanded form, a learning object might embody a structure similar to that set out in Table 1.

ACKNOWLEDGMENT

The National Science Foundation provided support for the research and development reported here. Thanks to Stephen Weaver for assistance in developing Flash applications to support learning activities in Geography 482, Lesson 2.

REFERENCES

Allen, B.S., Hoffman, R.P., Kompella, J., & Sticht, T.G. (1993). Computer-based mapping for curriculum development. In *Proceedings of selected Research and Development Presentations Technology sponsored by the Research and Theory Division*, New Orleans, LA. (Eric Document Reproduction Services No. ED 362 145).

Buendia-Garcia, F., & Diaz, P. (2003). A framework for the specification of the semantics and the dynamics of instructional applications. *Journal of Educational Multimedia and Hypermedia, 12*(4), 399-424.

Buzan, T., & Buzan, B. (1996). *The mind map book: How to use radiant thinking to maximize your brain's untapped potential.* New York: Plume.

College of Information Sciences and Technology. (no date). *CiteSeer.IST: Scientific Literature Digital Library.* University Park, PA: The Pennsylvania State University. Retrieved January 26, 2008, from http://citeseer.ist.psu.edu/

DiBiase, D. (2000). Is distance education a Faustian bargain? *Journal of Geography in Higher Education, 24*(1), 130-135.

DiBiase, D. (2004). The impact of increasing enrollment on faculty workload and student satisfaction over time. Journal of Asynchronous Learning Networks, 8(1), 45-60. Retrieved January 26, 2008, from http://www.aln.org/publications/jaln/v8n2/pdf/v8n2_dibiase.pdf

DiBiase, D., & Rademacher, H. J. (2005). Scaling up: How increasing enrollments affect faculty and students in an asynchronous online course in geographic information science. *Journal of Geography in Higher Education, 29*(1), 141-160.

DiBiase, D. (2005). Using concept mapping to design reusable learning objects for e-education in cartography and GIS. In *Proceedings, International Cartographic Association*, A Coruña, Spain.

DiBiase, D. (2008). *Nature of geographic information: An open geospatial textbook.* University Park, PA: The Pennsylvania State University. Retrieved January 26, 2008, from http://natureofgeoinfo.com

DLESE. (no date). *Digital library for earth systems education.* Retrieved January 26, 2008, from http://www.dlese.org/library/index.jsp

Edmondson, K. M. (1993). *Concept mapping for the development of medical curricula.* Paper presented at the Annual Conference of the American Educational Research Association, Atlanta, GA. (Eric Document Reproduction Services No. ED 360 322).

Gahegan, M., Agrawal, R., & DiBiase, D. (2007). Building rich, semantic descriptions of learning activities to facilitate reuse in digital libraries. *International Journal on Digital Libraries, 7*(1-2), 81-97.

GeoVISTA Center. (2005). *ConceptVista: Ontology management.* University Park, PA: The Pennsylvania State University. Retrieved January 26, 2008, from http://www.geovista.psu.edu/ConceptVISTA/index.jsp

Henze, N., Dolog, P., & Nejdl, W. (2004). Reasoning and ontologies for personalized e-learning in the Semantic *Web. Educational Technology & Society, 7*(4), 82-97.

L'Allier, J. J. (1997). *Frame of reference: NETg's map to its products, their structures and core beliefs.* Intermedia. Retrieved July 18, 2006, from http://www.im.com.tr/framerefer.htm

Martin, D. J. (1994). Concept mapping as an aid to lesson planning: A longitudinal study. *Journal of Elementary Science Education, 6*(2), 11-30.

McClellan, J. H., Harvel, L. D., Velmurugan, R., Borkar, M., & Scheibe, C. (2004). CNT: Concept-map based navigation and discovery in a repository of learning content. In *34th Annual ASEE/ISEE Frontiers in Education Conference,* Savannah, GA (Session F1F, pp. 13-18). Retrieved May 5, 2008, from http://ieeexplore.ieee.org/iel5/9652/30543/01408581.pdf?tp=&isnumber=&arnumber=1408581

Navarro, L. I., Such, M. M., Martin, D. M., Sancho, C. P., & Peco, P. P. (2005). Concept maps and learning objects. In *Proceedings of the Fifth IEEE International Conference on Advanced Learning Technologies* (ICALT'05) (pp. 263-265).

Novak, J. D. (1990). Concept mapping: A useful tool for science education. *Journal of Research in Science Teaching, 10*, 923-949.

Polsani, P. R. (2003). Use and abuse of reusable learning objects. *Journal of Digital Information, 3*(4). Retrieved July 18, 2006, from http://jodi.tamu.edu/Articles/v03/i04/Polsani/

Qin, J., & Finneran, C. (2002). Ontological representation for learning objects. In *Proceedings of the Workshop on Document Search Interface Design and Intelligent Access in Large-Scale Collections,* JCDL'02, Portland, OR, USA.

Rodríguez-Artacho, M., & Verdejo Maíllo, M. F. (2004). Modeling educational content: The cognitive approach of the PALO language. *Educational Technology & Society, 7*(3), 124-137.

Wiley, D. A. (2002). Connecting learning objects to instructional design theory: A definition, a metaphor, and a taxonomy. In D. A. Wiley (Ed.), *The instructional use of learning objects.* Bloomington, IN: Agency for Instructional Technology and Association for Educational Communications and Technology. Retrieved April 30, 2005, from http://www.reusability.org/read/

Chapter XI
Semantic Tools to Support the Construction and Use of Concept–Based Learning Spaces

Terence R. Smith
University of California at Santa Barbara, USA

Marcia Lei Zeng
Kent State University, USA

ABSTRACT

We describe a digital learning environment (DLE) organized around sets of concepts that represent a specific domain of knowledge. A prototype DLE developed by the Alexandria Project currently supports teaching at the University of California at Santa Barbara. Its distinguishing strength is an underlying abstract model of key aspects of any concept and its relationship to other concepts. Similar models of concepts are evolving simultaneously in a variety of disciplines. Our strongly-structured model (SSM) of concepts is based on the viewpoint that scientific concepts and their interrelationships provide the most powerful level of granularity with which to support effective access and use of knowledge in DLEs. The SSM integrates a taxonomy (or thesaurus), metadata (or attribute-value pairs), domain-specific mark-up languages, and specific models for learning scientific concepts. It is focused on attributes of concepts that include objective representations, operational semantics, use, and interrelationships to other concepts. The DLE integrates various semantic tools facilitating the creation, merging, and use of heterogeneous learning materials from distributed sources, as well as their access in terms of our SSM of concepts by both instructors and students. Evidence indicates that undergraduate instructional activities are enhanced with the use of such integrated semantic tools.

INTRODUCTION

In this chapter we describe the design and implementation of a digital learning environment (DLE) that may be organized around sets of concepts selected by an instructor to represent a specific domain of knowledge. While such DLEs are applicable to the whole range of scientific and engineering disciplines, the operational DLE described in this chapter supports a core introductory course in physical geography that is currently being taught at the University of California, Santa Barbara (UCSB). This DLE was designed and implemented as part of UCSB's Alexandria Digital Earth Project (ADEPT). The main goal of ADEPT has been the design, implementation, testing, and application of DLEs that take advantage of digital library (DL) collections and services in the construction and use of learning materials, particularly for undergraduate classes (ADEPT, 2003). A further goal for the development of the ADEPT DLE has been to foster the development of a student's understanding of specific domains of knowledge by organizing and accessing a large array of learning materials drawn from the collections of a DL.

DIGITAL LEARNING ENVIRONMENTS (DLE) FOR TEACHING SCIENCE

The Rationale

A necessary condition for understanding any domain of science is that students possess sufficient familiarity with the concepts, as well as with the interrelationships between concepts, that are employed in representing the phenomena of that domain. Many approaches to teaching and learning at the undergraduate level, however, typically present the concepts that are employed in representing the phenomena in ways that are implicit and relatively superficial.

Traditional courses in physical geography, for example, typically confront students with large arrays of scientific concepts that are introduced sequentially in textbooks and usually "modeled" as lists of keyword terms at the end of chapters and/or described in terms of definitions given in glossaries at the end of the book. Students typically have significant problems in developing and using integrated cognitive representations of these concepts, their interrelationships, and their applications to scientific understanding. Their confusion is well summarized in a common question about the keyword lists at the end of each chapter: "Do we have to remember all of the words in these lists?" In particular, this question indicates a lack of understanding of either the role that concepts play in the representation and organization of scientific knowledge or of the interrelationships between different concepts.

It is clear that students in traditional course environments are rarely presented with much of the information about concepts that is critical to an understanding of their role in representing scientific knowledge. Such information includes, for example, the discovery, the various syntactic representations, the semantics, the manipulation, and the interrelationships of concepts. As a result, students too often emerge from a course of study with: (1) limited notions of the nature of concepts; (2) large, but poorly-structured, memorized sets of terms denoting concepts; (3) incompletely structured associations of relationships between concepts; and (4) partial knowledge of how to use concepts in creating representations of phenomena or how to create new concepts as the need arises in some domain of science.

In relation to learning, there is a growing consensus that science education should be an activity in which students learn to think like scientists rather than solely memorize information. This strongly suggests the importance of students developing an understanding of scientific concepts and their role in the scientific approach, including their representation, creation, use, and evaluation. The National Science Education Standard (NRC, 1996), which makes frequent reference to concepts, calls for students to engage in scientific reasoning in which they become familiar with modes of scientific inquiry, rules of evidence, ways of formulating questions, and ways of proposing explanations. The success of such activities depends on the depth of understanding of the concepts underlying the practice of science.

There are three generally accepted principles concerning students' acquisition of a deep understanding of scientific knowledge (Mayer, 1991, 2001; Bruer, 1993) that we may interpret in terms of concepts. These principles, derived from research in learning and education, are:

- **Domain-specific learning:** Students learn best when cognitive skills are taught in the context of a specific domain of knowledge as opposed to general contexts. An interpretation of this principle in terms of concepts suggests the importance of explicitly supporting students' acquisition of a deep understanding of the concepts, concept interrelationships, and reasoning skills related to developing and using scientific representations of phenomena in any domain of scientific knowledge.
- **Case-based learning:** Students learn best when cognitive skills are learned in the process of solving authentic problems rather than when pieces of information are presented as isolated facts to be learned. A concept-based interpretation suggests the importance of constructing a learning environment in which specific problems are presented in ways to motivate the use of sets of concepts and interrelationships germane to the problems.
- **Scaffolded learning:** Students learn best when the task difficulty is adjusted to meet their capabilities. This interpretation suggests providing students with subsets of

concepts and concept representations appropriate to their level of knowledge.

From these principles, we may infer the value of learning environments that have an explicit focus on concepts and their interrelationships. Such environments do, we believe, lead students to a deeper understanding of: (1) the nature, structure, and classes of the concepts that, together with the interrelationships between them, provide a basis for representing scientific knowledge; (2) the scientific roles of various classes of concepts across the spectrum of scientific activities; and (3) the global structure of various domains of scientific knowledge as the underlying framework of concepts.

Technologies that are being used to develop digital libraries (DLs) and digitally-based learning environments offer great promise for building environments that are focused on organizing, accessing, and applying scientific knowledge explicitly based on the sets of concepts that underlie various domains of scientific knowledge, as well as on providing a deeper understanding of the role of concepts and their interrelationships in scientific understanding. With current digital library technology, it is possible to develop digital learning environments (DLEs) in which we may integrate:

- Pedagogically useful models of concepts and their interrelationships;
- Domain-specific knowledge bases (KBs) of such representations;
- Associated DL collections of "illustrative materials" concerning different aspects and attributes of the concepts; and
- Services supporting the creation, modification, viewing, and use of concepts for various purposes in learning contexts.

From the instructor's viewpoint, such learning environments facilitate the efficient reuse and re-purposing of the KBs of concepts and the associated collections and services in creating new instructional materials.

Supporting DLEs with Digital Library Technologies

While a DL is often regarded as no more than a digitized collection with information management tools, it is in fact an information environment that integrates collections, services, and people in support of the full life cycle of creation, dissemination, use, and preservation of data, information, and knowledge (Duguid & Atkins, 1997). A particular strength of DL technology lies in its ability to incorporate services supporting content-based organization of, access to, and use of its collections in learning applications. Many applications of DL technology, such as the National Science Foundation's program for constructing a national DL to support education in science, engineering, mathematics, and technology (NSF, 2003), depend for their success on this ability.

Despite this potential, however, a significant portion of current DL development activities that support learning is merely focused on providing fast and efficient access to educational materials in terms of traditional "information containers," such as electronic versions of books, journal articles, dissertations, images, and videos. The richness and diversity of the learning materials that may be accessed from both local and networked sources, however, makes it difficult for instructors to create and maintain an environment in which knowledge is organized and integrated based on content (e.g., controllable and consistent sets of scientific concepts) rather than merely the whole container (e.g., physically identifiable and accessible files of a whole book or chapter) of the content. This is, in fact, one of the weaknesses of using Web portals to provide access to un-integrated collections of learning materials.

When services that support content-based organization are focused on support for learn-

ing, DLs evolve naturally into what we may term "digital learning environments" (DLEs). DLEs offer two advantages over the learning environments that are based on traditional information containers. The first is the ease with which rich and diverse learning materials may be accessed from both local and networked sources, including DLs. The second is the capability of knowledge representation with Web technologies for integrating such materials at fundamental levels of knowledge granularity, such as the level of scientific concepts.

The availability of established and emerging DL technologies for providing more fundamental access to learning materials, as well as our experiences in having taught a wide variety of concept-rich undergraduate courses in geography, computer science, and library and information sciences for many years, has led us to believe that a valuable approach to accessing and presenting learning materials for undergraduate students studying scientific and engineering disciplines may be based on organizing the learning materials according to sets of objective scientific concepts and their interrelationships. Objective definable concepts provide the fundamental "granules" of knowledge from which scientists create their symbolic representations of phenomena. Such granularity is characteristically reflected in the manner in which scientific knowledge is organized and represented in learning materials such as lectures, textbooks, and laboratory manuals. These materials are typically structured around a sequence of some selected set of concepts that is augmented with additional information, although a "formal acknowledgement" of such organization is typically relegated to lists of key terms or glossaries in textbooks.

Knowledge bases of concept representations provide an excellent basis for organizing the DL collections concerning particular domains of knowledge as per their basic concepts and relationships. Not only do such KBs facilitate the creation of different conceptual organizations of knowledge about a domain, but they also support a greater focus on critical aspects of concepts. They may, for example, be employed in creating tailored courses of instruction that take the form of a "trajectory" through the space of concepts underlying some domain of knowledge. This approach contrasts strongly with the traditional organization of, and access to, scientific knowledge in such standard containers of this knowledge as books.

The approach taken in this chapter regarding the question of organizing DLEs is based on the premise that "scientific concepts" and "relationships between concepts" provide a powerful, and perhaps the only, level of granularity with which to support effective access and use for learning. As we show below, DLEs are capable of supporting the construction of KBs of concepts and the services that support the integration of diverse information sources. We have designed and implemented semantic tools for use in environments in which heterogeneous learning materials from many distributed sources may be explicitly synthesized, integrated, accessed, and organized on the basis of strongly structured models (SSMs) of concepts. These SSMs of concepts, which we describe more fully below, are focused on such attributes of a concept as its objective representations, operational semantics, use, and interrelationships with other concepts, all of which play important roles in constructing representations of phenomena that further the understanding of scientific domains of knowledge.

THE ADEPT DIGITAL LEARNING ENVIRONMENT AND ITS SERVICES

The DLE developed as part of UCSB's Alexandria Digital Earth Project (ADEPT) is essentially a set of KBs and DL services organized around our SSM of concepts. It was designed to provide a set of services that support instructors in organizing learning materials based on SSMs of key concepts, and that support students in accessing and using learning materials in terms of concepts. For developmental concreteness, the ADEPT DLE was initially designed and tested in teaching a relatively large undergraduate level class in physical geography during the period of 2002-2008 at UCSB. The domain of physical geography is a valuable test case, since it employs concepts from a large number of scientific fields, in addition to having a large and well-developed collection of its own concepts. We note, however, that the abstract model of the ADEPT DLE is applicable to any domain of science or engineering.

Figure 1. A Classroom presentation. Left to right: knowledge window, lecture window, and collection window

DLE in Classroom Settings

In current applications of the DLE in classroom settings, items from each of three basic collections (knowledge bases of concepts, a DL collection of lecture presentations, and a DL collection of learning objects) are separately projected onto three different screens (termed from left to right the ``knowledge window,'' the "lecture window," and the "collection window"). An edited photograph of the three screens taken during a class presentation is shown in Figure 1.

The lecture window (centre) controls the content of the knowledge window (left) and collection window (right) through icons and links pre-built into the lecture, although direct search over the knowledge bases and the learning object collections may be directed from these windows if desired. It should be noted that although the three screen setting seems ideal, it is feasible to present this material if only one or two screens are available. The instructor can, for example, navigate among three opened browser windows on one screen.

Materials from the Supporting ADEPT Collections

The three screens represent materials from the supporting ADEPT collections. The first is the collections of learning objects from the DL that may be used for exemplifying and illustrating concepts in terms of their representation, meaning, use, and interrelationships. The collection window (i.e., the right screen in Figure 1) is used in presenting such material and is controlled through icons and links in the lecture window. For example, by clicking on the icons listed under a concept (see Figure 2), corresponding objects such as images, maps, figures, videos, and simulation models that exemplify the phenomena and the associated concepts are projected on the collection window. This functionality can usually be supported by any DL.

The second collection of materials from supporting ADEPT collections consists of re-usable presentation materials such as lectures or laboratory sessions. They are composed by integrating items from the KBs and the learning object

Figure 2. A lecture window

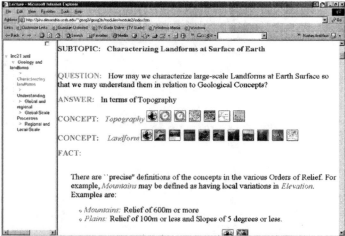

Table 1. The structure in each lecture (as showing in the lecture window)

Teaching Objective	Lecture Component
Identification of Scientific Phenomena	TOPIC; SUBTOPIC; SUB-SUBTOPIC; OBSERVATIONAL PROCEDURE; EXAMPLE
Representation of Scientific Phenomena	FACT; CONCEPT; THEORY
Understanding of Scientific Phenomena	QUESTION/ANSWER; PROBLEM/SOLUTION; HYPOTHESIS/EVALUATION; STATEMENT/DERIVATION; PREDICATION/TEST; COMMENT

collections and may be organized as trajectories through some relevant space of concepts. The lecture presentations shown in the lecture window (centre screen in Figure 1) form the backbone of any class (Figure 2).

Each lecture may be structured according to the teaching objectives, as shown in Table 1.

The third type of materials from the supporting ADEPT collection is one or more knowledge bases (KBs) containing collections of SSMs of relevant scientific concepts that are necessary and/or sufficient for representing knowledge in the given domain of science. During a presentation, the conceptualization may be shown as a dynamic graph on the left screen, or knowledge window (see Figure 1). As the instructor teaches the concepts, relationships between the concept being studied and a set of related concepts are automatically centered on the screen (see Figure 3). Figure 3 shows the output of the Java applet that represents the conceptualizations of the concept mass movement from the physical geography class. This representation shows four levels of conceptualization and conceptual structures that may be used in "explaining" the concept mass movement. For example, one may note that the figure includes the hierarchical structuring of the four concepts: MassMovement← Mechanics ← ForceBalance ←Stability.

Graphic displays of "concept spaces" visually indicate the relationships and important subsets of concepts, particularly subsets that constitute ontological commitments for representing a given phenomena. These displays are intended to provide students with large-scale (and even global) views of the structure of concept spaces. The conceptualization operation allows users (instructors or students) to associate a given concept occurring in a presentation (such as stream velocity) with other concepts occurring in the presentation (such as depth, slope, roughness), providing a network structure among the concepts.

A concept map is naturally interrelated with all the other elements in the SSMs of concepts. The knowledge bases of SSMs enable the generation of dynamic and scalable concept-centered maps and presentations. This mechanism also solves problems of concept map generation because it would be impractical to rely on the preparation of many maps in advance, even though most concept maps described in the literature or on the Web are "hand-built."

The ADEPT DLE supports students' ability to "zoom-in" to relevant, and more detailed, components of some concept spaces (see Figure 4), observing and studying the knowledge structure of a particular concept, while not losing the context of the concept, including its characteristics and relationships. The conceptualization allows students to "zoom-out" to reveal large-portions of the concept space and navigate to other concepts they wish to explore (see Figure 3). Such conceptual-

Figure 3. Visual presentation of the concept space: The concept of mass movement from the physical geography course (Source: Smith, Zeng, and ADEPT Team, 2002a)

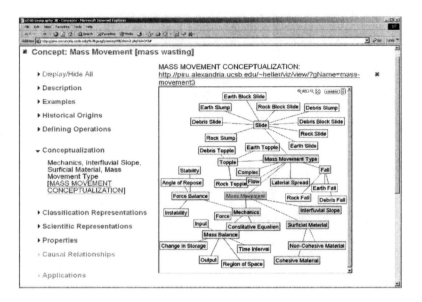

Figure 4. Partial display of the information about a concept in the KB

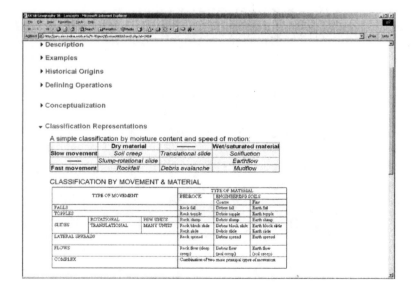

izations typically expand into a network and have values in providing students with incrementally constructed but global views of the conceptual structure of the learning materials.

In summary, the models of knowledge organization and representation developed by ADEPT provide a foundation for DLEs, while current computational technologies facilitate the implementation of these models.

Some Implementation Details for the ADEPT DLE

With the understanding that these details represent little more than transient technical solutions, we briefly describe a current implementation of the ADEPT DLE.

The DLE is designed to have thin client access through a Web browser that supports the creation, access, and display of concepts. Representations of the SSMs may be accessed in two forms: visual and textual. The visual forms are intended to show the interrelationships among concepts and to provide "global" views of the structure of the concept spaces. In particular, we have focused on providing visualizations based on the conceptualization element of the SSMs. Textual forms allow instructors and students to browse through the contents of the elements of some SSMs of concepts. Visual representations of the concept spaces have been implemented in two clients, an OpenGL graph visualization tool and a lightweight Java applet (Ancona & Smith, 2002). The Java Graph Applet has the advantage of a live database connection through a PHP script on the server. The visualizations currently center on a specific user-selected concept and illustrate a view of the KB from that concept. The Java Graph client allows for fine-tuning of a graph with the ability to change the number of relationship levels that are visible, and to hide individual nodes in order to focus on the precise topic at hand. Conversely, many parameters can be passed to the applet via a PHP script, allowing for links to dynamically generated graphs centered on any concept in the KB. It also has the ability to save a graph to a database so it can easily be retrieved via a URL as part of a lecture.

Basic input and editing features for SSMs have been implemented. The system currently supports listing/searching of concepts though a concepts control page, where users can search for a concept to begin editing, viewing, or creating a new concept. Since a concept consists of several separate parts, both the concept entry/edit page and concept viewing page use a JavaScript collapsing menu system, which allows users to view any combination of the concept parts.

The operational KB of approximately 1,200 concepts from the domain of physical geography was created, using these input tools, by a small (5-person) group of "experts." Each expert entered a set of concepts, using reference materials to support the process. Our experience is that a typical concept takes around 0.5-1.5 hours to create, although efficiency increases rapidly with experience. Weekly reviews were held to discuss created concept models and to provide uniformity over the KB. The design, implementation, and evaluation of ADEPT DLE are reported in a number of conference papers (Janee & Frew, 2002; Smith et al., 2003).

A current implementation of DLE is being employed operationally in undergraduate classroom settings. We view this implementation as one possibility in a large class of tailorable DLEs in which logically distinct DL collections and associated services are integrated in terms of KBs of SSMs of scientific concepts. The focus is on the preparation and presentation of materials supporting lecture-type modes in the domain of elementary physical geography/geology. However, other modes of presentation, including laboratory sessions, have also been developed. The DLE is designed to be generally applicable to all domains of science.

THE ADEPT STRONGLY STRUCTURED MODEL (SSM) OF CONCEPTS

ADEPT has developed SSMs of concepts for scientific domains with a frame-based knowledge representation system with slots and attribute-value fillers. We now describe the ADEPT SSM of concepts and a use-based classification of concepts that have proved of value in helping students understand the different roles that concepts play in representing some domains of knowledge. The references from the work of various scientific groups in constructing detailed, objective models of the concepts underlying their domains, and from the theoretical work of Gärdenfors (2000) and others will be discussed in a later section.

Elements of ADEPT SSMs of Scientific Concepts

The ADEPT SSM of concepts significantly extends the typical thesauri-like definitions of concepts that have traditionally been used to access information in bibliographical reference databases. Such definitions usually limit the presentation of attributes of concepts within two general types of relationships (hierarchical and associative) (Smith et al., 2002a, 2002b). SSMs have similarities with the applications of "ontologies" that involve not only the term representations of concepts and their interrelations, but also the values of their properties. It should be noted that our knowledge bases are not structured to function as an ontology for reasoning or retrieval purposes.

Figure 5. Elements of ADEPT SSMs of scientific concepts

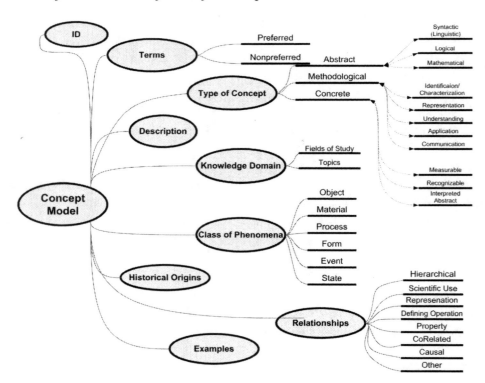

The current ADEPT SSM of a concept is aimed at integrating all of the information about a concept that is required for undergraduate levels of education. Figure 5 describes the main elements of the SSMs and the following text describes the elements in further detail.

Elements of the ADEPT SSMs of Scientific Concepts include:

- **Domain context:** The domain of knowledge to which the concept is relevant. For example, the concept water, having distinct connotations for physical geography, chemistry, agriculture, and meteorology, will have different SSMs for different contexts. Furthermore, many of the terms that are used to represent concepts may be quite ambiguous if the context is not prescribed. For example, transport in the context of physical geography typically denotes the transport of earth materials by some fluid medium.

- **Terms:** Simple (linguistic) representations that do little more than denote a concept, although they may be employed in reasoning about hierarchical and partitive relationships.

- **Descriptions:** Natural language representations of the concept. Many of those employed in the current KB are drawn from glossaries of books (with citations) and similar materials.

- **Class:** A class of a concept relates to some chosen classification, such as our classification into abstract, methodological, and concrete concepts, and associated subclasses (discussed in the next section.)

- **Historical origin(s):** History and evolution of a concept.

- **Example(s):** One or at most a few representations of prototypical examples of the concept. For example, photographic and image representations of landforms are used extensively in the physical geography KB.

- **Defining operations:** Objective descriptions of activities that define the semantics of the concept.

- **Hierarchical relations:** Specific relations between the terms denoting concepts, such as ISA and PARTITIVE relationships. While such terms provide bases for inference (for example: a ISA b AND b ISA c IMPLIES a ISA c) these terms and relationships are essentially lacking strong scientific semantics;

- **Conceptualizations:** A presentation of relationships between the term representing a concept and a set of other concepts may be viewed as "weak" representations of a concept. This component of concept definition is intended to provide answers to the question: "What other concepts in the KB do you require in order to explain/understand a given concept?" For example, an answer to this question for the concept of average stream velocity might be roughness, water surface slope, and flow depth, since these are the concepts that enter into the most basic scientific representations of the concept. We note that conceptualizations typically expand in a hierarchical manner, as in the case: Polygon → LineSegment → Point → RealNumber, and may be graphically represented as a "concept map" (see Figure 3).

- **Scientific classification:** A moderately strong scientific representation of a concept as some (generally complex-shaped) region in a space whose dimensions may be given numerical scales. A good example is the classification of igneous rocks as regions in some space characterized by chemical and mineral composition. Scientific classifications may be viewed as providing simple representations of a concept in terms of other concepts.

- **Scientific representations:** Representations of a concept that are given in scientific "languages" and from which useful information may be derived with various deductive and inductive procedures. Important classes of such representations include data, graphical, mathematical, and computational representations. For example, a frequently used and powerful scientific representation of the concept of average stream velocity is the Chezy Equation ($V = CS1/2R1/2$, where C is a constant, R is the hydraulic radius, and D is the flow depth).

- **Properties:** Attributes that "characterize" concepts and whose scientific representations may be manipulated to obtain a characteristic value of a concept. For example, properties of a drainage basin that may be computed from its representation in terms of a digital elevation model (DEM) include its area and its (Strahler) stream order.
- **Causal relations:** Term and scientific representations of causal relationships between one concept and others.
- **Co-relations:** Co-relational, and often statistical, rather than causal relationships between one concept and others.
- **Applications:** The applications of a concept for a specific scientific concept.

Figure 4 is an example of the display of these elements in the concept KB.

It is suggested that generalizations and refinements of similar models could be used for developing SSMs of scientific concepts that are acceptable to scientific communities. For example, to indicate the degree to which our general model of concepts is consistent with the SSMs in other domains, we mapped the Chemical Abstracts Service (CAS) model for chemical substance concepts into our general model (Smith et al., 2002a). The result is that they are consistent in their basic structures and functions of the elements while having domain-specific attributes.

A Use-Based Classification of Concepts

For the value spaces of "type of concept," we developed a use-based classification of scientific concepts, informally based on a National Research Council publication regarding scientific education (NRC, 1996). This classification provides one basis for characterizing concepts in terms of objective scientific operations specifying operational semantics. The basic form of our classification of scientific concepts is shown in Figure 5 under "Type of Concept." At the highest level, the classification represents operationally interpretable concepts as belonging to one of three classes:

- **Abstract concepts:** have operational semantics defined in terms of syntactic (or computational) manipulations of symbolic representation. Three possible subclasses include, but are not limited to:
  - o Syntactic (linguistic) concepts;
  - o Logical concepts; and
  - o Mathematical concepts.
- **Methodological concepts:** have semantics defined in terms of various classes of scientifically well-defined operations that may be carried out in relation to them. Possible subclasses include, but are not limited to, concepts relating to procedures for:
  - o Identification/characterization;
  - o Representation;
  - o Understanding;
  - o Application; and
  - o Communication.
- **Concrete concepts:** have semantics defined in terms of scientifically well-defined operations that provide the concept with an interpretation. The concept of river discharge, for example, has a characterizing set of operations that defines the concept in terms of various sets of measurement procedures that may be carried over to real-world contexts. Such procedures determine

the amount of water passing through some cross-section of a river during a given interval of time. These concepts are, by and large, the class of concepts used in model and theory construction and include the important subclasses of:

- Measurable concepts;
- Recognizable concepts; and
- Interpreted abstract concepts.

The classification recognizes that the concepts employed by scientists cover a broad range of contexts, as illustrated by such concepts as polygon, experiment, dataset, multiple linear regression, hypothesis, momentum, hydraulic geometry, and heat diffusion equation.

We believe it is important for students to understand the main context of scientific activity in which a given concept finds application. Hence, one application of the classification lies in providing students with a "model" of scientific activities.

In the ADEPT SSMs, the types of concepts also cover non-trivial relationships between the three top-level classes. For example, the classical heat diffusion equation is a concrete concept in the sense that it is a (implicit) representation of the heat diffusion process using variables that are measurable. It may, however, also be viewed as a certain mathematical concept, namely a specific partial differential equation whose terms have been given an interpretation in terms of measurable concepts, such as temperature and spatial location.

This classification of concepts permits the representation of relatively high-level concepts, such as topics, that generalize various (sub)classes of concepts. For example, an abstract mathematical concept, such as an arithmetic equation, may be defined in part by the objective syntactic operations that may be carried out on a symbolic representation of an equation; or a measurable concept, such as the average velocity of a river, may be defined in terms of the objective opera-

tions that may be used to determine the average velocity of a river. We classify concepts that are interpretable within the span of this basis as operationally interpretable concepts. It follows that there are many concepts that cannot be so classified including such un-interpretable concepts.

Theoretical Foundations and SSMs in Other Disciplines

A large literature extends over many academic fields, including psychology, education, and cognitive science, concerning the development, use, and evaluation of the conceptual bases for various domains of scientific knowledge. Within the context of human knowledge in general, Binwal and Lalhmachhuana (2001) define "knowledge representation as a systematic way of codifying human knowledge" and affirm that any knowledge representation system requires an ontological commitment as to what set of concepts should be used in representing some aspect of the world. They further argue that "a central part of knowledge representation consists of elaborating: (1) a set of abstract objects; (2) concepts and other entities; and (3) the relations that may hold between them." They note that structured object systems, such as semantic nets and frames, provide bases for especially useful knowledge representation structures characterized by attribute-value pairs containing both procedural and declarative representations.

In a similar vein, Gärdenfors (2000) argues that the central question for any theory of knowledge representation is how concepts should be modeled. Insights such as these provided us with a rational basis for organizing the services and knowledge bases of the ADEPT DLE according to SSMs of concepts. In constructing a convincing theory of cognitive (or non-objective) representations of concepts, Gärdenfors (2000) explicitly lays the groundwork for a theory of objective (including scientific) representations of concepts and their interrelationships. He describes three interrelated classes of theories concerning the cognitive representation of concepts, namely: (1) low-level connectionist/associationist/procedural approaches; (2) high-level symbolic approaches; and (3) bridging-level conceptual approaches.

Gärdenfors gives priority to the third class of conceptual representation, for which he develops a representational theory based on geometrical structures in well-defined conceptual spaces. It should be noted that his main contribution to the theory of objective representations of concepts and interrelationships is apparent when noting that: (1) the three interrelated classes of theories of cognitive representations of concepts have natural and direct counterparts in theories of scientific (or objective) representation of concepts; (2) his representational theory of concepts at the conceptual level translates almost verbatim to the language of scientific representations; and (3) the intermedi-

ate class of SSMs of concepts at the conceptual level are highly expressive, semantically-rich, and very useful in applications. The specific design to represent the structure of knowledge in the knowledge base and lecture materials also reflects this approach. SSMs of concepts introduced in the previous section may be interpreted as special cases of Gärdenfors' theory.

Various scientific disciplines, such as materials science and chemistry, have developed SSMs of their concepts in order to organize their knowledge for ease of access and reference. The MatML Working Group of the National Institute of Standards and Technology (NIST) has constructed a model for representing concepts relating to materials science (MatML, 2001) and a schema for materials property data (MatML, 2004). At the highest level, the MatML_Doc element (root element) contains one or more material

Figure 6. Sub-elements of materials from MatML (XML for materials property data) Composite based on MatXML schema version 3.1 http://www.matml.org/schema.htm

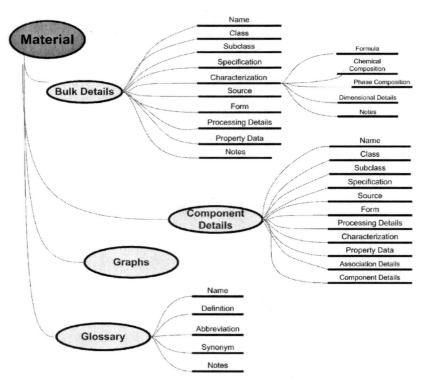

elements, each of which describes a material and its properties. The information contained by the material element is compartmentalized into four major elements:

1. BulkDetails element contains a description of the bulk material;
2. ComponentDetails element contains a description of each component of the bulk material;
3. Graphs element encodes two dimensional graphics; and
4. Glossary.

The elements included in the Version 3.0 are presented in Figure 6.

Recording, storing, and retrieving information on chemical substances have been critical in the progress of chemistry (Weisgerber, 1997). The chemistry field has developed an authoritative knowledge representation structure for representing concepts relating to chemical substances. The Chemical Abstract Service (CAS), a division of the American Chemical Society, produces the largest and most comprehensive databases of chemical information, including Chemical Abstracts (CA) and the CAS Registry (CAS, 2008). CAS extended nomenclature principles of the International Union of Pure and Applied Chemistry (IUPAC) to develop unique names of any substance. When a chemical substance is first encountered in the literature processed by CAS, its molecular structure diagram, systematic chemical name, molecular formula, and other identifying information are added to the CAS Chemical Registry. CAS' SSMs for chemical substance concepts include:

*   **Note:** Illustrative structural diagrams: identifying the molecular skeleton of a ring

Figure 7. CAS' SSMs for chemical substance concepts; Composite based on CAS. (2008). The CAS registry. Retrieved March 5, 2008, from http://www.cas.org/EO/regsys.html

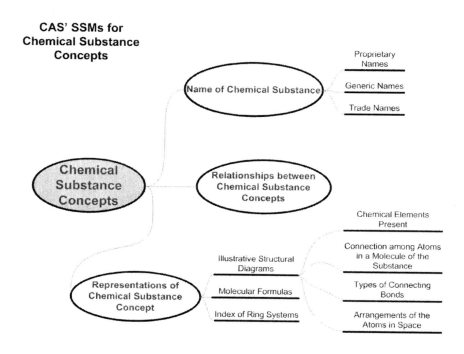

system or stereoparent for convenience in interpreting associated entries in the Chemical Substance Index.

- **Molecular formulas:** Representing invariant properties of a chemical substance and derivable from its molecular structure.
- **Index of Ring Systems:** Listing names of cyclic skeletons contained in organic chemical compounds in the order determined by their ring analysis.

The General Subject Index of CA includes general concepts relating to chemical substances, such as classes of substances, physical representations, chemical concepts, phenomena, biochemical and biological subjects (other than specific biochemicals), and concepts relating to animals and plants (especially their common and scientific names) (CAS, 1999).

The cases presented above have emphasized objective representations, operational semantics, use, and interrelationships of concepts. Such representations are strongly-structured models of concepts, since all of these play important roles in constructing representations of phenomena that further the understanding of scientific domains of knowledge.

EVALUATING THE EFFICACY OF THE ADEPT DLE

As noted above, the ADEPT DLE has been in operational use as an environment for teaching an introductory course in physical geography over the period 2002-2008. During that time, we have made both formal and informal evaluations of the efficacy of using the ADEPT SSM of concepts as an organizing knowledge structure for a DLE. The results of these evaluations indicate that (1) students attain a deeper understanding of the field of physical geography with the use of the ADEPT than with approaches based on traditional

information containers, and (2) instructors benefit greatly from the DLE services that support the construction of KBs of concepts, collections of course materials organized in terms of concepts, and presentations constructed around webs of interrelated concepts. We must point out, however, that major and long-continuing studies are needed to generate scientifically supported results concerning the relative efficacy of such systems for student learning.

Formal Evaluations of the Efficacy of the ADEPT DLE

The education and evaluation component of the ADEPT project addresses two general research questions: (1) how to design a useful DL for undergraduate education in geography, and (2) how to design a DL that university faculty will choose to employ. A series of evaluations of ADEPT project was conducted by the Education and Evaluation (E & E) team of the ADEPT project, led by University California Los Angles (UCLA), during a five-year period (1999-2005). The E & E team studied the design, deployment, and adoption of prototype ADEPT learning environments at two university campuses: UCSB and UCLA. Research methods included classroom observations, interviews with faculty, students, teaching assistants, and developers; analysis of teaching materials (lectures, assignments, exams); and analysis of available metadata standards (Borgman et al., 2000; Borgman, 2006). The evaluation provided valuable feedback and helped improve usability and functionality (Champeny et al., 2004). An evaluation conducted during the spring 2003 academic term showed that students' learning outcomes significantly improved in scientific reasoning ability when compared to fall 2002 results. These latter results were generated from teaching the course with traditional approaches and information containers (but with the same instructor). In particular, students performed significantly better on tests of graph comprehension and hypothesis generation (Champeny et al., 2004). Furthermore, participating students found classroom presentations with the DLE useful for understanding concepts (D'Avolio et al., 2004).

While students enjoy the new environment, instructors hold concerns about the reliability of the technology since the "live" presentation of lecture material depends on the reliability of the underlying electronic services and the ability of technicians to keep the systems in operation. The E & E report concluded that "ADEPT is likely to make the greatest contributions to education by building tools and services that facilitate inquiry learning with primary sources. The geography faculty in our study has the greatest need for primary sources and wishes to make more use of them in their research and their teaching" (Borgman et al., 2005, p. 654).

Current Experiences in Using the ADEPT DLE in an Operational Setting

One of the authors (Smith) has used the ADEPT DLE in teaching the introductory course in physical geography, typically twice per year, over the whole period 2002-2008. His informal observation over that period is that the single most important contribution of the DLE for student understanding is its use of the structured concept maps for organizing the course materials. A key question put to students is, "What set of concepts would you employ in describing your understanding of some phenomenon in response to a friend's question?" It is clear from the students' responses to the repeated use of this question during the teaching of the course, as well as to the repeated use of the question, "What set of concepts could we use in representing and understanding this phenomenon?" that students benefit greatly from the explicit organization of materials around webs of concepts. As noted above, a great deal of further research is required to fully address the question of the advantages of using DLEs organized in terms of SSEs of concepts.

Some Final Comments

Our experiments in developing and employing a DLE in undergraduate classes have reinforced the obvious notion that one needs to be able to integrate heterogeneous DL materials at a conceptual level for much instructional work. Just having Web portals and search engines as the navigation and access tools to poorly integrated digital collections of learning materials is insufficient. In our case, the three sets of collections, (a set of knowledge bases for over 1,200 scientific concepts, a learning object collection of approximately 2,000 objects, and a collection of reusable presentation materials), are integrated and made usable with the ADEPT DLE model and architecture.

Building ADEPT DLE has allowed us to extend conventional knowledge organization systems (KOS) and structures, such as thesauri, gazetteers, and ontologies, and integrate their constructs into more informative representations of concepts. Instructional activities are greatly facilitated with the use of such integrated knowledge organization structures as taxonomies, thesauri, metadata, and domain-specific mark-up languages (Hill et al., 2002). Combining these with our general models of scientific concepts has allowed us to create knowledge bases of information about concepts that may be used in organizing learning materials for access and presentation. Conventional KOS elements can be integrated with other scientific knowledge in an interactive manner, particularly when such knowledge is represented as the content of traditional information containers. In contrast, poorly integrated and elementary knowledge structures cannot easily provide access to knowledge concerning many of the attributes of concepts that make them useful in scientific modeling activities.

ACKNOWLEDGMENT

The work described here was partially supported by the NSF-DARPA-NASA Digital Libraries Initiative and the NSF NSDL initiative under NSF IR94-11330, NSF IIS-9817432, DUE-0121578, and UCAR S02-36644. The Alexandria Digital Earth Prototype (ADEPT) Project Team, University of California, Santa Barbara. Web site: http://www.alexandria.ucsb.edu/

REFERENCES

ADEPT. (2003). *Virtual learning environment.* Retrieved February 15, 2007, from http://www.alexandria.ucsb.edu/research/learning/index.htm

Ancona, D., & Smith, T. (2002). *Visual explorations for the Alexandria Digital Earth Prototype.* Paper presented at the Second International Workshop on Visual Interfaces to Digital Libraries, The ACM+IEEE Joint Conference on Digital Libraries (JCDL), Portland, OR. Retrieved February 15, 2007, from http://vw.indiana.edu/visual02/Ancona.pdf

Binwal, J. C., & Lalhmachhuana. (2001). Knowledge representation: Concept, techniques, and the analytico-synthetic paradigm. *Knowledge Organization, 28*(1), 5-16.

Borgman, C. L., Gilliland-Swetland, A. J., Leazer, G. H., Mayer, R., Gwynn, D., Gazan, R., & Mautone, P. (2000). Evaluating digital libraries for teaching and learning in undergraduate education: A case study of the Alexandria Digital Earth Prototype (ADEPT). *Library Trends, 49*(2), 228-250.

Borgman, C. L., Smart, L. J., Millwood, K. A., Finley, J. R., Champeny, L., Gilliland, A. J., & Leazer, G. H. (2005). Comparing faculty information seeking in teaching and research: Implications for the design of digital libraries. *Journal of the American Society for Information Science and Technology, 56*(6), 636-657.

Borgman, C. L. (2006). What can studies of e-learning teach us about collaboration in e-research? Some findings from digital library studies. *Computer Supported Cooperative Work, 15,* 359-383.

Bruer, J. T. (1993). *Schools for thought: A science of leaning in the classroom.* Cambridge, MA: MIT Press.

CAS. (1999). *1999 CA index guide.* Columbus, OH: Chemical Abstracts Services.

CAS. (2008). *The CAS registry.* Retrieved March 5, 2008, from http://www.cas.org/EO/regsys.html

Champeny, L., Borgman, C. L., Leazer, G. H., Gilliland-Swetland, A. J., Millwood, K.A., D'Avolio, L., Finley, J. R., Smart, L. J., Mautone, P. D., Mayer, R. E., & Johnson, R. A. (2004). Developing a digital learning environment: An evaluation of design and implementation processes. In *ACM/IEEE-CS Joint Conference on Digital Libraries* (JCDL 2004), Tucson, AZ. New York: ACM Press.

D'Avolio, L. W., Borgman, C. L., Champeny, L., Leazer, G. H., Gilliland, A. J., & Millwood, K. A. (2005). From prototype to deployable system: Framing the adoption of digital library services. In A. Grove (Ed.), *Proceedings 68th Annual Meeting of the American Society for Information Science and Technology (ASIST)* 42, Charlotte, NC, US. Retrieved from http://eprints.rclis.org/archive/00005052/

Duguid, P., & Atkins, D. E. (Eds). (1997). *Report of the Santa Fe Planning Workshop on Distributed Knowledge Work Environments.* Retrieved from http://www.si.umich.edu/SantaFe/

Gärdenfors, P. (2000). *Conceptual spaces: The geometry of thought.* Cambridge, MA: MIT Press.

Hill, L., Buchel O., Janee, G., & Zeng, M. L. (2002). Integration of knowledge organization systems into digital library architectures: Position paper. In J.-E. Mai, C. Beghtol, J. Furner, & B. Kwasnik (Eds.), *Advances in classification research. Proceedings of the 13th ASIS&T SIG/CR Workshop* (Vol. 13, pp. 62-68). Medford, NJ: Information Today. Retrieved February 15, 2007, from http://alexandria.sdc.ucsb.edu/~lhill/paper_drafts/KOSpaper7-2-final.doc

Janee, G., & Frew, J. (2002). The ADEPT digital library architecture. In *Proceedings of the Second ACM/IEEE-CS Joint Conference on Digital Libraries (JCDL)* (pp. 342-350). New York: ACM Press.

MatML. (2001). *MatML overview.* National Institute of Standards and Technology. Retrieved March 28, 2008, from http://www.matml.org/

MatML. (2004). *MatML schema, version 3.1.* MatML Schema Development Working Group. Retrieved March 28, 2008, from http://www.matml.org/schema.htm

Mayer, R. E. (1991). *The promise of educational psychology: Learning in the content areas.* Upper Saddle River, NJ: Merrill Prentice Hall.

Mayer, R. E. (2001). *Teaching for meaningful learning.* Upper Saddle River, NJ: Merrill Prentice Hall.

National Research Council (NRC). (1996). *National science education standard.* Washington, DC: National Academy Press.

NSF. (2003). *National science, technology, engineering, and mathematics education digital library (NSDL) - program solicitation.* NSF 03-530. Retrieved February 15, 2007, from http://www.nsf.gov/pubs/2003/nsf03530/nsf03530.htm

Smith, T. R., Zeng, M. L., & ADEPT Knowledge Team. (2002a). *Structured models of scientific concepts as a basis for organizing, accessing, and using learning materials* (Technical Report 2002-04). Santa Barbara, CA: University of California Santa Barbara, Department of Computer Science.

Smith, T. R., Zeng, M. L., & ADEPT Knowledge Team. (2002b). Structured models of scientific concepts for organizing learning materials. In M. J. Lopez-Huertas & F. J. Munoz-Fernandez (Eds.), *Challenges in knowledge representation and organization for the 21st century: Integration of knowledge across boundaries: Proceedings of the seventh international ISKO conference,* Granada, Spain (pp. 232-239).

Smith, T. R., Ancona, D., Buchel, O., Freeston, M., Heller, W., Nottrott, R., Tierney, T., & Ushakov, A. (2003). The ADEPT concept-based digital learning environment. In *Proceedings of the 7th European Conference on Research and Advanced Technology for Digital Libraries (ECDL 2003),* Trondheim, Norway (pp. 300-312). Berlin: Springer.

Weisgerber, D. W. (1997). Chemical Abstracts Service Chemical Registry System: history, scope, and impacts. *Journal of the American Society for Information Science, 148*(4), 349-360.

Chapter XII
Simple Geography–Related Multimedia

Richard Treves
University of Southampton, UK

David Martin
University of Southampton, UK

ABSTRACT

Teaching geography at university level involves students in study of complex diagrams and maps. These can be made easier to understand if split into parts. This chapter reports the work of a team writing a series of courses in geographic information systems (GIS) and their solution to the problem, which involved authoring simple multimedia animations using Microsoft PowerPoint™ software. The animations were authored by those writing the courses with little input from the multimedia Web specialist supporting the team. The techniques that the team used to produce the animations are explained, as are the nine points of best practice that were developed and how the animations were used with other non-animated content. Three sub-categories of these animations are described and explained and the issues of maintenance and reuse of the animated content is considered.

INTRODUCTION

A major challenge facing the teaching of geography or geology is students' interpretation of complex maps and diagrams. Laurillard (1997) states that this is because the "[graphical] representation and the content are unfamiliar" to a novice user. These graphics are often used to communicate important multidimensional and locationally specific concepts that are extremely difficult to present verbally. Taking an example: a professional geographer will inspect a contour map and immediately be able to understand the aspects of the landscape that it represents: they can "read" maps. A novice geographer would not be able to interpret a map in the same way, for example, they may be able to understand what a contour line symbolizes but the complex pattern of contours on the map may overwhelm their ability to interpret the physical landscape it represents. One solution to this issue is to break the map or diagram down into component parts or concepts and present them in a way that builds back up to the original complexity of the diagram. In a contour map example with a dipping plane dissected by a valley ending at a cliff, the problem could be split into three stages: in the first a contour map of a dipping plane is presented, then a stage showing the same plane dissected by a valley and the final stage would show the patterns of contours with all the elements present. This sequence of diagrams would be annotated by labels and notes. In this chapter we present our experience in the development of complex maps or diagrams (hereafter referred to as "complex diagrams" for the sake of brevity) for delivery to students engaged in independent learning. Our work was initially implemented as part of a fully online master's-level program and we have subsequently utilized the approach in self-paced practical sessions with traditional face-to-face students.

The traditional approach to breaking down a complex diagram would involve an expert geographer acting as author (hereafter referred to only as "author") who analyses the diagram and then splits it into logical parts, annotates it, and presents it using a static diagram form. A second, alternative approach would have the author and a learning technologist (a multimedia Web specialist) working together to produce an animation of the diagram. The problem with this approach is that learning technologist time is expensive and highly constrained in most higher education environments, in addition communication between the author and learning technologist could slow production speeds compared with just having the author work on the material alone. In this chapter we advocate a third approach: by keeping the animation simple the authors can produce anima-

tion-based content for themselves. For reasons that will become apparent we term these animations "animated slide sets". We argue that this is an educational improvement on the first static text and diagrams approach, whilst avoiding the problems of the second approach.

The teaching materials described in this chapter have been written by a multi-author team comprising several academic staff and a learning technologist. The initial work was undertaken to support an online distance learning MSc in GIS, delivered collaboratively by the Universities of Leeds and Southampton. In the case of our actual team the learning technologist had subject specialist skills, blurring the distinction between author and learning technologist. We think such blurring of roles should not affect any project that aimed to reproduce our type of work flow, all that is required is that members of the team maintain a sense of what role they are taking when undertaking team tasks. Our learning materials were written as standard HTML Web pages in learning object format (described in more detail by Wright et al., under review) and are each between approximately 500 and 2,500 words in length. Within each object there are typically one or more animated slide sets, activities and a reference section. Wright et al. (under review) also addresses such issues as learning object reuse, storage, editorial control, and intellectual property so we have avoided discussion of such issues here.

The remainder of this chapter is arranged in six sections. Firstly, we present a definition of animated slide sets, the particular type of animation we developed for our courses, and we review a range of work that deals with the presentation of complex concepts through complex diagrams, particularly focusing on the role of animation and the issue of cognitive load. We then present a worked example of an animated slide set, to which the remaining discussion will make reference. In the third section we address software and technical issues before turning to a consideration of design principles that we used when producing animated slide sets. In the penultimate section we consider issues of maintenance and reusability before concluding with some observations on the lessons learned from our experiences.

COMPLEX DIAGRAMS FOR COMPLEX CONCEPTS

In this section we will define some terms that we will use in the rest of the chapter, including animated slide sets, which are a subset of animations as a whole. We also review various theoretical perspectives on student learning and discuss how they relate to the techniques we developed in order to address the complex diagram problem.

"Animation" is a widely used term with multiple meanings; Mayer and Moreno (2002, p. 88) define it more specifically as being a "simulated motion picture depicting movement of drawn (or simulated) objects". This differs from video, which shows movement of real objects. Our advocated approach is to produce animated slide sets with characteristics as defined by Mayer and Moreno but with additional characteristics:

- Graphical elements are grouped together onto "slides," commonly each slide is built up from a number of simpler graphical elements.
- Using sequential navigation: "next" and "previous" buttons control movement rather than "play" and "pause."
- Graphical elements are usually simple cartoon-like drawings rather than actual images.
- Using labels and text linked to the graphical elements, no audio is used.

Animated slide sets have much in common with presentation material produced by programs such as Microsoft PowerPoint™. The key difference is that notes and annotations tend to be more detailed in animated slide sets since there is no equivalent of the presenter delivering an audio commentary.

Having described animated slide sets we still have not answered the basic question: why should we choose to use an animation to help students understand complex diagrams? Mayer and Moreno (2003) believe that animations are effective at explaining physical processes that can be illustrated graphically such as how a human breathes or a car brake operates. Nicholson (2002) states that 73% of earth science students surveyed approved of the use of lectures that used animations generated by PowerPoint. They felt that the animations helped them understand complex concepts, which include maps and tables. Laurillard (1997) directly addresses our complex diagram problem. She explores three case studies, one involves categorizing paintings from an art course, another involves a two-dimensional (2D) representation of gravity from a physics problem, and the final case study involves geological maps. Her suggested solutions to these problems all use multimedia with adaptive learning made possible by the interactivity and the graphical elements of the animation. She states that this is the important advantage of using multimedia. The geological case study is an example of the complex diagram problem; as with our approach, Laurillard begins by suggesting that the map be broken down into parts. However, she also recommends that the animation produced should operate as a model: the student inputs geological parameters and the multimedia application would output the corresponding geological map. In the paper she also discusses the differences between traditional teacher/student interactions and the value of interactions with multimedia materials. Her conclusion is that human interaction has a special value but that interaction with multimedia content can produce important adaptive learning in a student.

Our aims in producing animated slide sets were similar to Laurillard's in that we were not trying to replace the tutor in our teaching, we agree that tutor-student interactions are especially valuable to students. Like her, we attempted to improve learning by splitting complex diagrams into parts and, through the use of limited inter-activity (students control the speed with which they move through our animations), make the learning as easy as possible for students. However, we felt the cost of learning technologist time that would be needed to produce such "model-based" multimedia was not cost effective.

Mayer and Moreno (2003) explain how animations should be designed within the framework of cognitive load theory. This theory was first proposed by Sweller (1988), the work a student does to understand a concept is split into different parts: intrinsic cognitive load is held to be the mental effort required to understand a concept. For a given topic it is immutable, thus the concept of how to obey traffic lights is relatively easy to understand and has low intrinsic cognitive load. By comparison, the electronics and programming that control the traffic lights are much more complex to understand and would involve a higher intrinsic cognitive load. Extrinsic cognitive load is variable and describes the extra work necessary to understand a concept beyond the intrinsic load. For example in a diagram explaining the sequence of colors seen in traffic lights, if the label text was to be printed upside down there is more work for the student to understand the diagram than if the text was printed the correct way up; intrinsic load would remain the same whilst extrinsic load would be higher. Germane cognitive load is the work needed to create new schemata to understand a new concept. A schemata is a mental construction, such as a stereotype, against which similar concepts can be understood. For example, a driver who has learnt that the concept of orange representing "warning" in a traffic light can quickly grasp that flashing orange on a motorway sign also denotes a warning. A driver who has never seen a traffic light would find it more difficult to learn that orange denotes "warning." In designing learning materials authors should try and decrease extrinsic cognitive load; however, to give the student mental scaffolding to process similar concepts in the future an author may design material in a way that increases germane cognitive load.

Mayer and Moreno (2003) proposed "segmenting" a presentation as a technique to reduce extrinsic cognitive load. In our context that translates to mean that animated slide sets can help reduce the extrinsic cognitive load needed to understand complex diagrams or maps by splitting them up. This is not the only way that animated slide sets can reduce extrinsic cognitive load: the animation means that elements that are related can appear at the same time, which instantly links them in the mind of the student. For example, a graphical element may have an explanatory comment, these two elements are linked when they appear at the same time in an animation but if a student was viewing a complex diagram there would be extrinsic cognitive load related to "hunting down" the comment that went with the graphical element. If a complex diagram is replaced by an animated slide set the actual decrease of extrinsic cognitive load is small, but when a number of such static diagrams are replaced then the cognitive load savings become significant.

Cognitive load theory is not the only conceptual insight that has relevance to the issues in this chapter: we can also draw on theories of memory capacity, dual coding, and change blindness. Miller (1956) discussed the capacity of memory; his finding was that most people could hold between five and nine "chunks" of information in their short-term memory at any one time. A chunk is a concept that is a discrete packet of memory, thus letters such as I, B, and F would be three chunks, whereas put together in reverse order they become the acronym FBI and would be one chunk. By processing the chunks, the concept to be learnt is moved from short-term memory to long-term memory and if memory capacity is exceeded, the processing cannot take place. In our context, by splitting a complex diagram into layers we reduce the number of chunks needed to be processed by a student at any one time. Dual coding theory was first proposed by Paivio (1986). It suggests that language and visual information are processed in two separate channels and the brain has a certain capacity for processing each channel. To be most efficient, information should use both channels and should be complementary rather than conflicting. For example, a video clip of a presenter with an audio commentary accompanied by a graphic screen showing bullet points would overwhelm the visual channel. This is because two graphical forms of information are presented at the same time. It is better to have only one form of graphical presentation accompanying the audio commentary. This raises the question of whether animated slide sets should use written or audio commentary; we address this point in the section on design principles. Change blindness is an effect related to detecting small changes in an image. An observer views a computer screen, which displays a large number of similar, randomly arranged dots. If the computer adds a single dot to the view, most observers can reliably detect the addition. However, the observer's ability is severely affected if their view of the screen is interrupted by a blank screen for a period of more than 100ms at the instant when the new dot is added. Details of the effect are reported in Phillips (1974). By making elements appear instantly on screen, animated slide sets make use of this ability to detect change whereas the static diagram version of an animated slide set (as will be seen in Figure 2) in effect introduces a blank screen as the student moves her eyes from one diagram to the next, thus producing change blindness. This eye movement is more precisely known as a saccade and its effect on change blindness is discussed in McConkie et al. (1996).

We have considered the concept of an animated slide set in the light of several theories in order to determine how this type of animation could help solve the complex diagram problem. By relating animated slide sets to cognitive load theory and memory capacity we explained how splitting a complex diagram into parts could aid learning. The theories of cognitive load and change blindness

Figure 1. Part of the process of "Local Smoothing with Spatial Filters" in static form (Source available at http://tinyurl.com/2x9ooo)

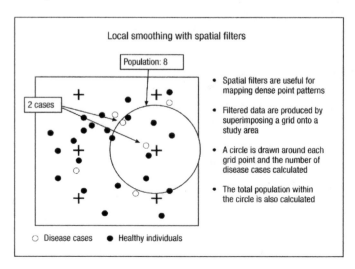

were used to discuss how the use of animation in our materials could be an improvement on the use of static diagrams. Finally, dual coding theory raised the question: should commentary be added via audio or text? In the following sections we develop these arguments from a general definition of animated slide sets into a series of best practices and principles, drawing on the literature presented in this section.

WORKED EXAMPLE

To explain the structure of an animated slide set in more detail we will explore part of an animated slide set from one of our courses.

Context of the Example

Techniques are available in geographic information systems (GIS) for smoothing spatial data. The example we discuss here applies to a data set which represents the locations of individuals' home addresses and whether or not they have a particular illness. To complete the smoothing process GIS software performs a number of sequential tasks on the data and it is important that students understand the steps the software is performing. In Figure 1 part of this process is illustrated.

Rationale for Using an Animated Slide Set

There are a number of different types of symbol and text visible in Figure 1:

a. Title
b. Key (bottom left)
c. A filled square illustrating the study area
d. Light fill circles representing the location of disease cases
e. Dark fill circles representing the location of healthy individuals

Figure 2. Screen shots making up the animated slide set for "Local Smoothing with Spatial Filters" (Source available at http://tinyurl.com/2x9ooo)

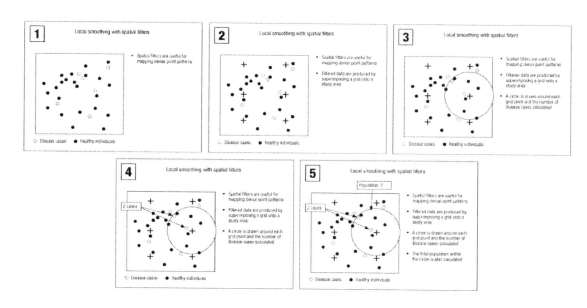

f.   Crosses representing the imposed grid
g.   A sampling area circle
h.   Labels marking cases
i.   Labels marking population
j.   Commentary

The items in this list can be thought of as "layers", a concept common in GIS analysis.

When the process is presented to students as shown in Figure 1 there is extrinsic cognitive load caused by the requirement to sort the multiple elements into logical relationships, for example, there is a need to decipher the multiple symbols before the graphic as a whole can be interpreted. In our materials we use an animated slide set to break down Figure 1 and so explain the process being performed by the GIS software. The animated slide set is illustrated by Figure 2, which shows the sequence of screen shots that appear as the student clicks a "next" button.

It can be seen that the first screen image introduces layers (a), (b) (c), (d) and (e), which together make up the base layer data. In the second screen image we add layer (f); in the third screen (g) is added on top of the layers visible in the second and the pattern is repeated for screens [4] and [5]. The commentary is split into bullet points on the right of the diagrams. An extra bullet point is added on all slides except the fourth. Presenting each bullet point and the layer to which it relates at the same time allows the student to associate each element with the relevant bullet point of the commentary without having to search to connect comment and graphical element.

Explaining an animated slide via Figure 2 suffers from the problem of translating an animation into a static figure and the reader is encouraged to follow the links in the links section below to view the actual animated slide set as they appear in our materials. Tufte (2001) states that graphical excellence in static diagrams is achieved when complex ideas are communicated with clarity, precision, and efficiency. This could be used to argue that despite being graphically complex,

Figure 1 would be sufficient for student understanding if time was spent improving its design. However, Figure 1 represents a subsection of the whole animated slide set, so a true graphical replacement of the animated slide set would involve a much larger number of screenshots. We think that this makes the animated slide set version the most effective in terms of cognitive load. We explore other advantages of animated slide sets over static equivalent diagrams in our discussion of spatial contiguity below.

SOFTWARE AND TECHNICAL ISSUES

A major problem with discussing software in an educational environment is that literature dates quickly because software develops at such a fast rate. We hope that the major point readers take from this chapter is the idea that, with an average set of information technology (IT) skills, and a limited amount of training, authors can produce quick, unpolished but effective multimedia.

Although our comments may date, we still think it is informative to explicitly discuss the software we have used because readers can attempt to reproduce our animated slide sets themselves. It also gives us an opportunity to explain the key characteristics that should be present in any suite of software for producing animated slide sets so the reader can produce animated slide sets with different software tools as and when they appear.

There are three meanings of "flash" used from this point forward, flash format (hereafter referred to simply as 'flash') refers to a Web-based standard, essentially it is a form of encoding that can be read and is widely used for rich Web applications and animations. To decode flash elements browsers require a flash plug-in. Finally, the Flash™ editor (Adobe, 2008) is a specialist program that can be used to produce flash format output.

Our content authors used PowerPoint software, to produce animated slide sets making special use of the "custom animation" feature. They used a template, the central feature of which was a consistent use of a white background, black lettering, and an orange border which allowed students to easily differentiate the animated slide set from the rest of the HTML learning object. Once produced in PowerPoint the animated slide set is passed over to the learning technologist who converts it to flash using a commercial add-on to PowerPoint called PowerCONVERTER (PresentationPro, 2008). The conversion is essentially a one-click operation for the learning technologist with a small amount of file manipulation to get the flash element into the relevant Web page. In the case of the GIS smoothing example discussed above, this process was free of problems, but in some cases the conversion from PowerPoint format to flash has produced errors. The conversion process is completely automated, so to avoid the errors the learning technologist has to alter the PowerPoint until it can be converted correctly: this can involve steps such as changing transitions from being custom animations to being complete changes of slide.

The combination of PowerPoint and PowerCONVERTER provides a suitable set of software for producing an animated slide set for a number of reasons. Firstly, a basic knowledge of PowerPoint is already common among content authors, which means that learning how to use custom animations in PowerPoint is usually all that is technically required in order to proceed to the production of animated slide sets. Secondly, the custom animation feature allows the author to make elements such as labels, lines, and circles appear and disappear as the student clicks. Nicholson (2002) outlines her use of this feature to annotate micrographs of rock in lectures, an example of a complex diagram, and suggests that the feature should be more widely used. Whilst use of the custom animation feature does not allow the detailed and rich animation controls available in

software such as the Flash editor, it is perfectly adequate to produce a custom slide set of the type shown in Figure 2. Finally, PowerCONVERTER allows a PowerPoint presentation to be converted to flash, which will download to a Web browser faster than the equivalent PowerPoint file. There are other advantages because of the way flash streams data to a user's computer but these are too specialist and minor in impact to discuss here.

The simplicity of the conversion process from PowerPoint via PowerCONVERTER makes it easy for an author, should they choose, to teach themselves how to process their own animated slide sets to the flash format. Once in this form, embedding the animated slide sets in HTML pages is a more technical task but one that could be avoided if the author has access to a VLE for delivering content such as Moodle (Moodle, 2008), which allows authors to insert flash content in a course without requiring them to understand HMTL code.

Within an animated slide object we adopted the practice of only having a simple sequence of slides and layers, this compares to a branched structure in which alternative sets of slides and layers are invoked as a result of the student answering questions. Branching can be educationally useful but has the disadvantage that it complicates authoring and is difficult to achieve with the software approach used here.

AUDIO NARRATION

It is technically possible to add audio narration to an animated slide set. The reasons why this should be done, how it should be done, and the skills required are explored in the design principles section below. Adding audio in our PowerPoint to PowerCONVERTER process is not possible, but other software tools could be used to do this such as the Flash editor and Camtasia (TechSmith, 2008). Even if appropriate software is selected to produce animated slide sets with audio narration and the skills needs of the author (explored in the design principles section) are overcome, there are still download size issues to consider. Adding audio increases the size of the flash element considerably. For a course aimed at students with access to a good broadband connection, such as traditional campus university students, audio would not be a problem. However, we note that in developing countries and in some rural areas of developed countries, especially for home study, broadband access is not widely available so for distance learning courses serving these areas text narration may be the best solution to adopt.

In selecting software to produce animated slide sets it should be noted that Web-based software is often free compared to PC-based software, which requires the purchase of a license. However, the "lock in" risk must be assessed in this case (Farrell & Shapiro, 1989): those running a Web-based service may decide to start charging or even stop allowing users access to their content if it serves their purpose. If content can be exported from the Web-based service in some way then this can ensure the lock in risk is neutralized.

EMBEDDING AND ALTERNATIVE FORMATS

As noted in the introduction, animated slide sets form part of the HTML learning objects that make up our MSc courses. We developed the practice of embedding the animated slide sets within the HTML pages rather than putting them in a pop up window that could be accessed via a text link. It is unclear whether our embedding practice is the best to avoid extrinsic cognitive load but we can report that it was a robust solution and students have not reported any major problems.

A request our students have repeatedly made to us is the ability to print out materials. We have reacted by converting our animated slide sets to PDF format, which students can access and print

out. However, if students do access our animated slide sets via the PDF format they lose the cognitive gains that the animations offer. We make this point to them in the introduction to the courses but they still request printer-friendly materials. This creates a dilemma for the authorship team: how do we compromise between giving the students what they request whilst providing them with what we believe is the best educational experience? We have no easy solution to this problem; it is similar to a discussion of whether distance learning in geography should be completely rejected on the grounds that geographical teaching is intrinsically about place (DiBiase, 2000). In that case students want to take distance learning courses in GIS but some geographers think they should only learn through face-to-face courses. Both problems involve the ethical dilemma of giving students what they want or what educators believe they should have. Returning to our dilemma, it is worth noting that developments in mobile technologies may make screen reading more attractive to students and that later generations of students may be happier with not printing out materials.

It would be possible to produce animated slide sets with software tools other than the ones we have used, for example Keynote software (Macintosh computer users only, see Apple Inc., 2008) producing flash output; the Flash editor producing flash output; Camtasia producing software which can output in MP4, flash or WMV output. However, at the time of writing we have not found another suite of software that combines the following key characteristics: (1) relatively easy for the author to learn; (2) produces simple animated effects; (3) converts the output to a Web-friendly format; (4) while maintaining all effects of the animation. Most options fail on counts (1) and (4). For example, learning to use the Flash editor would involve a big investment of time from authors and OpenOffice (2008) does not convert the custom animations in a PowerPoint presentation, but only the text and formatting. However, for a team that uses Macintosh computers rather than PCs,

Keynote software meets all the requirements and deserves further investigation. Further discussion of formats or software is felt to be beyond the scope of this chapter: it is a complex topic and our aim was to produce a software solution that was sufficient to produce effective animated slide sets rather than identify a (short-lived) optimal software solution.

In certain circumstances we have left animated slide sets in their native PowerPoint format. We believe this practice should be avoided if possible for a number of reasons:

- Flash animated slide sets download faster than PowerPoint equivalents and when they are embedded in HTML pages they download progressively, that is, when the HTML page is first opened an animated slide set off-screen is already downloading and so is ready to be viewed when the student reaches that part of the content.

- Embedding animated slide sets allows students to move seamlessly from supporting text to animated slide set without the hiatus of downloading material. This is not possible with a native PowerPoint file.

- Students who do not use PowerPoint (e.g., Macintosh users) will have to download an additional plug-in to view the animated slide set (the Flash plug-in is very common in browsers so does not present a similar problem).

- Students will need to be directed to only use the slide show feature otherwise they could inadvertently edit the animated slide set or view it without the required animation.

If the choice is between not using animated slide sets at all or using them in PowerPoint form, the gains in terms of cognitive load are probably worth the problems associated with downloading and plug-ins.

DESIGN PRINCIPLES FOR ANIMATED SLIDE SETS

In our discussion of complex diagrams for complex concepts section we considered the theoretical background to the use of animation in teaching. Now that we have described animated slide sets in greater detail it is useful to consider how the characteristics of animated slide sets compare to other studies of animations in teaching. Mayer and Moreno (2002) outline seven design principles for multimedia presentations involving animations. In this section we will compare their empirically derived principles with the best practice we have evolved in producing animated slide sets. For clarity, we summarize these principles in Table 1, which also contains a brief explanation of each principle and examples of its application in our materials.

It can be seen that our animated slides sets conform to four of these principles (multimedia, spatial contiguity, temporal contiguity, and coherence). We did not conform to the personalization principle because of concerns about how the animated slides sets would be perceived by our target students. In the two principles concerned with audio narration (modality and redundancy principles), we choose not to conform because of technical issues and the extra time that authors would need to invest in order to produce narrations. We feel the issues around the skills training deserve expansion in the following subsection because readers planning to produce animated slide sets may feel it worthwhile to add narration to animated slide sets.

Extended Discussion of the Modality Principle

Our animated slide sets use text commentary instead of audio narration primarily because of problems with bandwidth and the software that was available when we began developing

Table 1. Application of Mayer and Moreno's (2002) Design principles for multimedia presentations

Principle	Explanation	Description/Justification of use in animated slide sets
Multimedia principle	Animation and commentary more efficient when used in combination than singly	We always use both diagrammatic and textual labels or commentary
Spatial contiguity principle	Text should be close to the elements it describes	Labels always connected by lines or proximity to the features described
Temporal contiguity principle	Elements should appear at the same time as the commentary	Layers appear with associated bullet points which obeys the temporal but fails the spatial principle. Mayer and Moreno report that the temporal principle is stronger of the two
Coherence principle	Extraneous words, video and sounds such as music should be excluded from animations	Priniciple is obeyed. For example, we avoid complex transitions, our elements either appear or blend in
Modality principle	Students learn more deeply from audio narration and diagrams than from text labels and diagrams	Not obeyed. Advantages out weighed by the need to give extra training and increase in download size for students. See extended discussion below
Redundancy principle	Related to the Modality principle: animations are better when they use just audio narration rather than using audio and text narration	Does not apply since we do not use audio narration
Personalization Principle	Use of a conversational rather than formal tone in the animation. Characterised by use of personal pronouns such as 'you' and 'I'. Note Mayer and Moreno tested this principle on college level students	Not obeyed. Our students are at MSc level and tend to be busy professionals, a conversational tone in our materials was thought to be inappropriate. However, we have not tested this assumption

this technique as discussed with regard to audio narration above. However, there is also a skills issue with using audio narration. To master our PowerPoint process authors needed to learn how to use custom animations in PowerPoint and how to apply the principles described in this section. In a related video podcasting project we found that authors had to learn audio narration skills such as: avoiding sounding monotonous; cutting out naturally-occurring "erm" and "err" from the narration; tool manipulation and adopting the same tone of voice between recording sessions. Mastering skills related to manipulating audio materials as well as those needed to produce animated slide sets may be too onerous for an academic author with limited time. A compromise would be to use text only commentary as we have

described and accept the loss of teaching quality by not using audio narration.

Further to the empirically derived seven principles of animation, we propose two further aspects of best practice based on our own experience. They are "focus" and "active control" and we list and describe them in Table 2.

The focus principle is the more complex of these two principles to describe, since it involves consideration of some concepts that are difficult to define and also consideration of the context in which animated slide sets appear. We expand on this in the following subsection.

Table 2. Team-derived design principles for animated slide sets

Best Practice	Explanation	Description/Justification of use in animated slide sets
Focus Principle	Animated slide sets should encapsulate a whole concept whilst avoiding covering a concept that is too large. E.g. it would be ineffective to explain just half of the smoothing process discussed in the Working Example section, the process is easier to understand as a whole	Our materials are activity based, in each learning object we produce a student activity and use this to define the concept that needs to be explained. The length of the animated slide set should be long enough to cover the concept, no more and no less. Related to the coherence principle in table 1. See extended discussion below
Active Control principle	Slide navigation with 'next' and 'previous' buttons is better than movie clip controls with 'play' and 'pause' buttons. Using slide navigation nothing will happen on screen unless the student engages actively	We believe this encourages students to vary the speed of their clicking so that they properly absorb the information contained in each layer before proceeding to the next one. Downside is that having to click a mouse continually can be annoying to a student

The "Focus" Principle

This principle concerns the scope of the content presented in the animated slide set. Mayer and Moreno (2002) note that their animations were "short and focused" from 30 to 180 seconds in length although they do not discuss focus as a principle. We derived a rule of thumb that no animated slide set should exceed ten slides (where a slide consists of a diagram split into a number of layers) but that it was more important that the content presented should encapsulate a single concept than obey this rule. In practice this meant that sometimes an animated slide set with more than ten slides was used because to break it down would mean presenting a logical concept in two parts. It also meant that animated slide sets with more than ten slides were often split into two logical sub-concepts.

Defining what a concept is and what focus an animated slide set should have is difficult and necessarily imprecise. Wiley (2000) discusses issues around learning object focus relating to four theories of instructional design and does not come to a clear, simple conclusion. However, for our purposes it is sufficient to address the problem in terms of Kolb's learning cycle (based on Kolb, 1984). In this cycle the student is modeled as undergoing a cycle of stages: concrete experience is followed by reflective observation, then abstract conceptualization, and finally active experimentation before the cycle completes by returning to concrete experience.

Kolb stated that these stages should be present in any learning situation and that the stages in the cycle are emphasized more or less strongly by different types of learner. We contend that animated slide sets are best represented by the "abstract conceptualization" stage. After students have completed our animated slide sets we usually provide activities in the learning object that contains the animated slide set. An example activity would be a data set to analyze or a Web search to be performed. We believe that our activities provide the active experimentation and concrete experience stages of the cycle after which the student hopefully passes through the final reflective observation stage before returning to the start of the cycle. Within each of our learning objects we envisage students cycling through this cycle one or more times. This leads to our definition of the scope of an animated slide set as the concepts required to be covered so that the associated activity can be completed.

ANIMATED SLIDE SET TYPES

Our animated slide objects were all developed from the same PowerPoint template and followed design principles as outlined, however, they can be further split into three types by their purpose. These are layered maps/diagrams, conceptual diagrams, and image slideshows.

Layered Maps/Diagrams

This is the most important type and also the one we have discussed in this chapter up to this point. It was the most extensively used of the three types in our materials because teaching GIS involves many complex processes that are easy to explain using animated slide sets. Figure 1 is a previous screenshot example showing a layered map, Figure 3 shows a layered diagram, in this case a graph.

Conceptual Diagrams

In this type of animated slide object a concept is represented and explained spatially, for example a Venn diagram or the example shown by screenshot in Figure 4 where the relationships of public health agencies are illustrated graphically. Just as with the layered map type the cognitive load of the diagram is reduced by splitting it into layers and the other principles of animated slide set design are adhered to. All our animated slide set types attempt to explain "space," in the case of a map this is a representation of real space, in the case of Figure 3 it is the graph space. In the conceptual diagram it is relational space, for example Figure 4 shows relationships between the doctor who is informed by medical knowledge and is constrained by societal norms. This type of animated slide set could be usefully used by other disciplines beyond geography, as any area worthy of study will have complex abstract concepts that could be illustrated as a complex diagram.

Figure 3. A screenshot of a layered diagram animated slide object discussing animal movements; the actual object can be accessed from the links section

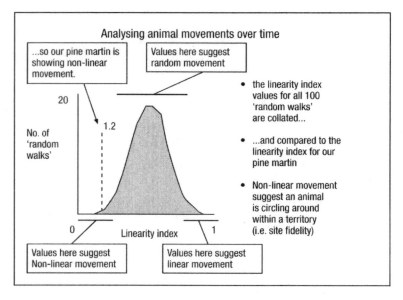

Figure 4. A screenshot of the final slide of a conceptual diagram type animated slide object (Source available at http://tinyurl.com/2cutuc and embedded within HTML page at http://tinyurl.com/2amjjb)

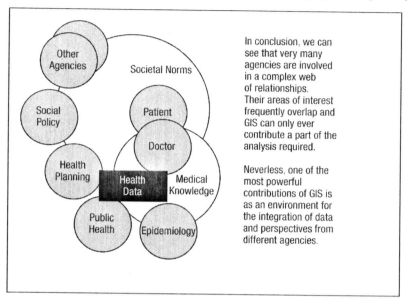

Systems diagrams are good examples of diagrams that could be usefully represented as conceptual animated slide sets. Open University (2008) provides an overview of systems diagrams. This site not only explains system diagrams, it also uses a form of presentation that is itself an animated slide set of the conceptual diagram type, the only difference being that voice narration is used.

Image Slideshow

In this type of animated slide set the principal purpose is to explain a concept through a series of images rather than maps or diagrams. In using this type of animated slide set care must be taken that the coherence principle is adhered to otherwise the images used may not add particular educational value. For example, in our MSc materials an image of a river drainage system was used in a slideshow about river networks. Just showing the image would probably fail the coherence principle but a layer was added that was used to illustrate the parallel nature of the system, a spatial fact that was important to emphasize. This form of animated slide set was often used in the introduction of a unit (a set of learning objects) to summarize the topics covered.

MAINTENANCE AND REUSE ISSUES

It is important to design materials that can be reused in other applications and maintained easily within the team environment. In our work flow the use of PowerPoint enables this, not because it is a particularly elegant way to produce simple multimedia but because it is well known by academic authors. To consider this further, we will compare our system of producing animated slide sets to an alternative technique where an author prepares a description of an animated slide set, which is then produced by a learning technologist using Camtasia (TechSmith, 2008), a specialist software tool for the production of multimedia content. Three maintenance and reuse scenarios will be examined, repurposing for face-to-face delivery after development for distance learning students, error correction, maintenance, and as the membership of a team changes.

Part way through producing our distance MSc courses we decided to reuse our materials with more traditional face-to-face MSc students. If we had been using the Camtasia route both author and learning technologist would have been involved in rewriting the materials so that they could be used in a face-to-face presentations, there would be a need to iterate between the two team members as the learning technologist produced versions of the animation that the author would then check. Because our materials were produced in PowerPoint, the authors could just customize the animated slide sets in this format for face-to-face delivery. No learning technologist time was required.

Despite extensive peer review and checking, our materials have sometimes produced errors that have needed to be fixed whilst the course was running. Using the Camtasia route both author and learning technologist would be involved in editing the animation to fix the error. This problem is mitigated with our PowerPoint workflow, as an interim solution can be published in PowerPoint format until a learning technologist becomes available to produce the flash final output.

In the present UK context, learning technologists are often prone to switch jobs due to concerns over short-term contracts. If animated materials have been created via the Camtasia route it may be difficult to find someone with the skills needed to use Camtasia. However, because the source files are in PowerPoint format for which multiple conversion tools exist (including PowerCON-VERTER) the materials are not locked into an unusable format.

CONCLUSION

We have defined animated slide sets, described an example, and examined best practices for their design. Within the design principles section we compared our practices to seven evidence-based best practices for using animation in learning materials outlined by Mayer and Moreno (2002). We went on to define the principles of focus and active control, which we have developed as our own best practices, but which do not map onto any of Mayer and Moreno's principles. We have argued that animated slide sets are effective in terms of cognitive load from the premise that they are similar to other animations and teaching situations that have been reported in the literature. Our conclusions would be more powerful if we had tested our animated slide sets directly, this has not been done but we have used them in multiple presentations of our courses with very few technical difficulties and with generally positive feedback from our students.

Animated slide sets were our solution to the university teaching problem of how best to support our students' learning of complex diagrams and maps in a format suitable for distance learning. In choosing a solution for the team there was a compromise to be made between the uses of simple or advanced multimedia technologies. A simple approach could have involved authors splitting complex diagrams up and producing sets of static diagrams. It has the central disadvantage that students will expend mental energy flipping between the separate parts of the diagram rather than grasping the concept it describes, in psychological terms such materials have high extrinsic cognitive load. Other more complex solutions involving interactive multimedia were available, for example the models advocated by Laurillard (1997), branched materials, or audio narrative. A complex multimedia solution will minimize extrinsic cognitive load but has the disadvantage of authorship team costs such as learning technologist time, speed of material production, maintenance, and reuse. Our solution was to choose a point on the continuum between complex and simple content. It was not just a simple case of compromise; there is a significant jump in costs when the learning technologist becomes involved in editing the animation rather than just converting it to a Web friendly format. Our strategy was therefore to involve the learning technologist only in conversion with the editing of the materials done entirely by the authors using PowerPoint. Authors were already familiar with PowerPoint so relatively little training was necessary to enable them to produce animated slide sets. They only needed to grasp the best practices outlined in our design principles discussion and learn how to operate the custom animation functionality of PowerPoint. This approach did involve costs in terms of staff training, but, as we have examined under the headings of complex diagrams for complex problems and in our consideration of design principles, the literature suggests that this approach offers an improvement in learning beyond the simple static diagram solution.

In our discussion of software and technical issues, we explained exactly how we produce our animated slide sets and implement them within the supporting course materials. This approach will date quickly but we hope that the principles of how we chose the software are clear to readers so that if they wish to produce their own animated slide sets they can successfully evaluate other types of software as they become available. The bulk of the discussion of animated slide sets has focused on the most common type we produced: layered maps/diagrams. We also produced other subtypes of animated slide sets: conceptual diagrams and image slideshows. In geography learning materials, layered maps and diagrams may be the most useful type to consider but outside of this topic area we think that the conceptual diagram type could be usefully implemented.

In our discussion we have framed the complex diagram problem as one of cognitive load, although we stress that this is only one factor amongst many that should be considered when courses are designed. Other factors include enabling communication, producing engaging content, and encouraging students to reflect on their learning. In this wider context we believe that setting activities where students produce their own animated slide sets suitable for their peers could be a very effective form of learning; it would encourage students to learn transferable skills of communication whilst also forcing them to engage with the topic area deeply in order to produce learning materials.

It seems to us that the topic we have examined here represents a general effective approach to the use of IT in learning situations. Much discussion in the literature, on blogs, and in the broadcast media focuses on the newest technologies and it appears to assume that the latest technologies necessarily offer effective solutions to learning problems. An example would be the current discussion of the educational value of virtual worlds such as second life (Weber, 2006). We would counter with the argument that, "The mere fact that something can be done does not mean it should be done" (Mowshowitz, 1997). Animated slide sets are far from the cutting edge of technology, but we contend they are an effective solution to presenting complex diagrams, be they maps, geographical diagrams, or conceptual relationship diagrams. Their development has been informed by considering the complex diagram problem from a holistic viewpoint, which includes consideration of factors such as the skill set of our authors, how teams operate in a university, designing for ease of maintenance, designing for reuse, and the relationship between the materials and how the course is delivered. We recommend academic teams consider a similarly holistic view of their overall learning problems before they define a format for content and start producing materials.

REFERENCES

Adobe. (2008). *Adobe Flash CS3 Professional.* Retrieved March 27, 2008, from http://www. adobe.com/products/flash/

Apple Inc. (2008). *Keynote: Cinema-quality presentations for everyone.* Retrieved March 27, 2008, from http://www.apple.com/iwork/keynote/

Dibiase, D. (2000). Is distance education a Faustian bargain? *Journal of Geography in Higher Education, 24*(1), 130-135.

Mowshowitz, A. (1997). Lessons from a cautionary tale. *Communications of the ACM 40*(5), 23-25. Retrieved March 27, 2008, from http://portal. acm.org/citation.cfm?id=253777&coll=portal&dl=ACM

Farrell, J., & Shapiro, C. (1989). Optimal contracts with lock-in. *American Economic Review, 79*(1), 51-68.

Kolb, D. A. (1984). *Experiential learning experience as a source of learning and development.* Englewood Cliffs, NJ: Prentice Hall.

Laurillard, D. (1997). Learning formal representations through multimedia. In F. Marton, D. J. Hounsell, & N. Entwistle (Eds.), *The experience of learning* (2nd ed.) (pp. 172-183). Scottish Academic Press.

Mayer, R. E., & Moreno, R. (2002). Animation as an aid to multimedia learning. *Educational Psychology Review, 14*(1), 87-100.

Mayer, R. E., & Moreno, R. (2003). Nine ways to reduce cognitive load in multimedia learning. *Educational Psychologist, 38*, 43-52.

McConkie, G. W., & Currie, C. B. (1996). Visual stability across saccades while viewing complex pictures. *Journal of Experimental Psychology: Human Perception and Performance, 22*(3), 563-581.

Miller, G. A. (1956). The magical number seven plus or minus two: Some limitations on our capacity for processing information. *Psychological Review, 63*, 81-97.

Moodle. (2008). *Moodle - a free, open-source content management system for online learning.* Retrieved March 26, 2008, from http://moodle. org/

Nicholson, D. (2002). Optimal use of MS PowerPoint for teaching in the GEES disciplines. *Planet, 4*, 7-9.

OpenOffice. (2008). *OpenOffice.org: the free and open productivity suite.* Retrieved March 27, 2008, from http://www.openoffice.org/

Open University (2008). T552 diagramming. Retrieved March 27, 2008, from http://systems. open.ac.uk/materials/t552/index.htm

Paivio, A. (1986). *Mental representations: A dual coding approach.* Oxford, UK: Oxford University Press.

Phillips, W. A. (1974). On the distinction between sensory storage and short-term visual memory. *Perception and Psychophysics, 16*, 283-290.

PresentationPro. (2008). *Convert your PowerPoint to Flash with PowerCONVERTER.* Retrieved March 26, 2008, from http://www.presentationpro. com/products/powerconverter.asp

Sweller, J. (1988). Cognitive load during problem solving. *Cognitive Science, 12*, 257-285.

TechSmith. (2008). *Camtasia screen recorder.* Retrieved March 26, 2008, from http://www. techsmith.com/camtasia.asp

Tufte, E. R. (2001). *The visual display of quantitative information* (2nd ed.). Cheshire, CT: Graphics Press.

Weber, A. (2006). NOAA comes to second life. *Second Life Insider.* Retrieved March 27, 2008, from http://www.secondlifeinsider. com/2006/08/18/noaa-comes-to-second-life/

Wiley, D. A. (2000). Connecting learning objects to instructional design theory: A definition, a metaphor, and a taxonomy. In D. A. Wiley (Ed.), *The instructional use of learning objects: Online version*. Retrieved March 27, 2008, from http://reusability.org/read/chapters/wiley.doc

Wright, J. A., Treves, R. W., & Martin, D. (under review). Challenges in the reuse of learning materials: Technical lessons from the delivery of an online GIS MSc module. *Journal of Geography in Higher Education*.

Chapter XIII
Evaluating the Geography E-Learning Materials and Activities:
Student and Staff Perspectives

Karen Fill
KataliSys Ltd, UK

Louise Mackay
University of Leeds, UK

ABSTRACT

This chapter is concerned with the evaluation of learning materials and activities developed as part of the DialogPLUS project. A range of evaluation activities was undertaken, focusing on the experiences of students, teaching staff, and the entire project team. Student evaluations included both quantitative and qualitative approaches, particularly using a questionnaire design drawing on a specific methodology and generic quality criteria, facilitating comparative analysis of results. Discussion of the student evaluations is focused on specific taught modules from both human and physical geography. Results of these evaluations were discussed with teaching staff and contributed to improvements in the various online resources. Both internal and external evaluators were involved in interviewing key project staff and their different perspectives are presented. The chapter concludes by reflecting on the effectiveness and impact of different DialogPLUS activities, highlighting the principal impacts of the project as perceived by the students and staff involved.

INTRODUCTION

Earlier chapters in this book describe and discuss some of the online materials and activities developed under the auspices of the DialogPLUS project, to enhance learning for geography students. Student-focused evaluation of these innovations involved quantitative and qualitative methods, including questionnaires, observation, interviews, and analysis of online discussion board activity. The questionnaire design drew on a specific methodology and generic quality criteria, facilitating comparative analysis of results. Teaching staff were invited to add any questions of particular interest and preliminary findings from the analysis were discussed with them. Their reflections informed the final evaluation reports, which in turn led to improvements in the resources. Additionally, towards the end of the three-year development phase, key project staff were interviewed about their experiences by both internal and external evaluators. This chapter describes and discusses the approaches to, and results of, both student and staff-focused

Table 1. DialogPLUS stakeholder groups, their interests and concerns

Stakeholders	Interests and concerns
Geographers at partner institutions	Relevance and value of the project to their local context. Access to digital library resources. Barriers and enablers to nugget development, usage and sharing. Effectiveness of nuggets in their local learning & teaching context. Collaborative nature of the project and how it has worked. Usability and effectiveness of the toolkit. Embedding of outputs / outcomes. Changes to professional practice resulting from involvement in the project.
Learners at partner institutions	The kind of skill or conceptual understanding needed to use the nuggets. Accessibility of resources / nuggets. Effectiveness of resources / nuggets. Impact on their learning processes and outcomes.
Computer scientists and educational technologists at partner institutions	Barriers and enablers to development of the toolkit. Usability and effectiveness of the toolkit. Barriers and enablers to developing systems for nugget sharing. Usability and effectiveness of solutions for nugget sharing. Convergence with emerging standards in learning design, interoperability, resource discovery and reuse.
Educationists and the evaluation team	Innovation in teaching and learning. Pedagogical soundness of the toolkit. Barriers and enablers to changing practice. Effectiveness of the evaluation methodology adopted. Evaluation findings and their relevance.
Project team	Ensuring that the project is successfully completed on time and to budget and meets the original aims and objectives of the proposal. Facilitating communication between partners. Monitoring of project activities against project plan. Collaborative nature of the project and how it has worked.
Institutional managers	Successful project completion. Usability and effectiveness of project outcomes.
Funding bodies	Value of the collaboration. Synergies between related JISC/NSF projects and programmes. Applicability and transferability of the outcomes to the wider community.
The Higher Education community	Project contribution in the areas of digital resources / repositories, distributed learning design, development and implementation, international collaboration, teaching and learning in Geography at tertiary level.

223

evaluation. It concludes with reflections on the lessons learned.

BACKGROUND

The education team at Southampton developed an overall evaluation strategy for DialogPLUS based on the principles of utilization-focused evaluation (Patton, 1986).

... in the real world of trade-offs and negotiations, too often what can be measured determines what is evaluated, rather than deciding first what is worth evaluating and then doing the best one can with methods. Relevance and utility are the driving force in utilization-focused evaluation; methods are employed in the service of relevance and use, not as their master. (Patton, 1986, p. 221)

The approach involves identifying key stakeholders and working with them to understand how they intend to use the outcomes of evaluation and which major questions are useful to answer. This information then informs the design of evaluation approaches and instruments. The full range of the

DialogPLUS stakeholders was identified early in the project (see Table 1).

Subsequently, the appointment by the UK funding body of external evaluators meant internal evaluation focused largely on the students' views of the resources, the impact on their learning, and on the project staff (see Table 2). The internal project evaluation activities were led by educationalists at the University of Southampton and used a mix of quantitative and qualitative methods.

STUDENT-FOCUSED EVALUATION

Student-focused evaluation of the DialogPLUS innovations involved both quantitative and qualitative methods.

Quantitative evaluation methods used with the students were based on the MECA-ODL methodology outlined below. Qualitative methods included observation of students using the resources, discussions with individual or groups of students, a nominal group technique (Harvey, 1998, pp. 44-45) and analysis of contributions to online discussions. Statistical analysis of com-

Table 2. Focus of the internal evaluation activities

Aspects to evaluate	Indicators
Extent of exchange of materials between the geographers.	Number of 'nuggets' exchanged. Usage by students. Staff views.
Impact of the project on local teaching environments (culture change).	Staff views.
Development of the academics involved with the project.	Staff views.
Resources newly available as a result of the project. (Includes the Toolkit as well as teaching & learning resources.)	Number of resources. Evaluation of resources.
Transferability of materials between institutional settings.	Staff views/experiences.
Evaluation of individual courses on which project has had an impact: student response.	Student views/evaluations. Assessment results. Staff views.
Enhanced student experience / learning outcomes.	Staff & student views & evaluations. Assessment results. Evaluators' views.

pleted questionnaires was thus supplemented with attempts to gain some insight from student behavior and commentary. When students were using the e-learning materials in timetabled lab sessions it was sometimes possible for the evaluator to observe and to question those who seemed to be struggling with or, indeed, obviously enjoying the activities. As Laurillard suggests (2002, p. 233), "*Designers learn more from watching a small number of students trying to learn from their materials than they ever do from questionnaire studies.*"

All findings were discussed in detail with the academic tutors. If necessary, and possible, further data was collected and analyzed for clarification or illumination. A brief summary of each evaluation was agreed and made available to external stakeholders. All the teaching and learning resources, plus the detailed and summary files were accessible to team members via the project Web site.

The MECA-ODL Methodology

The quantitative approach adopted for student-focused evaluation was based on the MECA-ODL methodology (Riddy & Fill, 2003, 2004a, 2004b). This was developed by a collaborative

European project that ran between 2001 and 2003 on quality in open and distance learning. The project partners were a mix of European universities and commercial training organizations. The project commenced with a review of other evaluation methodologies (Riddy & Fill, 2003), which compared the then small number of published e-learning evaluation approaches. The final output was a methodological guide and online evaluation tool with over 150 quality criteria covering the following phases of e-learning initiatives: conception, analysis, design, content, production, delivery, and evaluation. The tool and the methodological guide are still available on the Web: see MECA-ODL (2003).

The DialogPLUS educational research assistant proposed the use of a subset of the MECA-ODL quality criteria as the basis for student focused evaluation of the online learning activities and resources created during the project. This was accepted and supplemented with more qualitative approaches – interviews and focus groups. The subset was referred to as the "generic quality criteria."

Table 3. The generic quality criteria

Generic quality criteria for student evaluation
There was a full description of each online learning activity, including learning objectives.
The interface was easy to use.
Required tools were included (e.g. database, spreadsheet, bulletin board).
The content met the needs of my preferred learning style.
The content was relevant, appropriate and clear.
All embedded materials were easily accessible.
Mechanisms were provided for information and support.
Maximum response times to learner queries were defined.
The assessed elements of the activity were appropriate for the learning objectives.
The online learning activities improved my knowledge and skills in the subject.

Generic Quality Criteria

Ten of the user quality criteria proposed in the MECA-ODL methodology were utilised generically on all the student surveys conducted by the education research assistant for DialogPLUS evaluation. These generic quality criteria are shown in Table 3.

Students were asked to score their response to each statement as 0 (No), 1 (Somewhat), 2 (Yes), or N/A (not applicable). This range of responses was chosen, following feedback sessions on the MECA-ODL tool, which had suggested that the 5-point Likert scale originally suggested in the methodology was overly nuanced.

Students' scored responses were entered to spreadsheets and analyzed using standard statistical methods. Students were also asked to make comments if they wished and these were considered with other qualitative input.

There was a wealth of quantitative and qualitative data produced by these evaluation methods. Three specific examples are now described and discussed on the evaluation of earth observation, human geography and environmental geography modules (i.e., Chapters VII, IV, and VI).

Specific Student Evaluations

Evaluating the Leeds' Online Undergraduate Module in Earth Observation

In the first of our evaluation examples we look at the impact of delivering nugget material online and with distance tutoring to an undergraduate module in earth observation, described in Chapter VII. In 2004 a single module course was delivered to second year undergraduates within the School of Geography at the University of Leeds. The nugget/online resources were created to maintain on-campus lecture material from the previous years campus delivered course and to

allow appropriation of earth observation tutorials to the physical geography faculty. The course was designed to introduce the physical principles of earth observation, with associated image processing techniques, and was delivered as a series of ten online lectures and eight campus-led practicals. It was examined through summative assessment and by an end-of-year examination. The course was delivered through the University of Leeds virtual learning environment (VLE), Bodington Common, and supported by tutor e-mail contact and a moderated student message board. The course was managed and tutored by a School of Geography expert in earth observation. In its 2004 guise, the course was tutored at a distance, as the tutor was based in Tokyo, Japan.

The course was successfully run for fifty-six undergraduate students with no previous knowledge of earth observation or experience of online learning. The impact of the course on student learning, plus the tutor and faculty experience were evaluated using student feedback (individual questionnaire and group questioning), statistical analysis of class marks (t-test on class marks: between online and on-campus years and by gender), and by tutor interview. The overall results on the generic quality criteria are shown in Table 4.

The results of the evaluation were useful for understanding the student experience. There was a range of comments on the positive and negative aspects of the course. The students enjoyed the material content and the introduction of a new and interesting topic within their geography career. On a negative note, some students felt that they were not coping well with the course, had a poor understanding of the material, could not motivate themselves to work independently, and felt a lack of social contact with fellow students. However, the statistical analysis of overall marks obtained by students in 2004 and 2005 were encouraging and revealed:

Table 4. Student responses for Earth Observation Nuggets

Generic criteria	Percentage of the 35 student respondents giving each answer			
	No	Some-what	Yes	N/A
There was a full description of each lecture, including learning objectives.	0	3%	97%	0
The interface was easy to use.	0	23%	77%	0
Required tools were included (e.g. database, spreadsheet, note making, discussion board).	0	20%	80%	0
The content met the needs of my preferred learning style.	23%	51%	23%	0
The content was relevant, appropriate and clear.	3%	49%	46%	3%
All linked resources (e.g. image databases) were easily accessible.	3%	29%	69%	0
Mechanisms were provided for information and support.	11%	46%	40%	3%
My queries were answered in a timely manner.	17%	26%	43%	14%
The multiple choice quizzes were appropriate for the learning objectives.	9%	20%	69%	3%
The feedback from the quizzes was helpful to my learning.	9%	34%	51%	6%

- No statistically significant differences year on year for all students;
- No statistically significant differences by gender within either year or across the years;
- That overall results for female students in 2004 were slightly skewed towards the higher quartile of the class distribution;
- In 2005, overall results for female students were normally distributed.

Post evaluation the tutor reflected mostly on suggested changes to the delivery mode, specifically adapting future delivery of the course to incorporate more face-to-face contact in the form of trouble-shooting tutorials and seminars offered three to four times throughout the delivery of the course. The reflections and revisions for further delivery based on this course evaluation are discussed in Chapter VII. However, the course did not run again in e-learning mode during the DialogPLUS project so improvements based on the first evaluation and comparative analysis were not possible.

Evaluating Southampton's Census and Neighborhood Analysis Module

The second example considers one of the human geography modules described in Chapter IV. The evaluation activities undertaken at Southampton used the generic questionnaire during a focus session for students, conducted within a timetabled lecture slot towards the end of the unit delivery in academic year 2004/2005. This was attended by fifty-one of the sixty-three students taking the unit. At that point the CASWEB/CommonGIS tutorial was the most recently used of the online activities. Virtual observation of discussion board threads was also conducted throughout the unit.

Student responses to questions about the CASWEB/CommonGIS tutorial (Harris et al., 2002) were mixed, but not overwhelmingly negative, despite some initial technical problems they had encountered. Indeed, 79% recognized that it had improved their geodemographic knowledge and skills, at least somewhat. In the focus group session, CASWEB issues were in the middle of the list of worst aspects, disliked most by level three female students. It was noticeable that working with real census data, a core tenet of the unit and highly valued by the tutor, was last on the students' list of best aspects derived from the focus group session.

Overall, views on all the online activities were also mixed, with students acknowledging that they learnt from them but did not particularly enjoy them. However, they were very positive about the impact on their learning of the tutor's and other learners' contributions to the discussion boards. They were less positive about enjoying making such contributions themselves. Analysis of responses in the focus group session showed that female students, especially at level three, voted in large numbers for the discussion boards as the best aspect of the unit.

Although 65% responded that the blend of lectures and online activities was a good way to learn, only 19% positively enjoyed the online activities. Female students, especially those at

level three, enjoyed them least. The worst aspect nominated by the focus group was lack of face-to-face support. This seemed to be felt most keenly by level two female students. Level three male students felt strongly that the worst aspect was that the activities were really time consuming. However, the tutor was confident that the amount of effort required was in line with the credited hours for the unit. Further analysis of assessment scores for all components of the unit revealed no significant difference in results by student gender and/or level of study. The overall results on the generic quality criteria are shown in Table 5.

After the students had completed their responses to the questionnaire, they were asked to think about and write down, as individuals, the three best and three worst aspects of the unit. Once complete, they were asked to compare lists with one other student and agree on the three best and worst; then to repeat that process with another pair of students. This was an adaptation of the

Table 5. Student responses for the online Casweb/CommonGIS tutorial

Generic criteria	Percentage of the 51 student respondents giving each answer			
	No	Somewhat	Yes	N/A
There is a full description of the online activity, including learning objectives.	4%	47%	45%	4%
The interface is easy to use.	10%	43%	43%	4%
Required tools are included (e.g., database, spreadsheet, note making, bulletin board).	6%	33%	55%	6%
The content meets the needs of my preferred learning style.	25%	41%	27%	6%
The content is relevant, appropriate, and clear.	8%	43%	45%	4%
All embedded materials are easily accessible.	12%	57%	25%	6%
Mechanisms are provided for information and support.	10%	33%	51%	6%
Maximum response times to learner queries are defined.	29%	22%	22%	27%
The assessed elements of the online activities seem appropriate for the learning objectives.	4%	35%	41%	20%
The online activity improved my geodemographic knowledge and skills.	18%	22%	57%	4%

Table 6. Focus group votes on the census and neighbourhood analysis module

Focus group responses	Percentage of student votes
Best aspects	
Discussion board–problems/solutions–speed of response	61%
Self paced tasks	18%
Variety–mix of lectures & OLAs	10%
Improving IT skills	8%
Using real data	4%
Worst aspects	
Lack of face to face support	41%
Really time consuming	18%
Lack of instructions & support for CASWEB	12%
Navigation in CASWEB	8%
Forced to use ICT	6%
OLAs took time–not enough left for reading	6%
Release of materials/resources via Blackboard–overload/finding–not linked to unit timetable	4%
Reliance on WWW–availability/connections	4%
Loss of work	2%

nominal group technique (Harvey, 1998, pp. 44-45). Each group was then asked to volunteer one best and one worst aspect. These were listed and numbered on a PowerPoint™ slide. Once each group had spoken, any group items not already volunteered were solicited and added to the list. Finally, each student was asked to decide on one personal best and one worst aspect from the lists and to write the item numbers on the questionnaire before handing it in. The best and worst aspects reported by the groups are listed in Table 6, together with the number of individual votes for each item (in the final vote).

Again these results, and a wealth of student comments, were reported to the tutor. Reflections on the feedback are discussed in Chapter IV. The module was not run again during the DialogPLUS project so comparative evaluation was not possible.

Evaluating an Environmental Geography Activity

The third example pertains to the environmental geography module described in Chapter VI. The evaluation activities undertaken in academic year 2004/2005 were observation of students' use of the environmental indicators online activity, plus analysis of responses to the generic questionnaire. These are shown in Table 7.

Positive aspects from the observation were:

- Students seemed engaged with and interested in what they were doing;
- There were no technical problems;
- Students seemed confident that they would be able to complete the activity by the deadline.

Table 7. Student responses for the environmental indicators online activity

Generic criteria	Number & percentage of the 56 student respondents giving each answer			
	No	Somewhat	Yes	N/A
There was a full description of the learning activity, including learning objectives.	2%	9%	88%	2%
The interface was easy to use.	4%	30%	64%	2%
Required tools were included (e.g., database, spreadsheet, note making, bulletin board).	2%	7%	89%	2%
The content met the needs of my preferred learning style.	4%	45%	50%	2%
The content was relevant, appropriate, and clear.	0	52%	46%	2%
All embedded materials were easily accessible.	4%	43%	52%	2%
Mechanisms were provided for information and support.	2%	13%	84%	2%
Maximum response times to learner queries were defined.	7%	29%	55%	9%
The assessed elements of the activity were appropriate for the learning objectives.	0	25%	71%	4%
The activity improved my environmental indicator knowledge and skills.	0	25%	73%	2%

Negative aspects observed were:

• Unsatisfactory navigation (Next/Back only).

Student responses on the questionnaires were particularly positive with respect to:

• Description of content and learning objectives;
• Inclusion of required tools;
• Mechanisms for information and support;
• Appropriateness of assessments;
• Improvement of knowledge and skills.

The most negative response was to the suggestion that they might prefer to be assessed by essays rather than online learning activities, implying general acceptance of the online assessment components. There was a spread of views about the other aspects. More students were positive that the activity improved their environmental indicator knowledge and skills than were positive that they had learnt a lot from all the online learning activities. These evaluation processes and outcomes are described in more detail in Priest and Fill (2006). Tutor reflections on the results of the evaluation, and changes made as a result, are also discussed in Chapter VI.

Comparing the Results of Evaluations

One of the advantages of using generic evaluation questions should be the affordance of comparing student responses to different learning activities (Fill, 2005). The results of such comparisons can be used to encourage discussion with learning designers and teachers about both positive and negative aspects of specific approaches and outcomes. For

Figure 1. Chart comparing percentage of "Yes" responses across modules

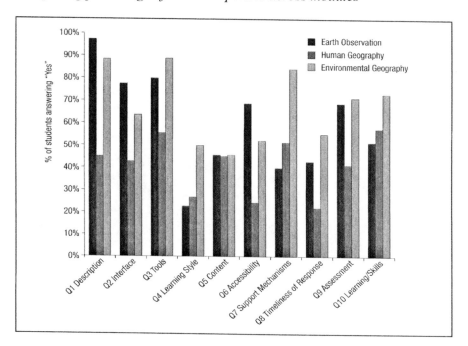

example, Figure 1 charts the percentage of "Yes" responses to each of the questions for the three activities described above.

From this we can see that, for example, a greater percentage of students were positive about the interface for the earth observation resources than for the other two. However, more students on the environmental geography course thought the online activities improved their learning and skills.

Ideally, such comparative results should not be used simplistically, but discussed with the tutors, learning technologists, the students, and other interested parties to try and understand differences between context, cohorts, subjects, and design approaches. The intention of such discussions should always be to learn from best practice and decide on any revisions necessary to improve the learning experience for future cohorts of students.

In the DialogPLUS project there was no provision made for the continuation of evaluation activities by the educational research assistant during the embedding phase. Ongoing evaluation and improvement was the province of the tutors and learning technologists. These are described in the geography exemplar chapters (Chapters IV to VII).

STAFF-FOCUSED EVALUATION

Methodology

In order to gather the staff views indicated in Figure 2, one-to-one, semi-structured interviews were conducted by the education researcher with sixteen key members of the DialogPLUS project team between November 2005 and January 2006. The guiding questions for the interviews are shown in Table 8.

Table 8. Questions used to guide the semi-structured interviews

Questions used
What has been your role in the DialogPLUS development phase?
Who have you worked closely with?
What has gone well? Why?
What has not gone well? Why?
What, if anything could have been done differently?
What significant outcomes are you aware of? How would you value them?
How have you disseminated the project/outcomes locally? Nationally? Internationally?
How has DialogPLUS affected you personally? Your colleagues? Your students?
What are your expectations of the embedding phase of the project? What will be your role in it, if any?
What do you think will be the lasting impact of DialogPLUS?
Is there anything else you'd like to say about the project processes or outcomes?

The eight interviews at Southampton and five at Leeds were conducted face-to-face, the two with Penn State University staff and one with a project team member at University of California Santa Barbara (UCSB) by telephone. All interviewees were asked, and gave permission, for the interviews to be recorded. Each interview was subsequently transcribed. The interviewees were asked to confirm, or correct, their transcripts and give permission for them to be analyzed and the findings reported. Approved transcripts were then analyzed using QSR NVivo™, a qualitative data analysis package.

Results

The sixteen interviewees were asked to describe their roles on the project and with whom they had worked most closely. From their descriptions it was evident that interviewees had played more than one role. These were coded as "attributes" as part of the analysis. Interviewee roles are summarized in Table 9. It is notable that nine interviewees had played some part in developing materials and eight had taught using project resources. The project manager/coordinator role includes people with overall responsibilities and those who played a pivotal role coordinating their local teams.

Collaboration, course innovations, and the high quality of the developed resources were mentioned most often in answer to the question about what had gone well. Table 10 shows the percentage of respondents by role who nominated these factors.

Interviewees stated that the collaboration was facilitated by

- Online and face-to-face meetings;
- Use of e-mail and the swiki;
- Innovative tools (ConceptVista and the Learning Activity Toolkit);
- Good people;
- Local technical support;
- Evaluation activities.

The outcomes from the collaboration that were mentioned most often were:

- Relationship building;
- Knowledge sharing;
- Personal learning;
- Course innovations;
- High quality learning and teaching resources;
- Sharing of some resources;

Table 9. Project roles undertaken by interviewees

Role	Primary role	Secondary role	No. of interviewees in this role
Materials developer	3	6	9
Teacher	5	3	8
Project manager/Coordinator	4	2	6
Technical developer	4	1	5
Technical support	0	2	2
Evaluation support	0	1	1
Researcher	0	1	1

Table 10. Nominated top three successes by role

Successes	Materials Developer (9)	Teacher (8)	Technical Developer (5)	Project Manager/ Coordinator (6)
Collaboration	100%	75%	80%	83%
Course innovations	78%	25%	60%	50%
Quality resources	44%	50%	40%	50%

Figure 2. Significant outcomes

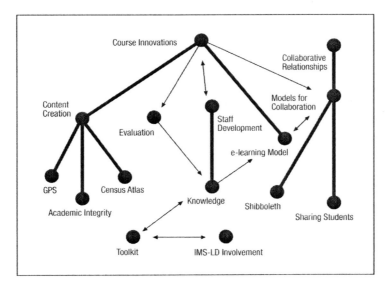

Table 11. Top three significant outcomes by role

Outcomes	Materials Developer (9)	Teacher (8)	Technical Developer (5)	Project Manager/ Coordinator (6)
Staff development/knowledge	44%	88%	100%	50%
Course innovations	44%	40%	20%	17%
IMS-LD	11%	0%	80%	17%

• Innovative tools (ConceptVista and the Learning Activity Toolkit).

The main weaknesses cited were the failure to create a shared digital library infrastructure or to demonstrate any significant reuse. Difficulties with access to externally maintained databases were mentioned by three interviewees as a contributory factor to the lower than expected level of sharing existing resources. Subsequently, many of the DialogPLUS resources have been submitted to the UK's Jorum repository (Jorum, 2008), and are freely available to the academic community. There was no striking consensus in answer to the question about what might have been done differently.

An NVivo™ model, based on interviewees' responses to the question about the projects' significant outcomes is shown in Figure 2. Staff development and increased knowledge were mentioned most often here, followed by course innovations and the involvement of DialogPLUS educational and technical specialists with the IMS Learning Design community of practice (IMS-LD). Table 11 shows the percentage of respondents by role who nominated these factors.

Most interviewees expected the embedding phase of the project to involve continued use of the resources that had been created and further improvements to teaching practices as a result of the lessons learned during the development phase. Some felt that ongoing local, national,

and international dissemination would have a real impact on e-learning strategies and policies. It was notable that the lasting impact was seen mainly in terms of the effect of the project on the participants, the changes in the geographers' teaching practice, and their hopes of continuing collaboration.

CONCLUSION

This chapter has described the approaches to, and some of the results of, student- and staff-focused evaluation of the learning and teaching innovations made during the DialogPLUS project.

Student responses were mixed, with a notable enthusiasm for online communications with responsive tutors, but an aversion to spending too much time themselves on the substantive learning activities. The majority of students questioned across all units that were evaluated did think the online activities improved their learning and relevant skills, and this was confirmed by teaching staff.

From the staff perspective, the quality of the collaboration between project team members, both within their own institutions and between the partner universities was a significant success factor. There were some failings, especially with reference to the amount of actual sharing and reuse of resources, and the absence of a digital library infrastructure as a mechanism for these aspects. While some of the created resources and course innovations were viewed as significant positive outcomes, staff were aware that they had a potentially limited lifespan. However, there was widespread recognition of the positive impact of the project on personal development, learning and professional practice. There was enthusiasm for the use of digital technologies to enhance student experiences and learning, improved understanding of how to develop, implement, support, and evaluate e-learning and appreciation of the effort involved to do this well. Staff also recognized that doing this well can be very time consuming. Above all else there was a desire to continue to collaborate that has borne fruit, including this book.

REFERENCES

Fill, K. (2005). *Student-focused evaluation of eLearning activities.* Paper presented at European Conference on Educational Research, Dublin, Ireland. Retrieved March 27, 2008, from http://www.leeds.ac.uk/educol/documents/143724.htm

Harris, J., Hayes, J., & Cole, K. (2002). Disseminating census area statistics over the Web. In P. Rees, D. Martin, & P. Williamson (Eds.), *The census data system* (pp.113-122). London: Wiley.

Harvey, J. (Ed.) (1998). *Evaluation cookbook.* Edinburgh, UK: Heriot-Watt University.

Jorum. (2008). *Jorum: Helping to build a community for sharing.* Retrieved March 29, 2008, from http://www.jorum.ac.uk/

Laurillard, D. (2002) *Rethinking university teaching: A conversational framework for the effective use of Learning technologies* (2nd ed.). London: RoutledgeFalmer.

MECA-ODL. (2003). *MECA-ODL: Methodology for the analysis of quality in ODL through the Internet.* Retrieved March 27, 2008, from http://www.adeit.uv.es/mecaodl/

Patton, M. Q. (1986). *Utilization-focused evaluation.* Thousand Oaks, CA: Sage Publications Inc.

Priest, S., & Fill, K. (2006). Online learning activities in second year environmental geography. In J. O'Donoghue (Ed.), *Technology supported learning and teaching: A staff perspective* (pp. 243-260). Hershey, PA: Information Science Publishing.

Riddy, P., & Fill, K. (2003). *Evaluating the quality of elearning resources.* Paper presented at the British Educational Research Association Annual Conference, Edinburgh, UK. Retrieved March 27, 2008, from http://www.leeds.ac.uk/educol/documents/00003331.htm

Riddy, P., & Fill, K. (2004a). *Evaluating eLearning resources.* Paper presented at In Networked Learning 2004. Retrieved March 27, 2008, from http://www.networkedlearningconference.org.uk/past/nlc2004/proceedings/individual_papers/riddy_fill.htm

Riddy, P., & Fill, K. (2004b). *Evaluating eLearning resources: Moving beyond the MECA ODL methodology.* Paper presented at ALT-C 11th International Conference, Exeter, UK.

Chapter XIV
Reflections, Lessons Learnt, and Conclusions

Louise Mackay
University of Leeds, UK

David Martin
University of Southampton, UK

Philip Rees
University of Leeds, UK

Helen Durham
University of Leeds, UK

ABSTRACT

In this book we have illustrated the materials, software, and experience of developing and delivering geography e-learning courses and learning activities. In this chapter we summarize how the teaching of a variety of geography topics has benefited from the following set of activities: creating media-rich online materials that take full advantage of linking to digital libraries; developing and adapting online, collaborative, and design software; and internationalizing materials through geography teachers in different countries working together. We take a moment to reflect on the experience of material development and the prospects for facilitating exchange of resources and student access. We provide advice to the aspiring geography e-tutor and describe how to access the wealth of materials that have been introduced in the preceding chapters. We then explain how the materials created will continue to be relevant beyond this book. We envisage that teachers, including ourselves, will download and then adapt the materials, borrowing content, techniques for presentation, or learning style. There will be an ongoing process of teaching and review that incorporates tutor and student feedback. The material, its delivery, and its style will not remain static but we hope new developments will be shared via learning repositories. It is important to sustain good online resources. This can be achieved by readers updating the geography e-learning materials and depositing improved versions in the new UK academic learning material depository Jorum.

INTRODUCTION

The preceding chapters document the exposure of a diverse group of university geography teachers to a wide range of e-learning tools and approaches. The academic geographers among the contributors to this volume remained throughout the DialogPLUS project far from the view that just because a particular technology could be applied to the teaching situation, it should be done. Some of these teachers were sympathetic from the start, perceiving the possibility that e-learning might offer approaches to tackling recognized challenges in their traditional teaching, while others have become involved through interest in the first round experiments of their colleagues.

Our evaluation of the many project outcomes has necessarily employed a variety of approaches. These include careful scrutiny of the standard student course evaluations required by our universities and additional evaluation exercises conducted as part of the project and summarized in Chapter XIII. Standard evaluations are helpful in so far as they tell us about the specific success or otherwise of teaching innovations in a specific context, but they are inherently limited because a single cohort of students only ever experiences our teaching in one way. Students bring their own set of expectations and competencies and are not able to compare their experience with that of earlier cohorts who have been taught the same or similar courses, using different methods.

In addition to these student evaluations, interviews with geography academics and learning technologists have been an integral part of the project's progress and the reflection involved in preparation of this text provides a further important aspect of our evaluation. It is entirely possible for the teacher to misinterpret the experience of a specific student cohort, yet these reflections do have the great advantage of providing context and allowing comparisons to be made with the impact of other approaches, both within the same institution and elsewhere. In forming an overall evaluation, both elements are necessary. In this conclusion we therefore discuss a broad conception of the success of our materials and methods, the effectiveness of our experimentation in learning and teaching, and the success of the learning nuggets, which formed such an important vehicle for our work. In the remaining sections we seek to draw lessons in three domains: collaboration, learning design, and learning delivery.

LESSONS FOR COLLABORATION

A common characteristic of the work described in Chapters I-XIII of this book, dating from the earliest discussions of the DialogPLUS collaborators, has been the concept of the learning "nugget" (Chapter I). From the outset, the underlying reasoning behind the nugget was that it would comprise a recognizable component of learning, less formally defined than a "learning object" (Wiley, 2002). The learning nugget would be readily recognizable by geography teachers and would therefore form a natural basis for the sharing of learning content between individuals and courses. Certainly the nugget concept has proved very useful, providing a common language for teachers to communicate with one another about the elements comprising their courses and the technologies used to support them.

However, one of the original aims of using the exchangeable nugget as a simple vehicle for reuse of learning materials between courses has largely failed and we believe that this experience may have important consequences for those who anticipate that learning repositories (Rieh et al., 2007) have the potential to provide large amounts of learning content at much reduced effort. In the past, repository projects have been challenged by the relatively small likelihood of users finding material that matches closely to their teaching requirements, but within DialogPLUS there has been a good match between courses and collaborating lecturers. However, even within the context of the project, teachers needing to design their own courses have developed highly individualized approaches and materials that reflect their personal expertise in the fields being taught. This is perhaps unsurprising, given the research-driven nature of much of the teaching described here, but the lesson seems stark. E-learning materials, even when well designed and clearly documented, will rarely fit directly into the courses and curricula delivered by other individuals or at other institutions. Teachers crave personal ownership in the learning materials they use, often necessitating significant adaptation of externally sourced materials for local use. An open question remains as to whether there is a gradient here, whereby more foundational materials and those dealing with more procedural skills have a higher probability of reuse. The academic integrity work described in Chapter VIII would suggest that this might be the case, whereby several institutions enthusiastically accepted a learning nugget that was seen to have very broad application across students and courses. Nevertheless, substantial adaptation was required before local implementation due to the specific requirements of institutional learning regulations. Leung et al. (Chapter II) do demonstrate that there is still value to the learning nugget concept, as it has provided considerable flexibility for course teams within the same institution to rework their own materials for different student audiences and different modes of delivery.

A more successful avenue has been the collaborative design of learning materials, as exemplified by the joint development between the University of Leeds and Penn State of the GPS and AI nuggets. These initiatives appear to succeed due to the direct involvement of the teachers who will use the materials. The collaboration introduces new ideas and skills to the process of designing teaching content, yet the teacher who will deliver the material retains a high level of control over design and delivery – this again represents a logical advance over the creation of personal teaching materials. Many academic readers will be familiar with the collegial benefits of team-teaching on undergraduate courses and these joint e-learning developments bring some of the same benefits to a collaborating team spread across different institutions. The unanswered questions in this area relate to the size and composition of the collaboration. Clearly, attempts to simultaneously co-author material across large teams are likely to prove impractical. Solutions are unlikely to be reached which meet the needs of many institutions at the same time. Nevertheless, the suggestion that collaborating institutions might cooperate more actively on the creation of high-quality learning nuggets when incorporating (for example) new techniques, concepts, or case studies into existing programs seems an attractive one.

By contrast, our experiences described in Chapter II do show that it is possible to exchange entire modules, and that this can be made to work even when the collaboration is international. Indeed, the technical barriers to module sharing are relatively low compared with those involved in obtaining inter-institutional academic approval for use of external modules but once approved, there are genuine benefits to be reaped. It is at first sight somewhat counterintuitive that it should be easier to exchange modules than nuggets, but this is due to the fact that modules are of a sufficient size to have institutional visibility. Course specification standards (embodied in institutional quality handbooks) require clear statements of learning objectives, teaching methods and outcomes at the module level. This clear articulation is effectively the module-level "metadata," which provides a mechanism for one department to assess and accredit a proposed teaching input from elsewhere. There are also well-established practices of student exchange between institutions, which also operate in terms of one or more modules accredited by a recognized partner. While module-level exchange represents more student study time than the exchange of individual learning nuggets, it does not impact on individual teachers in such a personal way: they are not required to adapt their approach or style within a specified module.

Unfortunately, we have encountered some important limitations to international collaboration that relate to the nationally-specific nature of some data licensing agreements – and this has particular relevance to some key geographical datasets. While census mapping data in the USA, for example, are freely available to all users via the web, we have been unable to share some UK datasets with USA partners due to licensing conditions, which specifically restrict use to the UK academic community. This applies both to census data purchased by the UK Economic and Social Research Council and mapping data from one of the UK's national mapping agency, Ordnance Survey Great Britain, available to UK academics through the Digimap scheme. We are thus unable to build shareable learning materials based on these data products because students outside the UK would not be able to make use of them. This issue has been considered in some depth in Chapter 4, but license restrictions on datasets, media or software tools is a relevant consideration for all would-be authors of shareable learning resources.

We believe that we have also learned practical lessons in inter-institutional collaboration that others would do well to note. Our project evaluations show that relationship building and knowledge sharing are among the most important benefits identified by staff taking part in the DialogPLUS project.

The project benefited greatly from the use of regular Access Grid meetings whereby team members in different locations were presented with a regular need to discuss progress with colleagues. Access Grid is a combined hardware and software package that enables multi-institution video-conferencing. To use Access Grid each collaborating institution needs to set up a suitably equipped room for the purpose (though the room can be used for normal meetings as well). Access Grid software enables the meeting convener to share presentations and files onscreen. The use of Access Grid technology permitted multiple sites to participate in fairly natural spoken discussion while seeing and hearing one another in real time. Access Grid offered the potential to share and demonstrate learning materials during the meetings, although in reality most content sharing occurred outside these occasions. The need to book an Access Grid facility for the same time slot at each institution incurs some additional financial and logistical costs. Sometimes the high capacity Internet connections needed are not available, for example.

Within the project, we have also made use of the desktop conferencing tool HorizonWimbaTM (http://www.wimba.com/), which allows interactive discussion through audio and video channels and presentation sharing from participants' own PCs. Other video- and Web-based conferencing technologies are likely to provide similar benefits to collaborating teachers, but our observation is that these media only begin to work best once the team members have met physically and begun to understand the context of one another's teaching. In addition to these real-time communication tools, much use was made of a swiki site, allowing collaborative authoring of project reports and sharing of documentation. "A wiki is a collection of web pages designed to enable anyone who accesses it to contribute or modify content, using a simplified markup language" (Wikipedia, 2008).

A swiki is a wiki written using the Squeak software. Again, these tools appear to be most effective once a relationship is established between team members at different sites and the collaboration has been "primed" by personal contacts.

One of the editors notes that the experience of sitting alongside a colleague from a different institution while moderating his student discussion list not only yielded practical benefits for their own teaching, but also opened a channel of shared understanding as the basis for subsequent discussion and development. While the practices of peer observation, staff appraisal, and departmental teaching strategy are widespread within institutions, the specialist teacher will often find greater commonality of practical teaching challenges with cognate subject specialists in other institutions and this is undoubtedly a benefit of the type of collaboration fostered by DialogPLUS. The geography lecturer seeking to find support for their development of e-learning innovation in their teaching is therefore well advised to seek involvement in inter-institutional projects and networks in order to find those with the closest teaching portfolios and the most relevant e-learning experience.

LESSONS FOR DESIGN

In those areas of the DialogPLUS project where new learning materials have been created – either by individuals or collaborating groups, the issue of learning design continually rises to prominence. The variation in styles and design philosophy between individual teachers has already been identified as a powerful force in determining the acceptability of externally written learning materials. The learning design toolkit described in Chapter IX serves to formalize many of these design considerations. It allows teachers to reflect on their pedagogic style and to select methods and tools, which may be appropriate. The benefits of such a tool appear to be different to differ-

ent types of teacher. The DialogPLUS project partners were mostly experienced teachers, who overwhelmingly cited the benefits of exposure to others' ideas and to new ways of thinking about their teaching. Use of a learning design toolkit could provide these benefits to an experienced teacher about to embark on writing a new module or undertaking substantial revision to existing material. However, the toolkit may play a very different role in the training and development of new lecturers, who do not have many pre-existing reference points and for whom it may serve to open up a broader perspective on the very many approaches available. The learning design tools discussed here, including the use of concept mapping, mind mapping, and linear thinking certainly assist with course design and make for future portability of materials, but these are not sufficient in themselves to ensure reuse. They can be powerfully used by collaborating authorship teams to design materials which have wide application, based on clearly specified models of student learning. Chapter X discusses the uses of concept mapping while Chapters III and XI have examples of its application.

Not only have our learning nuggets proved important for reuse and collaboration, but they also hold lessons for learning design, particularly the issue of granularity. Where materials have been produced from essentially small components it becomes much easier for them to be reused by the original authors and others. Most of our design challenges have not concerned file formats or transfer between VLEs – these materials rarely use strongly proprietary formats. This design issue therefore has important impacts on the durability of learning materials–large and indivisible nuggets are simply much harder to revise and reuse.

In general, our experience suggests that few of the most important learning design decisions are about e-learning technology per se, although it does offer the teacher a significantly increased array of tools and methods. Indeed, our endeavors to introduce a range of e-learning applications into our various teaching situations has more frequently challenged us to reconsider the fundamentals: the constructive alignment (Biggs, 2003) of learning objectives, delivery methods, the role of group and individual work, independent learning, interactivity and discussion, assessment, evaluation, and feedback. We have found a rich vein of teaching enhancement through the impetus of introducing e-learning innovations to courses and sharing these experiences with cognate partners and this is especially the case when taking previous face-to-face materials and putting them online. A particular area of challenge is to recognize that online students have far greater opportunity for independent learning and this places far higher responsibilities on the teacher to ensure that all instructions and supporting materials are robust and consistent. Some of our most frustrating experiences have been to discover that university teaching committees are often more ready to insist that rigorous design principles be applied to collaborative learning and e-learning proposals than they are to apply the same measures to existing face-to-face teaching practices! Martin and Treves (2007) review the broader process of institutional embedding of e-learning practices, which surround DialogPLUS and related initiatives.

Most of the teaching staff involved in DialogPLUS have now developed roles as "blended teachers," whereby a proportion of their teaching is delivered online. This not only equates to the assembly of online materials but, importantly, to the management of discussion lists and other electronic media for communication with students. A few among us have become e-tutors, whose primary contact with students is electronic, and this is a particular necessity when supporting distance learning students on online programs. This necessitates very different modes of working. The most obvious one, which has been often cited as an awful warning by our colleagues, is the need for continuous availability. This does not mean being continuously logged in to monitor student discussion forums, but rather a commitment to set out and maintain regular communication and to tackle teaching issues outside the normal university timetable and without lengthy interruptions. This is not without its benefits – it increases the capacity of the teacher to maintain student contact while travelling during term-time, to work from different locations, and generally increase flexibility within the working day. As so often with face-to-face teaching, it is not technical skills but factors such as clarity of communication, statement of expectations, fostering of student community, and discernment of individuals' needs that make for a successful e-tutor. There are likely to be distinct and, as yet, unresearched teacher characteristics that affect e-learning effectiveness, particularly concerning ability and willingness to recast their learning design in ways best suited to online or blended delivery.

LESSONS FOR DELIVERY

The e-learning materials described in the chapters of this book have been delivered to students through an extremely wide range of mechanisms. Our learning nuggets may be classified in various ways but, more importantly, it is the learning design and the willingness of staff to directly engage with it rather than the details of the institutional virtual learning environment that appear most important. Delivery is necessarily part of the course design and the main requirement of the delivery mechanisms is that they be reliable and appropriate to the rest of the learning.

E-learning materials ("nuggets") can be delivered in a variety of ways. At one extreme is fully automated e-learning where the computer delivers and manages both content and assessment. This mode was rare in the practice of DialogPLUS collaborators. In many cases the materials were part of an online distance learning module in which the materials could be downloaded and worked through by the student. The most successful examples required the student to interact with the materials (e.g., in digital libraries), respond to questions, and carry out analysis. Very frequently the materials at undergraduate level were part of a blended learning strategy that combined lectures and tutorials with lab assignments based on e-learning materials.

One important issue here is that regardless of course-specific learning objectives, student expectations regarding e-learning necessarily need to inform learning design. Our experiences include delivery of online modules to two very different types of student. While the distant student, often a mature professional with clearly articulated personal development objectives, will usually have engaged very thoughtfully in our online master's level programs (DiBiase, 2000), this is not the case for conventional face-to-face undergraduate students. While the latter are increasingly at ease with the Web as an information resource and a medium for social networking, they are not well developed as independent learners and still have a deep-seated expectation of regular face-to-face lecturer contact. Indeed, we readily recognize that the social aspects of learning in a traditional university rely very much on intensive personal interaction. Our delivery of wholly online learning in this setting is treated with some caution by face-to-face undergraduate students whose expectation is of weekly personal contact with a lecturer. On some other modules, our face-to-face students reported that online learning activities were time-consuming, yet these were often activities that replaced scheduled classes, transferring more of the responsibility for time organization to the students. By contrast, the introduction of judiciously placed and sized e-learning activities within face-to-face courses, such as the enhancement of practical sessions (e.g., Chapter III) has been well received and appears to deliver real advantages. There is an undoubted personal development benefit to students being able to engage appropriately with online resources and discussions, but the means

of determining the right proportion within any student's learning experience needs much more research. A final observation on this issue is that it could be argued that a small exposure to fully online learning in the undergraduate curriculum is valuable for providing a taster experience of what is provided in distance master's programs and in work-based training programs.

Our evaluations have not shown statistically significant changes in student assessments when similar material is taught using conventional or e-learning approaches. However, students do not often identify the same materials as most useful and most enjoyable! There are multiple challenges in interpreting these results, including the effects of normative marking and the inevitable differences between course delivery and student cohorts from year to year. Despite the many teaching delivery situations which formed part of DialogPLUS we have not been able to directly replicate the same student learning and assessment using conventional and e-learning methods in order to produce a satisfying "natural experiment" revealing the simple before-and-after effects of replacing an existing module with an e-learning equivalent.

CONCLUDING REMARKS

We trust that the e-learning examples presented in the preceding pages will not be interpreted as mere technological "boosterism." Indeed, many of the chapter authors reflect on the changing nature of the teachers' workload with increased use of e-learning. These materials take a long time to prepare and, although there may be a reduction in face-to-face teaching hours, there are new challenges of availability and flexibility in monitoring online discussion and responding to the needs of increasingly independent learners. There are however, real benefits to learners and teachers from the need for rigorous learning design, simply through the effect that preparing online materials forces many assumptions and instructions to be made explicit that remain implicit (or, indeed, unclear) in face-to-face settings. Five years from the commencement of the DialogPLUS project we are also able to see concrete examples of the reuse of learning materials–both within and beyond the original authorship teams. Our greatest evidence of reuse is as part of different or revised courses within the original institutions and this perhaps lends support to the current interest in developing institutional (rather than community) repositories. In the light of our experiences it does not seem likely that effective reuse of learning materials is likely to result from discovery and adaptation of resources within repositories without some degree of inter-institutional collaboration between educators–something to which we remain strongly committed.

In conclusion, we offer the DialogPLUS team's experiences in the belief that there is something here of value to each of the different types of reader identified in Chapter I: directors of learning/teaching programs, lecturers, drop-in tutors, research students aspiring to a career in research and teaching, and master's students in both geography and education. There are many research questions outstanding here. We have very gratefully recognized the role of the educational researchers and learning technologists in our teams, but we propose that real educational innovation is only possible when the academic teacher is willing to engage seriously with the entire learning process, encompassing both design and delivery, and to engage with e-learning not as an end in itself but as a source of new and reinvigorated solutions to longstanding teaching challenges. We hope that the preceding chapters will also convince many of the benefits of collaboration, recognizing not only the need for supporting technical specialists but also the insights that come from shared experience.

REFERENCES

Biggs, J. (2003). *Teaching for quality learning at university* (2nd ed.). Buckingham: The Society for Research into Higher Education and Open University Press.

DiBiase, D. (2000). Is distance education a Faustian bargain? *Journal of Geography in Higher Education, 24*(1), 130-135.

Martin, D., & Treves, R. (2007). Embedding e-learning in geographical practice. *British Journal of Educational Technology, 38*, 773-783.

Rieh, S. Y., Markey, K., Yakel, E., St. Jean, B., & Kim, J. (2007). Perceived values and benefits of institutional repositories: A perspective of digital curation. In *An International Symposium on Digital Curation* (DigCCurr 2007), Chapel Hill, NC. Retrieved April 28, 2008, from http://www.ils.unc.edu/digccurr2007/papers/rieh_paper_6-2.pdf

Wikipedia. (2008). *Wiki.* Wikipedia: The Free Encylopedia. Retrieved April 28, 2008, from: http://en.wikipedia.org/wiki/Wiki

Wiley, D. A. (2002). Connecting learning objects to instructional design theory: A definition, a metaphor, and a taxonomy. In D. A. Wiley (Ed.), *The instructional use of learning objects.* Bloomington, Ind.: Association for Educational Communications & Technology. Retrieved December 10, 2007, from http://reusability.org/read/chapters/wiley.doc

Section IV
Additional Selected Readings

Chapter XV
Online Learning Activities in Second Year Environmental Geography

Sally Priest
University of Southampton, UK

Karen Fill
University of Southampton, UK

ABSTRACT

This chapter discusses the design, technical development, delivery, and evaluation of two online learning activities in environmental geography. A "blended" approach was adopted in order to best integrate the new materials within the existing unit. The primary aim of these online activities was to provide students with opportunities to develop and demonstrate valuable practical skills, while increasing their understanding of environmental management. A purpose-built system was created in order to overcome initial technological challenges. The online activities have already been delivered successfully to a large number of students over two academic years. Evaluation and staff reflection highlight the benefits and limitations of the new activities, and the chapter concludes with recommendations for others wishing to adopt a similar approach.

INTRODUCTION

In the academic year 2003-2004, staff in the School of Geography at the University of Southampton created two online learning activities for students on a level two unit entitled *Physical* *Geography in Environmental Management*. The activities introduced the concepts of managing and querying environmental data, using and developing environmental indicators, analysis, reflection, and decision support. As they worked through the learning activities, the students had access to a wide range of Web-based resources plus repurposed data from the Environment Agency's River Habitat Survey (Environment Agency, 1998). Their responses to both formative and summative assessments were captured online. Some elements were computer assessed and others marked by the unit tutors. The staff and students were all campus based. These online activities complemented lectures and other face-to-face sessions on a unit that has been taught, in one guise or another, for 20 years at the University of Southampton.

This chapter describes the development and implementation of these online learning activities in terms of the pedagogic opportunities and technical challenges encountered and overcome. It reviews the learning outcomes achieved by the students and discusses their evaluation of the resources. Tutors' reflections on the impact of this innovation are included. The chapter concludes with recommendations, both specific to this unit and for those working in the wider field of technology-supported learning.

The innovations and evaluation were undertaken under the auspices of the *Digital Libraries in Support of Innovative Approaches to Learning and Teaching in Geography* project (DialogPLUS, 2004).

BACKGROUND

Overview of the Unit as Traditionally Taught

Physical Geography in Environmental Management is primarily lecture based; however, the unit is pioneering in many respects. Since its inception, it has been a test-bed for new pedagogic approaches within the School of Geography. The unit is taken by postgraduates and undergraduates, as well as geography specialists and non-geographers. This varied mix of students makes it ideally suited for evolving new and innovative

approaches. Indeed, the unit was an early adopter of the MicroCosm® open hypermedia system (Clark, Ball, & Sadler, 1995); the first in the School to use PowerPoint and subsequently Web-based resources; and the School's own virtual learning environment (VLE) was initially developed to house its resources.

The associated practical elements have developed from paper-based, via early computer techniques to support learning, to the first stages of Web-enabled education. Early attempts at e-learning focussed primarily on the delivery of resources across the Web and the use of simple computer models, rather than engaging students with any meaningful interaction, other than choice of options and parameters for modeling. The opportunity, within the DialogPLUS project, to address this perceived need for engagement, coupled with increasing numbers of students, provided the impetus for further change.

The Pedagogic Approach

A learning environment is a place where people can draw upon resources to make sense out of things and construct meaningful solutions to problems. (Wilson, 1996, p. 3)

For more than 20 years, tutors on this unit have adopted a constructivist perspective, progressively evolving an approach that embeds learning in "realistic and relevant contexts" (Honebein, 1996, p. 11). Practical elements give the students opportunities to learn in different ways, developing a variety of skills. The previous practicals involved electronic resources developed in MicroCosm® (see www.vmsi-microcosm.com). Advances in technology, primarily Web-based delivery, offered the potential for more experiential student learning, without a major change of rationale. This remains the activity-based enhancement of student learning in an alternative environment to lectures.

The mix of traditional lectures and online activities employed on the unit for the last two academic years conforms to the emerging 21st century paradigm (Kerres & de Witt, 2003) of "blended learning." Oliver (2004) is critical of this term, arguing that it is either inconsistent or redundant (p. 6) and that, because it originates from the world of corporate training, its use in higher education is somehow belittling (p. 7). However, increasingly lecturers are employing a mix of traditional (off-line) and online approaches and they and their students seem comfortable with the term and the approach. Indeed, Childs (2004) found students enthusiastic about blended learning, which they understood to mean a combination of listening to lectures, finding and using Web-based resources, and interacting with particular software to solve problems.

It is certainly crucial to specify and address clear learning outcomes, to ensure that there are opportunities for learners to engage with core concepts, and to provide what Weller, Pegler, and Mason (2003) refer to as an "overarching narrative" that directs and frames interaction with resources and constructive learning. These have been guiding principles in the creation of the online learning activities reported here. Within an activity, the students find a mix of media — text, image, and data — and of task types — assimilative, analytic, communicative, and reflective (Laurillard, 2002). This mix is intended to address the needs of different learning preferences and to engage all students in tasks that facilitate their individual construction of meaning.

In order to integrate the online elements fully within the overall approach to teaching and learning, the rest of the unit was examined and other aspects were redeveloped. For example, lecture content was modified to ensure that the role and purpose of the online activities blended appropriately with the more established elements. The timing of online activities and associated assessment was also addressed, as this is crucial to

learning (Perkin, 1999). It was important that the students were not overloaded and had sufficient time to concentrate on the set tasks. Therefore, the online activities were introduced quite early within the unit so that they did not impinge on the later coursework essay, or on examination preparation. Modification to the unit involved a slight overall reduction in content to facilitate independent work and in-depth study of the remaining topics.

TECHNICAL CHALLENGES

System Requirements

The technical environment was one of a number of different challenges addressed when developing the e-learning activities. The University of Southampton's preferred VLE is Blackboard™. However, this was unsuitable for the proposed online activities for a number of reasons. The pedagogic rationale behind the integration and adoption of these new online elements was the interactivity afforded by modern technology and the possibilities for introducing students to new and interesting datasets. The then current version of Blackboard™ did not offer the level of interactivity required to achieve the desired learning outcomes, it did not permit the students to enter "free" answers for automatic marking, and feedback options were restricted to "immediate" or "not at all". Therefore, it was decided to author the material independently of the established VLE, and to create content within dynamic Web pages authored in PHP (see www.php.net).

Despite the bespoke system overcoming many of the constraints of the Blackboard™ VLE, a number of outstanding issues remained. It was intended that these e-learning elements formed part of the summative assessment for the unit. This necessitated system recognition of student users, storage of online submissions, and recording of individual marks. This was achieved by using password protected log-ins and an underlying student database.

Although formative elements of the online activities were computer assessed, the summative components were marked by tutors. Thus, it was also important that submissions could be viewed and marked easily, and feedback entered. This was facilitated by the creation of a password protected tutor interface which allowed access to individual student entries. Although there are plans to implement electronic feedback, in the first instance the tutor screens were also designed to be printed out and handed back to the students.

Access to Datasets

A primary reason for developing the e-learning practicals was to allow students to gain firsthand experience of using data in a similar manner to a professional environmental manager. The chosen dataset was the River Habitat Survey (RHS) with entries for more than 14,000 UK river sites collected, authenticated, and managed by the

Environment Agency, who kindly permitted us to extract and reconfigure data for educational purposes. For ease of online access, the reconfigured dataset was mounted within dynamic Web pages. The technical design was complicated by the need for students to perform queries on the data in much the same way as if it were in the original database. Although time consuming and challenging, the repurposing of the RHS dataset significantly enhanced the students' experience, as they had the opportunity to manipulate real data. An unexpected result is that the repurposed RHS data Web pages have been suitable for use within other geography teaching at Southampton.

Web Design

There was initial debate about whether the Web pages should look different from those already authored for the School of Geography, or whether consistency was important for the students. A further complication was that these online activities were some of the first outputs of DialogPLUS, so it was important to highlight digital library aspects

Figure 1. Example of the layout and design of the Web pages

and consider any design implications that might impact more widely on that project.

After consultation with a number of people (i.e., educationalists, Web developers, geography tutors) about these design implications, it was deemed most important that the Web pages were identifiable to the students and seamless with those already being used within the School of Geography's undergraduate Web site. These had already addressed accessibility issues (see www.w3.org/WAI/), and consistency would reassure students that the online activities were endorsed by the School of Geography and build trust in what, to many, was a new learning experience (Graham, 1999; Carter, 2002).

In order to address these issues and as a way of separating content, tasks, and external resources, the main body of the screen was split into three areas: navigation, content, and digital library (see the example in Figure 1). From a pedagogic viewpoint, this layout reinforced the content of the task for students by placing it in the centre of the screen. The digital library resources held separately on the right then offered opportunities for wider learning and/or access to data to complete specific tasks. Incorporating all this via links within the main content might have overloaded the students and diverted them from following the narrative explaining the tasks and rationale.

THE ONLINE LEARNING ACTIVITIES

To assist the students, the structure of both activities was kept the same. Each had an *introduction, aims and learning outcomes, practical content,* and sections describing *expectations* and *assessment criteria.* This consistency was important as the majority of students had little experience with online learning. The potential nervousness of an inexperienced user group, plus that of the development team about how the system would perform, prompted the decision to hold associated "clinic" sessions which students were encouraged to attend if they had any problems or queries. These were timetabled in a university computer lab and staffed by a tutor and/or a postgraduate teaching assistant. Students could undertake the activities during these sessions as well as in their own time. Access to an activity was available at all times from campus or personal computers, so support was also offered via an online discussion board. The tutor monitored this regularly and responded swiftly to queries. This is discussed in more detail below.

Learning Outcomes

In order to ensure that core concepts were established, specific learning outcomes for the

Figure 2. Example of learning outcomes

"Getting your facts right":
Managing and querying environmental data
Learning Outcomes:
1. Appreciate that information is a basis for decisions and interpretations.
2. Be aware that information can be created out of data.
3. Understand the nature of database/ spreadsheet as data storage mechanisms.
4. Develop skills for the retrieving and manipulating data from database (query) and spreadsheet (statistic).
5. Appreciate the importance of the role of data quality and metadata.
6. Achieve an understanding of the River Habitat Survey as management information.

online activities were decided at the outset and informed the design of both content and learning environment (see Figure 2). These emphasise the importance, use, and interpretation of data in environmental management. They also provide the basis for examining the success of these learning activities below.

Practical Content

This section explores in more depth the different components of the online activities and their importance and role within student learning. The content of the activities was designed around a narrative to guide students through the use of resources and from task to task (see Figures 3 and 4).

The complete list of tasks and associated skills for the first activity is shown in Figure 5. Some tasks were formative, however all ultimately contributed to the assessed elements.

While these online activities are mainly linear (Plowman, 1996), it is appreciated that not all students learn best in this manner (Honey & Mumford, 1992). Some flexibility is possible as resources can be visited and tasks completed in any order. Significant aspects are the external

resources, which include papers in academic journals, external governmental Web sites, and Environmental Agency reports. This variety of sources was intended to encourage students to read broadly and gain experience of interpreting the different data types that underpin environmental management decisions. However, linking to external resources creates a maintenance issue, as availability depends upon stability on the host site.

Assessment Rationale

Consistent with the pedagogic motivations for implementing these activities and adding different learning experiences to the unit, the assessed elements also aimed to broaden the measures of student attainment. The online activities could have been purely formative, but in our experience, it can be very difficult to motivate students to complete optional exercises. To reinforce the importance of practical environmental decision-making, some components of the online activities had to be summatively assessed. Other assessments on the unit are more "traditional," including a 2,000-word essay-style report investigating a management problem and an unseen written examination.

Figure 3. Introduction and first tasks in an online activity

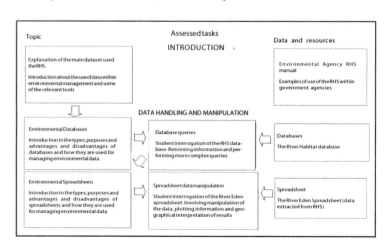

Figure 4. Assessed elements of an online activity

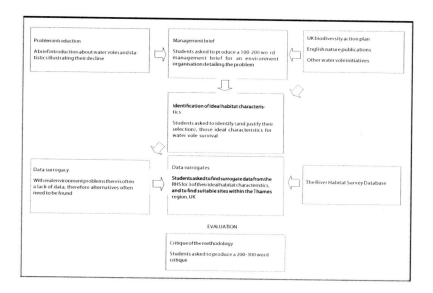

Figure 5. Indicative list of tasks performed in Online Activity 1

Task performed	Associated skill demonstrated
▪ Read explanatory text online	Research/interpretation
▪ View/read background resources	Research/interpretation/critical analysis
▪ Practice querying the database	Practical (data handling and manipulation)
▪ Use query to find answers to questions	Practical and interpretation
▪ Enter answers online	Completion of assessed activity
▪ Read explanatory text online	Research/interpretation
▪ View spreadsheet extract	Practical (data handling and manipulation)
▪ View map	Practical (data handling and manipulation)
▪ Explore spreadsheet	Practical (data handling and manipulation)
▪ Sort spreadsheet data	Practical (data handling and manipulation)
▪ Calculate and enter answers online	Practical and interpretation
▪ Plot & interpret graphs	Practical and interpretation
▪ Enter values read off the graphs	Completion of assessed activity
▪ Read explanatory text online	Research/interpretation
▪ Research/ reflect on a specified problem using external resources	Research/interpretation/critical analysis
▪ Write a 300 word management brief (in online box)	Interpretation/critical analysis/ability to summarise/ clear expression
▪ List 5 factors & assess their significance using a drop down list.	Interpretation of data and resources/ environmental decision-making/ writing and justification skills
▪ Use database to identify 3 attributes that will act as surrogates for the previous attributes	Practical querying skills/ data interpretation/ decision-making
▪ Enter 3 attributes, their RHS categories & reason for choosing them, site refs.	Writing and justification skills
▪ Use database to identify sites.	
▪ Enter list of suitable sites	Practical (data handling and manipulation)
▪ Write a 500 word critique of methodology/techniques used (in online box)	Completion of assessed activity
▪ Submit work online	Interpretation/critical analysis/writing and summarising skills
	Completion of assessed activity

In the first year, it was decided that the online activities would only comprise 10% of the overall assessment. This was felt sufficient to induce students to complete the tasks, in addition to any intrinsic motivation (Biggs, 2003). It was also conservative enough to provide a "factor of safety" if the technology failed. Although the system had been tested and seemed robust, it was impossible to test for large numbers of concurrent users, a situation likely to arise with 130 to 150 students on the unit. Therefore, it was considered inappropriate to place too great a weighting on the system that they were effectively "testing" for the first time. If one or both of the online activities did irretrievably fail, the situation could be resolved much more easily if only 5 to 10% of overall assessment was affected. This low weighting did cause some problems related to levels of study time and assessment, however, it was warranted to shield the students from potential technological problems. Following the success in the first year, the weighting was increased to 25%. Student feedback in year two appeared to endorse this allocation, with 75% of those who returned an evaluation form (n=56) agreeing that the maximum mark of 12.5% for each assessed online activity was fair.

The assessed tasks were a mixture of short answers and longer questions. The shorter ones enabled students to demonstrate that they could perform queries and find specific information. The longer ones, which involved entering written answers, aimed to assess whether students had understood the process of decision-making and all the problems involved. The majority of the shorter answer components were marked automatically. This assisted the speed, validity, and impartiality of marking and permitted feedback to the students at an earlier stage than would otherwise be possible, while also reducing tutor workload on a very popular unit.

Technological limitations are related to computer marking of written answers. Indeed, since the questions aimed to test the students' ability to make and justify sound environmental decisions, there is much ambiguity related to a "correct" answer. Therefore, although it was possible to develop a broad mark scheme from which to grade students' work, a great deal of judgement was still required to assess whether answers were appropriate. Certainly, some student answers were surprising in their level of ingenuity and thoughtfulness, and automated marking on these more innovative or subjective topics would entail a great deal of moderation. It is therefore likely that this component of the assessment will continue to be marked in the traditional way.

There were also system constraints. The boxes designed to hold student answers only permitted textual content. Although it would have been possible to develop them to hold images or graphs, it was felt that the time taken to do this would be counterproductive. Students were asked to produce graphs in a spreadsheet, thus practising these valuable skills. However, the most important learning outcome was not to produce attractive charts, but rather to interpret relationships in the data thus portrayed. Therefore, a detailed text-based description of the relationships between the data, and the significance of them, demonstrated to the marker not only that a student had undertaken the activity, but that they had also understood the results.

Students' Use of the Online Learning Activities

Students were observed doing the online activities during the timetabled clinics. Although no technical problems occurred, there was some dissatisfaction voiced with respect to the linear navigation (Next/Back only). As explained, tutors felt it was desirable to lead the learners logically through the activities, but many students wanted to be able to look ahead, jump back, or dip in and out of the resources. To facilitate this, a menu or tabbed navigation could easily be provided in the future.

Most students worked quietly as individuals, but it was noticeable that some worked in groups of three or four, discussing the resources and possible answers to the formative questions. While this usefully develops team skills required in their future working lives, the underlying approach was cognitive, rather than social and constructivist, and the tutors felt that such collaboration posed challenges for differentiating between students and (importantly) correctly rewarding individual effort and performance. This raises a number of important points about validating and assessing online coursework. Students seemed engaged and were overheard discussing key concepts, such as "bank erosion," reassuring each other about data selections and analysis. Many students made handwritten notes as they looked at the Web-based resources; under the current conditions in university labs, users cannot bookmark Web pages to return to later. This may change over the next few years with a proposed learning portal and greater customisation. Some kept several resource windows open (minimised) as they paged through the online activity, consulting them from time to time. Some students used MS Word to compose text answers, others typed directly into the Web form boxes. The embedded discussion forums were mainly used by a small number of students to pose questions to the tutor. This is discussed in this chapter's section on staff perspectives.

EVALUATION

Student Evaluation of the Resources

In the second year of implementation, students were asked to complete a questionnaire about the online learning activities. Completed forms were received from 56 (43%) of the 130 students who took the unit. Although a moderate sample, their responses were particularly positive with respect to description of content and learning objectives; the inclusion of required tools to access and manipulate online resources; the embedded mechanisms for information and support; the appropriateness of assessments; and, most encouragingly, the contribution of these activities to improving their knowledge and skills. The most negative response was the suggestion that they might prefer to be assessed by essays rather than online activities, suggesting general acceptance of the assessment components of the online activities. Only six students commented that the time estimation for completing the activities was too short. This was a marked improvement over the previous year, when the majority of negative comments were concerned with the length of time taken to complete the activities.

Twenty-six respondents contributed qualitative comments. Typical general comments indicated that this was an interesting new experience but not particularly enjoyable. Negative comments were about computer failures and limited navigation. Overall, both quantitative and qualitative feedback indicated strong endorsement of the online activities' contribution to improving skills.

On the scored questions, there were no significantly different responses from male and female students. The eight post-graduate respondents were significantly ($p < 0.0001$) more positive than the undergraduates about the ease of using the activities and the contribution that another, non-assessed, online activity had made to their submitted work.

Achieved Learning Outcomes

Students generally performed the set tasks well and attainment was high, the average grade being 72%, with a range of marks between 50 and 94%. In most cases, the results of the online activities illustrate that the learning outcomes were met. The majority of students correctly demonstrated that they had gained data manipulation and retrieval skills. They were then able to use them within other areas of the unit. Other written coursework, in particular, has revealed an increased understanding of the difficulties of making environmental decisions and how they can be overcome.

There is also anecdotal evidence to suggest that the computerised elements have contributed to other areas of student learning. Students completed these elements prior to starting their final year dissertation. A number chose to use some of the data and ideas within their dissertation and also on other units in their degree programme.

STAFF PERSPECTIVES

This section draws on both formal and informal student feedback and staff reflections over the two years of development and implementation.

Development Challenges

None of the technical challenges proved insurmountable. After a handful of minor teething problems, the dynamic Web page system has been extremely robust and has run, with slight content changes, for two academic sessions. The interactive elements are central to the student learning and have justified the design and creation of the mode of delivery. Many of the generic elements, such as the student login system and the tutor reporting structure, have been repurposed for other online activities on other units.

The development of content was stressful and time consuming, not only initially but also closer to delivery. However, time spent establishing unambiguous learning outcomes and tasks that allow students to achieve them were clearly important to our success. The upfront effort was also consolidated in the second year, and the subsequent delivery was therefore less demanding.

Implementation of the Online Activities

The role of the tutor in online learning activities is both similar and different to that in more

traditional teaching and learning (Bennett & Marsh, 2001). In practice, balancing the level of support was tricky; too much assistance might be counterproductive to student learning, but too little might be worse. Two support mechanisms were implemented: a discussion forum and the clinics described. The forum allowed students to submit queries at any time, enabled the tutor and/or other students to reply, and everyone to see the contributions. It had been hoped that students would discuss higher order ideas, in addition to asking specific questions. In year one, this did not happen and the forum was only used by a small number of students to pose questions to the tutor. In the second year, while postings were still mainly questions rather than discussion, students did volunteer answers to each other.

Similarly, the clinics did not always work as anticipated (i.e., that students would come to ask questions arising from work they had already done online). Rather, they chose to use them to do the work, raising queries dynamically. They also discussed things with each other at these times which obviated the need for the online forum.

Physical Geography in Environmental Management attracts large numbers of students, between 120 and 150 each year. Therefore, adding elements to the unit always raises issues of tutor workload, particularly with regards to marking. Computer-based practicals are not new on this unit, so although development time and resources were required, implementing "blended learning" was not onerous, and there was no significant increase in marking and administration. The computerised marking of some tasks did not reduce tutor effort significantly. Short textual answers still involved human marking, and where automatic marking was used, it required a thoughtful approach and substantial development time.

Student Attitudes to E-Learning

Many of the students seemed initially nervous about completing and submitting work online.

System stability and mechanisms for saving work concerned many. This eased after successful submission of the first activity. Another uncertainty related to their actual ability to complete the work, with some students concerned that they did not have sufficient computer skills. Most of these anxieties disappeared following encouragement to start the online activities, with the reassurance that the "clinics" and discussion forum could be used for support if needed.

A handful of students still struggled a little with the more practical elements of the tasks and became quite upset and annoyed about having to do them. This was tackled by working with these students and reinforcing the teaching and learning rationale. Explaining that they stood to gain important practical skills and would use them again in the future did seem to have a positive affect. Despite initial misgivings about working and submitting online, most students did well. Confidence gained from the first activity was apparent when observing their work on the second. Attendance at clinics was reduced and students appeared to be much more comfortable with the whole concept of online learning.

There is no denying that assessment is important to students. Feedback about the timing of the unit assessment has been mixed. Informal student comments indicated that some students did not appreciate that there were a number of smaller pieces of work and felt as though they were constantly working, rather than having periods where little was being undertaken. Some felt more under pressure and overworked. The majority, however, were positive about the spread assignments and being able to complete part of the overall assessment earlier in the unit. Most also welcomed the alternative means of assessment. Many find examinations very stressful and do not work best under these conditions. They enjoyed being challenged by new material and seemed to engage with the practical nature of the activities. This reinforces the rationale for broadening

assessment options and challenging the students in different ways.

Students also liked receiving their marks relatively soon after the submission deadlines. Activity one scripts were returned within two weeks, prior to activity two. Students seemed to appreciate this effort and were keen both to know their marks and to read the comments. They seemed to gain confidence from this feedback on their progress on the unit.

Some students perceived, wrongly, that work completed and submitted online did not warrant the same academic standards as more conventional assignments. For example, there was inappropriate referencing and poor English in textual answers. Following this discovery when marking activity one, tutors reinforced the need to demonstrate academic skills. This was less of a problem in the second year, as requirements were made clear upfront.

School Recommendations and Feedback

The School of Geography was very supportive of the use of new technology and the introduction of new methods of teaching. Staff were consulted about the online learning activities and, when asked for feedback, were honest, open and gave their time freely. For example, the mix of short answer questions and longer text-based questions meant that students who spent time and effort were rewarded with high marks for successfully completing the exercises. The average of 72% suggested that students did well on these types of assessment but raised a question. Objective, computerised marking can result in higher marks than some human markers might award, an issue debated at school level. Some staff expressed concern that both the tasks performed and the automated marking reduced the ability to distinguish between students.

The main purpose of the assessment was to ascertain whether students understood the role of data within environmental management and that they could take on the role of a manager and make justifiable environmental decisions. Marks were given if students were able to perform a particular task and demonstrate a certain skill. There was very little scope for differentiation, they had either gained the skill or not. Only in the more judgemental elements, was it possible to differentiate and these were human-marked.

Perceived Enhancement of the Student Experience

Overall, the development of these activities has enhanced the learning experience and provided an added dimension to the unit. Students have encountered some of the difficulties, conflicts, and frustrations involved when making decisions with limited data. In achieving the learning outcomes, they have gained relevant practical skills, and it has afforded those who struggle with traditional essays and examinations increased opportunities to do well.

Students need to be enthused and engaged with materials in order to promote effective learning (Laurillard, 2002). Our online activities have successfully done this. As well as investigating the River Habitat Survey data, students have benefited from reading the linked resources, such as professional or government reports that they might not have previously accessed. Coupled with more academic journal articles, exposure to these alternative sources of information helps to provide a more solid grounding in the subject.

Race (1996) states that feedback on progression is one of the four prerequisites for successful learning. The timing and relevance of student feedback is essential for student improvement (Laurillard, 2002; Mutch, 2003). This was certainly true, with both students and tutors benefiting from the early opportunity to gauge progress. Staff were reassured that students were beginning to grasp some

of the key issues in environmental management, while students gained the essential encouragement that they were on the correct track (Lisewski & Settle, 1996) and could continue to progress.

RECOMMENDATIONS

After two years of developing, implementing, supporting, and evaluating these online learning activities, the resulting recommendations fall into four categories related to technology, content, staff, and students.

Technology Related

Technology-supported teaching and learning can enhance student experience and attainment, particularly online activities that permit engagement with real data as students investigate and apply concepts introduced in off-line teaching sessions. This is the type of blended learning the authors of this chapter recommend for campus-based students in higher education. It also addresses the disadvantages of earlier computer-assisted courseware described by Spellman (2000, p. 74) which simulated "reality."

It is important to work within system constraints, as well as trying to push the boundaries. This implies being aware of limitations, as well as the benefits and opportunities, recognising the need for technical support, and planning for it in terms of availability, deadlines, and possible costs. It is essential to remain flexible, particularly early on, in order to deal with any unexpected technical issues.

Systems should be tested where possible for ease of use by the intended learners. Consistency with institutional or departmental learning environment serves to reassure. Seamless integration of online resources to support, but not impinge on the learning, is vital. Therefore, an uncomplicated Web design, checked for accessibility, is advocated.

To minimise risks from technological failure, start with low weighting for online assessments. For small-scale implementations, it is not always possible to engage in the full risk analysis and management recommended by Zakrzewski and Steven (2000), but it is vital to have contingency plans in the event of system malfunction (Harwood, 2005) and not to react in an ad hoc manner.

Content Related

There was a tendency to produce, or link to, more content than was truly required, resulting in tasks that took much longer than anticipated, or even occurred in testing. Recommendations here include careful selection of further resources that students might be directed to and early testing by both staff and students. The repurposing of datasets can be difficult and time consuming, but, once completed, the format may serve to support many different learning activities.

Careful consideration should be given to the type of activities developed. There were a number of other options explored that proved unworkable because of the high level of interactivity required and associated technological development to make them worthwhile for students. It is vital to align content and learning outcomes. As Honebein suggested, "interpretation of the goals and subsequent translation into learning activities is the real art in the design of constructivist learning environments" (1996, p. 18).

Staff Related

When developing activities of this nature, ensure that other teaching staff understand the role and also the limitations of the technology involved. Regardless of how nervous staff are, during the early stages of implementing innovative learning activities, it is important not to pass this on to the students. Try to instil confidence and offer thoughtful support. Both staff and students

should have a grasp on why the activities are being done online.

Staff should be aware of the time and resources required to design, implement, and maintain interactive online activities. The costs versus the benefits need to be carefully considered. In this case, the upfront effort was justified due to the large number of students on the unit and the skills and understanding that the tasks developed.

Student Related

Students can become very agitated about online assessments. Explain the rationale for using the technology and offer both online and face-to-face support. Formative assessments allow them to become familiar with the format and requirements and may lead to better summative results (Pattinson, 2004). Guidelines on the time expected to be spent on online activities should be reviewed, based on actual experience, and students advised to consult tutors if they are taking considerably longer. If students report that something is not working, do not always assume that the technology has failed. It is important to check student understanding of the subject, the task at hand, and their sequence of actions.

Campus-based students may not make much use of online discussion boards, unless there is real value in doing so. Marking forum contributions can distort the process. It is recommended that tutors monitor online discussion carefully, reply swiftly to queries, and praise student contributions that contribute to the learning of others.

CONCLUSION

This chapter has outlined the development, implementation, and evaluation of two online learning activities in geography. Despite initial technological problems and constraints, both staff and students have benefited from their development. The activities have provided students with the opportunity to experience some of the real difficulties and conflicts of professional environmental managers, while also developing practical skills. Staff have benefited from the opportunity to incorporate new and interesting data, ideas, and concepts into the existing unit, while gaining from the ability of the system to undertake some of the administrative and marking duties. The recommendations presented here provide both ideas and words of caution to those wishing to blend online activities with traditional teaching and learning on their units.

ACKNOWLEDGMENT

Thanks are due to UK's Joint Information Systems Committee and USA's National Science Foundation for funding and supporting the DialogPLUS project; and UK's Environment Agency for permitting the use of the River Habitat Survey dataset.

REFERENCES

Bennett, S., & Marsh, D. (2001). Are we expecting online tutors to run before they can walk? *Innovations in Education and Teaching International, 39*(1), 14-20.

Biggs, J. (2003). *Teaching for quality learning at university* (2nd ed.). Bury St Edmunds, UK: The Society for Research into Higher Education and the Open University Press.

Carter, J. (2002). A framework for the development of multimedia systems for use in engineering education. *Computers & Education, 39,* 111-128.

Childs, M. (2004). Is there an e-pedagogy of resource-and-problem based learning? Using digital resources in three case studies. *Interactions, 8*(2). Retrieved March 30, 2005, from www2. warwick.ac.uk/services/cap/resources/interactions/archive/issue24/childs/

Clark, M.J., Ball, J.H., & Sadler, J.D. (1995). Multimedia delivery of coastal zone management training. *Innovations in Education and Training International, 32*(3), 229-238.

DialogPLUS (2004). DialogPlus: Digital libraries in support of innovative approaches to learning and teaching in geography. *Proceedings of the Fourth ACM/IEEE-CS Joint Conference on Digital Libraries (JCDL 2004).*

Environment Agency. (1998). *River Habitat Quality: The physical characteristics of rivers and streams in the UK and the Isle of Man.* (RHS Report No. 2). Bristol, UK: Environment Agency.

Graham, L. (1999). *The principles of interactive design.* Devon: Delmar Publishing.

Harwood, I.A. (2005). When summative computer-aided assessments go wrong: Disaster recovery after a major failure. *British Journal of Educational Technology, 36, Special issue on Thwarted Innovation in e-Learning.*

Honebein, P. (1996). Seven goals for the design of constructivist learning environments. In B. Wilson (Ed.), *Constructivist learning environments* (pp. 17-24). NJ: Educational Technology Publications.

Honey, P. & Mumford, A. (1992). *The manual of learning styles* (3rd ed.). Maidenhead: Peter Honey.

Kerres, M., & de Witt, C. (2003). A didactical framework for the design of blended learning arrangements. *Journal of Educational Media, 28*(2-3), 101-113.

Laurillard, D. (2002). *Rethinking university teaching: A framework for the effective use of educational technology* (2nd ed.). London: RoutledgeFalmer.

Lisewski, B. & Settle, C. (1996). Integrating multimedia resource-based learning into the curriculum. In S. Brown & B. Smith (Eds.), *Resource-based learning* (pp.109-119). London: Kogan Page.

Mutch, A. (2003). Exploring the practice of feedback to students. *Active Learning in Higher Education, 4*(1), 24-38.

Oliver, M. (2004). *Against the term blended learning. OU Knowledge Network.* Retrieved March 30, 2005, from http://kn.open.ac.uk/public/document.cfm?docid=5053

Pattinson, S. (2004). *The use of CAA for formative and summative assessment – Student views and outcomes.* Paper presented at the CAA Conference. Retrieved March 22, 2005, from www.caaconference.com/

Perkin, M. (1999). Validating formative and summative assessment. In S. Brown, J. Bull, & P. Race (Eds.), *Computer-assisted assessment in higher education* (pp. 29-37). London: Kogan Page.

Plowman, L. (1996). Narrative, linearity and interactivity: Making sense of interactive multimedia. *British Journal of Educational Technology, 27*(2), 92-105.

Race, P. (1996). Helping students to learn from resources. In S. Brown & B. Smith (Eds.), *Resource-based learning* (pp. 22-37). London: Kogan.

Spellman, G. (2000). Evaluation of CAL in higher education geography. *Journal of Computer Assisted Learning, 16,* 72-82.

Weller, M., Pegler, C., & Mason, R. (2003). *Working with learning objects— Some pedagogical suggestions.* Paper presented at the Association for Learning Technology conference. Retrieved March 4, 2005, from http://iet.open.ac.uk/pp/m.j.weller/pub/altc.doc

Wilson, B.G. (Ed.). (1996). *Constructivist learning environments: Case studies in instructional design.* Englewood Cliffs, NJ: Educational Technology Publications, Inc.

Zakrzewski, S., & Steven, C. (2000). A model for computer-based assessment: The Catherine Wheel principle. *Assessment & Evaluation in Higher Education, 25*(2) 201-215.

Chapter XVI
Learning Geography with the G–Portal Digital Library

Dion Hoe-Lian Goh
Nanyang Technological University, Singapore

Yin-Leng Theng
Nanyang Technological University, Singapore

Jun Zhang
Nanyang Technological University, Singapore

Chew Hung Chang
Nanyang Technological University, Singapore

Kalyani Chatterjea
Nanyang Technological University, Singapore

INTRODUCTION

With the rapid growth of digital information, there is increasing recognition that digital libraries (DL) will play important roles in education, research, and work. DLs have correspondingly evolved from being static repositories of information, in which access is limited to searching and browsing, to those that offer a greater array of services for accessing, interacting and manipulating content (Agosti, Ferro, Frommholz, & Thiel, 2004; Goh, Fu, & Foo, 2002).

Within the classroom environment, DLs have the potential to be useful tools for active learning in which activities are characterized by active engagement, problem-solving, inquiry, and collaboration with others, so that each student constructs meaning and hence knowledge of the information gained (Richardson, 1997). Consider, for example, a group of high school students working on a class project. Typical activities would require these students to acquire content from the teacher, gathering reference materials from the library or other sources, such as the Web, compiling and making sense of all the available information, synthesizing content, writing the project report and submitting the completed project for grading. Here, DL services could be designed to support these activities. An integrated work environment could allow students to collaboratively retrieve and store personal and group information objects relevant to the task at hand. Such a DL would therefore depart from the traditional role of facilitating access to digital content, and instead become an integral part of the learning process.

While there is much work in making such DLs a reality, many systems still offer basic levels of support for educational services, and users typically encounter one or more of the following problems:

- Content access is a separate task from other applications. Although advanced features for searching and browsing are available, DLs provide, at best, limited support for sharing content among other applications that support learning (Ancona, Frew, Janée, & Valentine, 2005). Exceptions are query and data dissemination services through protocols such as Z39.50 (Lynch, 1997) and OAI (Lagoze & Van de Sompel, 2001), but these are usually between other DLs instead of with integrated learning environments.
- DLs are not designed to cater to the needs of different learning activities. Instead, they excel at tasks such as cataloging/classifying content and metadata, searching, and browsing. Thus, activities such as laboratory experiments and field studies that need to use the services of a DL must be tailored to its capabilities.
- DLs are often not designed to meet the learning needs of individuals or groups. They are rather created as a generic collection of services for their target user populations. Support for groups within these target

populations requiring specialized services or content are typically lacking.

- Single-user delivery of information. In DLs, that support personalization, content is accessed and manipulated individually via personalized workspaces. One side effect of this feature is that users are often unable to share their findings with others. Thus, while individual learning can be supported, collaborative group-based learning is more difficult.

In the remainder of this chapter, we describe G-Portal, a DL of geospatial and georeferenced resources. G-Portal is designed to address the shortcomings previously mentioned to support collaborative learning among its users. This is achieved through personalized project spaces, in which individuals or groups gather and organize collections of resources drawn from the DL's holdings that are relevant to specific learning tasks. In addition, G-Portal provides facilities for classification and visualization of resources, spatial searching, annotations and resource sharing across projects.

G-PORTAL: AN OVERVIEW

G-Portal is a Web-based DL that supports a variety of services to access, manipulate, and manage geospatial and georeferenced resources (Lim et al., 2002). The resources in G-Portal are primarily of metadata records that describe and point to actual resources, such as Web pages, images, and other Web-accessible objects. Other types of information managed by G-Portal include semi-structured data records and annotations.

Since G-Portal focuses specifically on geography-related resources, they may be associated with spatial locations and plotted on a map. Consequently, all resources have an explicit inclusion of location in their metadata definitions. This location attribute, together with several other attributes such as ID, name, and source, constitute the core attributes that every resource must have. In addition, each type of resource may define attributes for descriptive purposes. The set of core attributes, together with the customized attributes, is defined by a resource schema (Lim et al., 2005). Examples of resource schemas include description of places of interest, examination questions, and user annotations.

G-Portal organizes resources into projects which are user-defined collections of resources relevant to a specific topic or learning activity. Within each project, resources are further grouped into layers for finer grained organization. Each layer serves as a category to store logically related resources. For example, a project studying flora and fauna in nature trails may include rivers, lakes and hills in a map layer, flora and fauna information in separate layers, and user annotations in another.

G-Portal offers three ways to accessing its resources. The map-based interface visualizes resources with location attributes on a map (see Figure 1). Navigation aids such as zooming and panning are supported. A layer selector feature

allows layers to be hidden or displayed and is useful in complex projects in which users would like to view resources belonging only to specific layers. This interface makes resources with known geographical locations easily and intuitively accessible and helps users discover the spatial relationships between resources.

For resources without location attributes, a classification interface is supported. The interface organizes resources based on a customizable taxonomy derived from common resource attributes and presents them in a tree view similar to the familiar Windows Explorer application. Here, resources are organized into categories, which are in turn organized into category trees. The map-based and classification interface are synchronized to allow seamless information access using either interface. For example, if resources are selected on the classification interface, they will also be highlighted on the map-based interface and vice versa.

Finally, a query interface (see Figure 2) that supports searches for resources based on keywords and spatial operators is also available. The query interface allows a bounding box to be drawn

Figure 1. G-Portal's map-based and classification interfaces

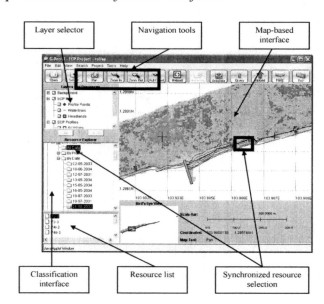

Figure 2. G-Portal's query interface

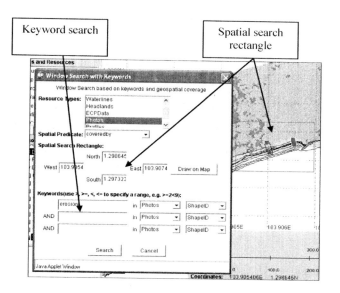

within the map-based interface and supports basic spatial query predicates such as containment and overlap. Search results are then shown in a results list window where resources can be selected for viewing. A similar synchronization mechanism between the results list window, the map-based and classification interfaces is implemented. Thus, when the user selects a resource in the results list window, the location of the resource on the map and the category that the resource is assigned to in the classification interface will also be highlighted.

G-Portal is implemented as a Java-based client/server system (Liu et al., 2004). The client is a Java applet that interacts with G-Portal's application server that functions as a gateway to the DL's content stored in two database servers. An XML database server is used to store resources in XML form while a relational database is used for storing spatial data and processing spatial queries. As designed, G-Portal can be accessed from any Java-enabled Web browser at http://www.ntu.edu.sg/project/g-portal/, making it possible for users to access the DL anywhere, anytime.

PERSONALIZED PROJECT SPACES

At first glance, G-Portal projects appear to be suitable for supporting learning activities since they allow users to manage and manipulate resources. However, a few limitations exist. First, projects are public and are created by G-Portal administrators to serve the needs of all G-Portal users. This idea is similar to current DLs that provide uniform access mechanisms (for example, query interfaces and classification schemes) to their entire collection of materials. Being publicly accessible, projects are not user customizable and as such, they do not cater towards supporting individual or group learning. Further, no provision is made for creating projects that only a certain user or group can access. G-Portal's personalized project space, however, is designed to address these issues so as to better support learning activities within the DL.

PERSONALIZED PROJECT CREATION AND MANAGEMENT

A personalized project is owned by a user or group of users and allows the creation of customized collections of resources and annotations relevant to a particular learning activity. A personalized project has the same basic features as any project in G-Portal but only its owner can modify its contents. Personalized projects can also be made public in which case all users are able to access it.

Once a personalized project is created, new layers are added to organize metadata records. G-Portal therefore provides features to create/delete layers. Metadata records are then organized into layers according to the needs of the user or group. In addition, metadata can be assigned spatial locations to be displayed in the map-based interface under different layers. For example, metadata of buildings and roads can be displayed in one layer, while that of parks and lakes can form another. Metadata can also be organized under one or more category hierarchies and made viewable via the classification interface.

SCHEMA AND RESOURCE MANAGEMENT

Every resource in G-Portal is created using a resource schema that serves as a template for describing resources. In a personalized project, schemas can be either predefined or user-defined to meet the needs of a learning activity for a user or group. To begin, a user first selects a schema for the new resource from an existing list. Alternatively, if none of the schemas are suitable, a new one can be created, as shown in Figure 3. Here, each schema is represented as a tree structure with nodes representing metadata elements or attributes, as well as data type and multiplicity constraints. This tree representation is used as schemas are XML-based and maps well to the hierarchical presentation scheme in the schema creation user interface. The ability to create new schemas is useful in individual or group learning since the resources needed are typically personal or highly contextualized such that general purpose schemas may not be suitable. However, due to

Figure 3. The schema editor

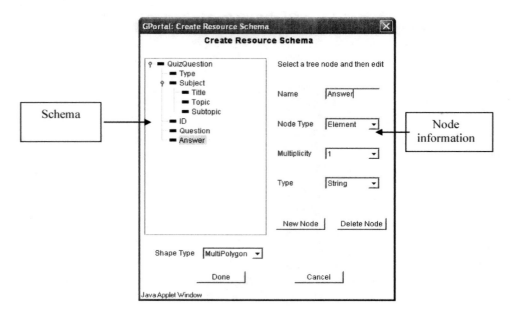

Figure 4. The resource editor

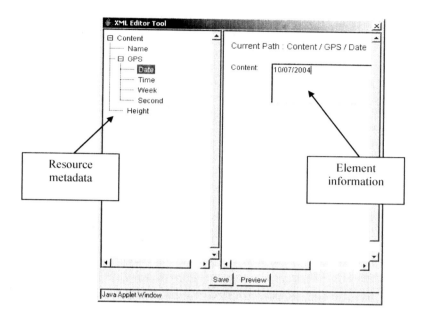

their personal usage nature, user-created schemas are only available to the current personalized project. Consequently, they are not subject to a formal review and registration process by G-Portal administrators.

Once schemas are defined, users can create new resources using G-Portal's built-in resource editor. Like schemas, metadata records are represented as a tree structure (see Figure 4) since they are XML-based. Element names are shown as tree nodes. When an element is selected, the right panel shows the attributes, values and/or the child elements of the selected element. The resource editor is schema guided. Before a resource can be created, its schema must be provided to the editor which parses it, and generates an empty resource and renders the user interface. The user then interactively uses the tree in the left panel of the editor to select elements and attributes, and fills in their values on the right panel. Once completed, the resource editor verifies that the entered values satisfy all constraints in the schema before saving. The use of XML schemas greatly reduces the chances of introducing errors thus making the resource creation easier.

G-Portal also supports the creation of annotations, which are a useful mechanism for learning and knowledge sharing (Marshall, 1997). Annotations are treated as a type of resource and can be visualized or queried via G-Portal's various interfaces. The creation of annotations follows a similar process with one exception. The user first selects the resources on the project to be annotated. Next, a suitable annotation schema is selected or created. The resource editor is then used to complete the annotation. The use of schemas provides flexibility for users in determining the content of annotations which in turn supports individual and group learning activities that typically require contextualized information.

RESOURCE SHARING

Active learning has a strong collaborative element and students often need to share information to fulfill a learning objective. G-Portal thus incorporates an easy-to-use mechanism for sharing resources across projects. For example, a teacher might maintain a public project containing information for an entire semester's course in geography. Students working on a particular learning activity may need to include existing resources in the public project into their personalized projects. Further, individual students working on independent projects may, at some point, come together as a group to work on a collaborative project which requires resources across these separate projects.

Sharing of resources in G-Portal is supported through an intuitive copy-and-paste function that is familiar with users of word processors and other office productivity software. Here, resources of interest are first selected from one or more accessible source projects. A copy function is activated and these resources are then pasted into the target personalized project. No records are duplicated and G-Portal simply provides links to them.

A SCENARIO OF USE

We present a typical scenario of use to highlight G-Portal's features and operation. Consider a field study of beach erosion at the East Coast Park (ECP) beach area in Singapore, a popular recreation spot. This beach is, however, prone to erosion as waves wash away large amounts of sand, which must then be regularly replenished with new sand. Suppose a high-school teacher plans a field study to the ECP beach and would like her students to investigate its erosion with the following objectives: (1) understand what beach erosion is and its causes; (2) assess the current state of the beach; (3) compute an estimated rate of erosion based on historical records; and (4) document the findings.

To help her students, the teacher creates a public master project that contains instructions for measuring beach profiles, project report format, the map of the beach to be accessed, the GPS positions of the beach profiles, and so on. An important resource type within this public project is the beach profile, each of which points

Figure 5. A sample beach profile with metadata

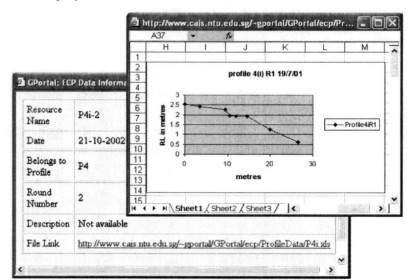

to a Web page containing its corresponding profile description. A profile description provides photos taken to help the student locate the profile, historical records, and other supporting information. Figure 5 shows a sample beach profile with its resource metadata.

Next, consider a student working on this field study. He first creates his personalized project in G-Portal and then adds the basic beach profile resources from the master project to his personalized project using the copy-and-paste facility. The field study instructions are also added for easy reference. When all measurements are collected, this new data is added to the personalized project. For example, the student could create a Web page to store each photograph taken at the beach, its corresponding beach profile data and an analysis of this data. The student then defines the corresponding metadata resource using a suitable schema to describe this Web page (see Figures 3 and 4). A link from the metadata resource to the Web page is also added and the latter now becomes accessible in the student's personalized project.

At a later time, the teacher visits each student's personalized project, perusing the resources and verifying their findings. This can be done either through the map-based or classification interfaces (see Figure 1). Alternatively, the teacher may want to locate a specific beach profile, in which case, the query interface would be used (see Figure 2). Due to the large number of beach profiles, each student may be required to analyze only a small number of profiles in their personalized projects. By making these projects publicly accessible, the findings of the entire class can be shared by everyone.

DISCUSSION AND CONCLUSION

G-Portal shares similar goals with existing digital libraries such as ADEPT (Borgman et al., 2004), DLESE (Sumner & Marlino, 2004) and CYCLADES (Avancini & Straccia, 2004). ADEPT supports the creation of personalized DLs of geospatial information and but owns its resources unlike in G-Portal where the development of the collection depends mainly on users' contributions as well as on the discovery and acquisition of external resources (such as geography-related Web sites). Our model is similar to DLESE although the latter does not support an interactive map-based interface or an environment for online learning. CYCLADES provides a suite of tools for personalizing information access and collaboration, but is not targeted towards education or the challenges of accessing and manipulating geospatial and georeferenced content.

DLs are beginning to play key roles in education, especially in the provision of information to learners. However, a crucial missing ingredient is the support for interactive and collaborative learning. The G-Portal DL represents a step in remedying this problem by offering a suite of tools for supporting geography and earth system science education. An important component in G-Portal is the personalized project space and this chapter has shown how its features can be used to support individual and group-based learning activities. This was illustrated with a field study example, a common type of learning activity in geography. With users having the flexibility to create, access and manipulate personalized content, we believe that a better integration between DL content and learning activities can be achieved.

REFERENCES

Agosti, M., Ferro, N., Frommholz, I., & Thiel, U. (2004). Annotations in digital libraries and collaboratories: Facets, models and usage. In R. Heery & L. Lyon (Eds.), *Proceedings of the Eighth European Conference on Digital Libraries* (Lecture Notes in Computer Science 3232, pp. 244-255). Berlin: Springer-Verlag.

Ancona, D., Frew, J., Janée, G., & Valentine, D. (2005). Accessing the Alexandria digital library from geographic information systems. In T. Sumner & F. Shipman (Eds.), *Proceedings of the Fifth ACM/IEEE-CS Joint Conference on Digital Llibraries* (pp. 74-75). New York: ACM Press.

Avancini, H., & Straccia, U. (2004). Personalization, collaboration, and recommendation in the digital library environment CYCLADES. In *Proceedings of IADIS conference on applied computing 2004* (pp. 67-74). Lisbon, Portugal: IADIS Press.

Borgman, C. L., Leazer, G. H., Swetland, A., Millwood, K., Champeny, L., Finley, J., et al. (2004). How geography professors select materials for classroom lectures: implications for the design of digital libraries. In E. P. Lim & M. Christel (Eds.), *Proceedings of the Fourth ACM/IEEE-CS Joint Conference on Digital Libraries* (pp. 179-185). New York: ACM Press.

Goh, D., Fu, L., & Foo, S. (2002). A work environment for a digital library of historical resources. In E. P. Lim, S. Foo, C. S. G. Khoo, H. Chen, E. A. Fox, S. R. Urs, et al. (Eds.), *Proceedings of the 5th International Conference on Asian Digital Libraries* (LNCS 2555, pp. 260-261). Berlin: Springer-Verlag.

Lagoze, C., & Van de Sompel, H. (2001). The open archives initiative: Building a low-barrier interoperability framework. In E. A. Fox & C. L. Borgman (Eds.), *Proceedings of the First ACM/IEEE-CS Joint Conference on Digital Libraries* (pp. 54-62). New York: ACM Press.

Lim, E. P., Goh, D. H., Liu, Z. H., Ng, W. K., Khoo, C., & Higgins, S. E. (2002). G-Portal: A map-based digital library for distributed geospatial and georeferenced resources. In G. Marchionini (Ed.), *Proceedings of the Second ACM/IEEE-CS Joint Conference on Digital Libraries* (pp. 351-358). New York: ACM Press.

Lim, E. P., Liu, Z., Goh, D. H., Theng, Y. L., & Ng, W. K. (2005). On organizing and accessing geospatial and georeferenced Web resources using the G-Portal system. *Information Processing and Management, 41*, 1277-1297.

Liu, Z., Yu, H., Lim, E. P., Ming, Y., Goh, D. H., Theng, Y. L., et al. (2004). A Java-based digital library portal for geography education. *Science of Computer Programming, 53*, 87-105.

Lynch, C. A. (1997, April). The Z39.50 information retrieval standard. *DLib Magazine.* Retrieved November 7, 2005, from http://www.dlib.org/dlib/april97/04lynch.html

Marshall, C. C. (1997). Annotation: From paper books to the digital library. In R. B. Allen & E. Rasmussen (Eds.), *Proceedings of the 2nd ACM International Conference on Digital Libraries* (pp. 131-140). New York: ACM Press.

Richardson, V. (1997). Constructivist teaching and teacher education: Theory and practice. In V. Richardson (Ed.), *Constructivist teacher education: Building new understandings* (pp. 3-14). Washington, DC: Falmer Press.

Sumner, T., & Marlino, M. (2004). Digital libraries and educational practice: A case for new models, In E. P. Lim & M. Christel (Eds.), *Proceedings of the 4th ACM/IEEE-CS Joint Conference on Digital Libraries* (pp. 170-178). New York: ACM Press.

KEY TERMS

Active Learning: A process of acquiring new ideas, skills and knowledge through problem-solving exercises, informal small groups, simulations, case studies, role playing, and other activities that require student engagement.

Annotation: A brief note that describes, comments, evaluates, explains or critiques the contents of a resource or a portion of a resource.

Digital Library: An information system that supports the access, retrieval, selection, organization and management of a focused collection of multimedia objects.

Field Study: Research carried out on location, undertaken outside the laboratory or place of learning, usually in a natural environment or among the general public.

Metadata: A set of attributes that describes the content, quality, condition, and other characteristics of a resource.

Metadata Schema: A metadata schema establishes and defines attributes and the rules governing the use of these attributes to describe a resource.

Personalization: In the context of a Web-based information system such as digital library or portal, personalization is the process of tailoring content to individual users' characteristics or preferences.

This work was previously published in Encyclopedia of Portal Technologies and Applications, edited by A. Tatnall, pp. 547-553, copyright 2007 by Information Science Reference, formerly known as Idea Group Reference (an imprint of IGI Global).

Chapter XVII
Collaborative Geographic Information Systems:
Origins, Boundaries, and Structures

Shivanand Balram
Simon Fraser University, Canada

Suzana Dragićević
Simon Fraser University, Canada

It is the theory that decides what can be observed.

Albert Einstein (1879-1955)

The scientists of today think deeply instead of clearly.

Nikola Tesla (1857-1943)

ABSTRACT

This chapter describes the origins, boundaries, and structures of collaborative geographic information systems (CGIS). A working definition is proposed, together with a discussion about the subtle collaborative vs. cooperative distinction, and culminating in a philosophical description of the research area. The literatures on planning and policy analysis, decision support systems, and geographic information systems (GIS) and science (GIScience) are used to construct a historical footprint. The conceptual linkages between GIScience, public participation GIS (PPGIS), participatory GIS (PGIS), and CGIS are also outlined. The conclusion is that collaborative GIS is centrally positioned on a participation spectrum that ranges from the individual to the general public, and that an important goal is to use argumentation, deliberation, and maps to clearly structure and reconcile differences between representative interest groups. Hence, collaborative GIS must give consideration to integrating experts with the general public in synchronous and asynchronous space-time interactions. Collaborative GIS provides a theoretical and application foundation to conceptualize a distributive turn *to planning, problem solving, and decision making.*

INTRODUCTION

Definitions within a community of practice have multiple benefits. Definitions reduce differences in semantics, and focus a community of practice towards goals that reinforce individual and collective efforts, make knowledge accessible to those at the edges of the community, and expand a study area by integrating related external concepts (Sager, 2000). Moreover, clearly defined concepts in a knowledge domain can better facilitate theory building. There are five types of definitions, and we have chosen to specify a *theoretical definition* for collaborative GIS since this type of definition aims to capture a commonality in the research area, and to relate that commonality to a broader intellectual framework (Sager, 2000). This chapter is organized as follows: firstly, a theoretical definition of collaborative GIS is presented; secondly, a historical footprint is established to reinforce the theoretical definition; and thirdly, the linkages between collaborative GIS and its broader conceptual framework are outlined.

What is Collaborative GIS?

There is a mutual influence between geographic information science and collaborative geographic information systems. GIScience is the rationale or science (axioms, theories, methods) that justifies the design and application of geographic information systems (Goodchild, 1992). Geographic information systems on the other hand are the physical designs and processes that integrate people and computer technology to manage, transform, and analyze spatially referenced data to solve ill-defined problems (Wright, Goodchild, & Proctor, 1997). Collaborative GIS are influenced by both GIS and GIScience. Hence, the name col-

laborative GIS will be used as systems, science, or both, depending on the context.

Collaborative GIS can be defined as *an eclectic integration of theories, tools, and technologies focusing on, but not limited to structuring human participation in group spatial decision processes.* In particular, the aim is to probe at the participant-technology-data nexus, and to describe, model, and simulate effects on the consensual process outcomes. The participants are typically a mixture of technical experts and the public, the technological tools being computers that are networked or distributed, and the data being spatially referenced maps and attributes. The outcomes do not result from implementing a task-oriented approach, but rather they emerge from a joint and structured exploration of ill-defined problems to benefit planning, problem solving, and decision making. In planning, the intention is to develop steps to achieve a desired outcome, while problem solving deals with the formulation of plans in new contexts. Decision making is the process of choosing among a set of alternatives.

Structuring is defined in the Webster Online Dictionary (http://www.m-w.com) as "the act of building, arrangement of parts, or relationship between parts of a construction." In this regard, structuring in collaborative GIS deals with the creation of process designs, how those designs enable the participant-technology-data interactions, and the relationships between the component parts of the designs. Hence, collaborative GIS is situated within the enhanced adaptive structuration theory 2 (EAST2) framework (Jankowski & Nyerges, 2001a). The framework outlines a detailed set of concepts and relationships linking the content, process, and outcome of collaborative spatial decision making. The content constructs of EAST2 examine the socioinstitutional, group participant, and GIS technology influences. The process constructs examine the social interactions between humans and computers. The outcome constructs address societal impacts of the decisions. Constructs five (group processes) and six (emergent influence) are important for collaborative GIS because they deal with "idea exchange as social interaction" and "emergence of socio-technical information influence" respectively. The interactions that occur in these constructs are more collaboration rather than cooperation.

Questions that engage collaborative GIS research activities include "What collaborative spatial decision making structures can generate meaningful outcomes? How can the attitudes and needs of participants be integrated into the group process? What are the effects of spatial data and cognitive overload on participation quality? How can prior solutions be integrated into the designs of collaborative spatial decision making systems? How can the outcomes of the process be evaluated and assessed for quality?"

The Collaboration vs. Cooperation Distinction

Some of the earliest works of educational psychologists attempted to distinguish collaboration from cooperation within teaching and learning contexts. The notion of "associated life" by John Dewey made the important recognition that human relationships are a key to welfare, achievement, and mastery (Dewey, 1916). This associative educational enterprise was the predecessor of the modern day interpretation and application of collaboration and cooperation in interactions that deal with groups of individuals (Bruffee, 1995).

The difference between collaboration and cooperation is subtle, but important. John Smith (1994) suggests that collaboration is an expectation of a common purpose, and this occurs at the implementation level with a close integration of component parts. On the other hand, coopera-

tion does not come with an expectation of close integration as individual tasks are combined at the hierarchically higher goal level. This means that for cooperative process, individuals can complete subtasks without being in close interaction with other supporting individuals. Bruffee (1995) points out that both collaboration and cooperation encourage group participation, but while cooperation guarantees accountability and risks maintaining authoritative structures, collaboration encourages self-governance and places guarantees of accountability at risk. Moreover, both collaboration and cooperation assume knowledge is socially constructed.

In the participatory GIS literature, collaboration and cooperation have been conceptualized in a hierarchical and cumulative arrangement consisting of four levels (Jankowski & Nyerges, 2001b). These participatory levels are communication, cooperation, coordination, and collaboration. Communication is meant to exchange ideas in social interactions, while cooperation uses the ideas generated from communication to develop an overall agreement, despite individuals may not interact with each other. Coordination occurs when there is a planned implementation of cooperative activity to reinforce collective group gains. Collaboration deals with a shared sense of meaning and achievement in the group process. The goal of collaborative GIS is to leverage collaboration towards a *collective* process. In collective participation, the participatory group, technology, and data operate as a single fused system.

Philosophical Orientation of Collaborative GIS

Understanding the philosophical orientation of a study area is important because it dictates what can be measured, and how measurements can be integrated and synthesized. A philosophical description can be characterized along four dimensions: ontology, epistemology, methodology, and praxeology. Based on a historical examination of collaborative GIS, a description of its philosophical dimensions is proposed in Table 1.

Ontology is about the essence of existence and its explicit specification when conceptualized concretely (Gruber, 1992). The ontology is usually organized into a hierarchy of top (general concepts), domain (specific knowledge domain), task (vocabulary), and application (context dependent) levels (Gómez-Pérez, Fernández-López, & Corcho, 2004; Torres-Fonseca & Egenhofer, 2000). For collaborative GIS, the ontology is relativist where the real word is socially and experimentally formed from multiple mental constructions.

Epistemology is the study of knowledge and its associations to truth and belief (Rescher, 2003). The interaction of the investigator and the investigated is a crucial consideration in the knowledge formation process. For collaborative GIS, the epistemology is subjectivist where the investigator and investigated are integrated as one entity.

Methodology is the study of methods, and seeks to examine how knowledge is obtained and verified (Fuller, 2002). The processing and assessment of mental constructions of reality are of importance. For collaborative GIS, the methodology is such that the processing is hermeneutic and the assessment is dialectic, with the outcome being a reduced set of consensus constructions.

Praxeology is the science of human action, and considers how that action can impact societal, human, and environmental situations (Oakeshott, 1975). Collaborative GIS has been applied extensively in the knowledge domains of geography and environmental studies. The three predominant action-oriented aims that can be synthesized for GIS applications are planning, problem solving, and decision making (Duckham, Goodchild, & Worboys, 2003).

Table 1. The philosophical orientation of collaborative GIS

Philosophical Dimension	Summary Description
Ontology	**RELATIVIST** In this interpretation, the real world exists in the form of multiple mental constructions that are based on social and experimental processes. These constructions are local and context specific because of the individual perspectives from which they are formed.
Epistemology	**SUBJECTIVIST** In this interpretation, the investigator and the investigated are combined into a single entity. Knowledge is created from the interaction processes between the investigator and the investigated.
Methodology	**HERMENEUTIC and DIALECTIC** The individual reality constructions are processed hermeneutically (interpreted based on experience and experiments) and assessed dialectically (synthesis of opposing assertions) for the purposes of achieving one or more consensus constructions.
Praxeology	**PLANNING, PROBLEM SOLVING, DECISIONS** The consensus constructions guide individual and collective action. The actions (with associated individual reflections) take the form of problem solving, planning and management, and decision making, with the aim to improve human, societal, and environmental conditions.

COLLABORATIVE GIS: ORIGINS AND BOUNDARIES

The origins of collaborative GIS are diverse and some level of aggregation is therefore necessary to clearly understand its origins and boundaries. A first strand of relevant knowledge is from the planning and policy analysis arena where environmental decisions are made. A second strand of knowledge is the aggregation of decision support systems, geographic information systems, and geographic information science. The key concepts from these strands of knowledge are chronologically presented in Figure 1 and summarized in Table 2.

The history shown in Figure 1 can be categorized into four cumulative and overlapping periods: argumentation, reasoning, representation, and synthesis, which can be mapped to data, information, knowledge, and intelligence (Klosterman, 2001). The argumentation period covered the 1950s and 1960s and focused on logical structures to construct lexical arguments, and to use those arguments in planning

and decisions. The reasoning and representation period covered the 1960s and 1970s, with much effort directed to showing relationships between arguments and processing those arguments with mathematical formalisms. During the 1970s and 1980s, the practical integration of planning concepts and computer-based decision making began to take hold. This was partly due to earlier progress made in decision support systems and geographic information systems. With the planning and computer integration solidifying, the 1980s and early 1990s were the synthesis years when the spotlight was turned towards groups and computer technology in decision interactions. The synthesis was further accelerated by the increasing importance of environmental matters during the time; integrated management using computer based data integration was seen as a promising way to manage the environment. With the emergence of Web GIS and supporting communication technologies during late 1990 and early 2000, the collaborative GIS focus is now converging towards a distributive paradigm, where systems and processes are aligned to incorporate a wider

Figure 1. A historical footprint of concepts related to collaborative GIS

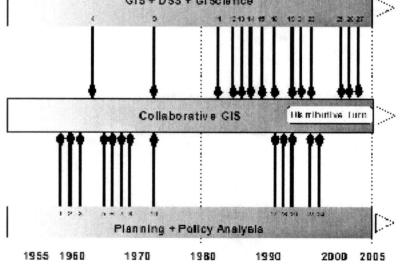

Table 2. Summary description of key concepts that influenced collaborative GIS (Note: The timelines represent the best estimate. There are time lags between when the concepts were formed and when they appeared in some published format.)

ID	Year	Concepts of influence	Summary Description
1	1958	Argumentation (Toulmin, 1958)	Sets out to establish a conclusion based on facts. The facts are connected to the conclusion by another argument called a warrant. The warrant is further supported by a backing. Together, these form an argumentation structure.
2	1960	Sketch planning and modeling (B. Harris, 1960)	Deals with the rapid and partial description of scenarios using computer-modeling methods. This was the precursor to present day planning support systems (PSS).
3	1960 - 1970	The Delphi process (Linstone & Turoff, 1975)	The Delphi process is used to explore consensus among decision-making groups. It consists of multiple iterations and feedbacks.
4	1963	Geographic information systems (Tomlinson, 1967)	A collection of computer tools and approaches to capture, manage, and transform spatially referenced data for planning and decisions.
5	1966	Mental maps (Gould, 1966)	Maps in the form of mental images are stored in our consciousness, and they seem to document spatial environmental relationships. Research was focused on clarifying the characteristics and uses of mental maps.
6	1968	Communicative rationality (Habermas, 1971)	A theory that assumes human rationality is a necessary consequence of successful communication. In the theory, implicit knowledge can become explicit through communication and discourse.
7	1969	Design with nature (McHarg, 1969)	Proposed a method for land use and human-settlements planning that involved manual inclusion and exclusion of map-based features. The layered analysis approach suggested here has been adopted by geographic information systems design.
8	1969	Ladder of citizen participation (Arnstein, 1969)	Clarified the levels of participation and nonparticipation using a ladder metaphor. The bottom rung corresponds to manipulation, and the top rung corresponds to control by citizens.
9	1971	Decision matrix framework (Gorry & Scott Morton, 1971)	Used a matrix to show the interaction between levels of management and decision-making structure at multiple levels. This was the precursor to decision support systems (DSS)
10	1971	Wicked problems (Rittel & Webber, 1973)	A class of problems for which no analytical solutions exist. These problems possess 10 characteristics. One characteristic is that a wicked problem has no definitive formulation.
11	1982	Human computer interaction (Badre & Shneiderman, 1982)	Deals with the design, evaluation, and implementation of interactive computer systems for use by humans.

continued on following page

Table 2. continued

ID	Year	Concepts of influence	Summary Description
12	1985	Group decision support systems (DeSanctis & Gallupe, 1985)	Proposed a system design where the purpose and configuration depended on the length and duration of the decision process, and on the physical proximity of the group members.
13	1985	Computer supported cooperative work (Bannon & Schmidt, 1989)	Addresses the design and deployment of computer technologies to support interactions between groups, teams, and organizations.
14	1985	Hypermaps (Laurini & Milleret-Raffort, 1990)	The spatial referencing of documents and cartographic products in a networked (Internet) environment.
15	1989	Multicriteria spatial analysis (Jankowski, 1989; Malczewski, 1996)	An approach integrating qualitative and quantitative information with MCE in a group spatial decision-making structure.
16	1992	Geographic information science (Goodchild, 1992)	The science that deals with geographic information technologies, designs, and their impacts on individuals and society.
17	1992	The argumentation turn in planning (Fischer & Forester, 1993)	An approach using argumentation to define problems, and structure viable solutions. Argumentation deals with rational persuasion towards changing the perspectives of others.
18	1992	The communicative turn in planning (Healey, 1992)	An approach that used communication to resolve disagreements and conflicts towards consensual solutions. A key goal is to improve local participation in policy processes.
19	1993	Virtual reality GIS (Faust, 1995)	A traditional GIS with a virtual reality interface and interaction method. The intention is to improve communication and collaboration in decision making and simulation contexts.
20	1993	Bioregional mapping (Aberley, 1993)	An approach using biophysical and cultural knowledge as a basis to construct maps of environmental places and spaces. The maps combine scientific and traditional information.
21	1993	Web geographic information system (Palo Alto Research Center, 1994)	Uses a distributed network (LAN, Internet, wireless) to share, process, and transform spatially referenced data.
22	1996	The deliberative turn in planning (Forester, 1999)	An approach where participants deliberate under conditions that support reasoned reflection. Deliberation is the process where individual reflection on issues can lead to a change in perspective.

continued on following page

Table 2. continued

23	1996	Collaborative spatial decision making (Nyerges & Jankowski, 1997)	A framework integrating aspects and concepts relevant to group, spatial, decision making.
24	1997	Ladder of empowerment (Rocha, 1997)	Clarifies various levels of empowerment by using a ladder metaphor. The bottom rung of the ladder is individual empowerment and the top rung is community empowerment.
25	2001	Geovisualization (MacEachren & Kraak, 2001)	Methods and techniques focusing on the novel display and integrated understanding of large volumes of spatial data.
26	2002	Geocollaboration (MacEachren, Brewer, Cai, & Chen, 2003)	A visual approach to collaboration using geospatial technologies in group processes.
27	2002	Agent interactions (Gimblett, 2002)	A paradigm where human entities are represented as agents in computer environments, and possible collaboration scenarios are explored through simulations.

cross section of participants in the planning and decision making process.

Table 2 provides a summary of key concepts that have influenced the evolution of collaborative GIS. The integration of these concepts provided the foundation for contemporary spatial group decision systems (Balram, 2005). An early form of collaborative spatial decision making was the Strabo technique, designed to elicit and forecast planning strategies based on a consensus of expert opinions (Luscombe & Peucker, 1975). The Strabo technique produced map and error summaries to aid decision makers in assessing a group's perspective about geographic planning problems. Technological limitation presented an immediate hurdle for the Strabo technique, with a critical challenge being how best to quickly obtain geographical summaries of expert feedback for input into the next iteration of the workshop group discussion. Nevertheless, the Strabo demonstrated the valuable contributions of expert groups in the spatial planning process.

The rapid advances in GIS software, hardware, and networking technologies have resulted in many new opportunities to integrate spatial mapping and analysis tools into group decision-making processes. In this respect, Armstrong (1994) argued for a greater integration of group mapping and visualization technologies into spatial decision making. Godschalk, McMahon, Kaplan, and Qin (1992) reported on a group design that allowed participants to manipulate criteria during the decision-making process. The key role of data in the decision-making process was also recognized, and collaborative multimedia technologies were used to make data more accessible (Shiffer, 1992). A loose-coupled electronic meeting and map overlay system was also designed for land-use planning applications (Faber, Watts, Hautaluoma, Knutson, Wallace, & Wallace, 1996). The issues of qualitative and quantitative data integration using multicriteria analysis was also at the forefront of research efforts in collaborative spatial decision making (Carver, 1991; Jankowski, 1989; Malczewski, 1996). These developments highlight stages in the evolution of a research area that would later come to benefit from a coordinated research direction.

The collaborative spatial decision making (CSDM) research initiative of the National Center of Geographic Information and Analysis (NCGIA), USA, and the first specialist meeting in September 1995 added focus to the research direction of CSDM by emphasizing the design of "highly interactive group-based decision making environments." The research thereafter reflected this new focus, and there now exists a well-established and growing body of literature on the theory and application of collaborative, spatial, decision making (Densham & Rushton, 1996; Feick & Hall, 1999; Horita, 2000; Jankowski, 1995; Jankowski & Nyerges, 2001b; Jankowski, Nyerges, Smith, Moore, & Horvath, 1997; Jiang & Chen, 2002; Klosterman, 1999; Kyem, 2000, 2004; Malczewski, 1996). However, the multitude of variables that are usually involved in the CSDM process makes it a challenge to conduct experimental studies and compare results across implementations. This was a driving factor in the development of the Enhanced Adaptive Structuration Theory 2 by Jankowski and Nyerges (2001a). The EAST2 framework outlined a detailed configuration of "concepts and relationships linking the content, process and outcome of collaborative spatial decision making." The content constructs examined the socioinstitutional, group participant, and GIS technology influences. The process constructs examined the social interactions between humans and computers, and focused on structuring the group decision-making process. The outcome constructs addressed societal impacts of the outcome decision.

Geographic data and the structure of the collaborative group process are two important micro-level factors that influence the group constructs of the EAST2 framework. Effective participation and decision making is dependent on access to scientific data and information (Craig, Harris, & Weiner, 2002; Jankowski & Nyerges, 2001b; Nyerges, Jankowski, & Drew, 2002; Sieber, 2000). During group deliberations, many alternative scenarios are generated as a result of the diversity in participant beliefs and interests, and as these scenarios become less distinct, more data and knowledge is required to develop informed solutions. But obtaining this knowledge is difficult, and when available, it is usually partial, transitory, and contested. New and synergistic opportunities for generating relevant knowledge are obtained by aggregating participant knowledge and spatial map data (Jankowski & Nyerges, 2001a). The merging of context-dependent participant knowledge and context-independent spatial data with digital maps and user-friendly exploration tools enhances critical thinking and creativity, producing a comprehensive understanding of values and change structures. The result is broader participant satisfaction, better management plans, and improved decision making (Geertman, 2002).

In recent times, a number of studies have reported on integrating digital map data into the group modeling and decision-making process (Fall, Daust, & Morgan, 2001; Horita, 2000). The general trend has been to use this data either to support existing arguments, or to choose among a predefined set of alternatives. When the data is not integrated into the decision-making process, two negative consequences occur. First, arguments and counterarguments among participants using independent data can lead to more confrontation, due to inherent differences in knowledge sources. Second, participants do not have the flexibility to define or explore common spatial scenarios and therefore, opportunities to develop new perspectives and understanding about an environmental situation are restricted. Despite these disadvantages, the use of prepackaged data in the process has persisted because of the perceived cognitive difficulties that digital map data and supporting technologies impose on participants. However, practical experience has shown that embedded digital-map technology can be modified to suit the needs of a targeted end-user, and that the technologically uninitiated is capable of adapting to new levels of sophistication in short time intervals (Mitcham, 1997; Talen, 1999).

The explicit integration of spatial map data and visual exploration tools into the group decision-making process can be achieved by embedding a collaborative geographic information system into the participatory structures of the process. A collaborative GIS is a tool and a system consisting of a networked collection of computer hardware, software, and user groups with the objective to capture, store, manipulate, visualize, and analyze geographically referenced data and knowledge, so as to provide new information in an institutional setting for solving unstructured planning problems (Armstrong, 1994). As a sociotechnical system, the collaborative GIS facilitates synchronous interactions, as stakeholder and scientific knowledge are combined using exploratory tools to share, annotate, analyze, and visualize numeric, text, and map data in search for solutions within shared geographic place and space (Faber et al., 1996). The collaborative GIS allows for elicitation of knowledge, simulation of data, scenario development, and encouraging spatial critical thinking about all issues. In order to best implement the collaborative GIS to articulate participant ideas, a careful structuring of the group decision-making process is needed for equitable and sufficient issues representation.

Structuring the group decision-making process can help focus the discussions so that constructive ideas are generated during argumentation. Usually, the structuring is conducted in stages involving shared understanding of the environmental situation, criteria identification and ranking, data and knowledge availability, and the generation of alternative scenarios (Godschalk et al., 1992). This is an effective way to integrate individual perspectives, resources, institutions, and organizations towards common solutions. A consequence of integration has been process structuring using top-down, bottom-up, and facilitator-based workshop settings, with advisory committees (Vasseur, LaFrance, Ansseau, Renaud, Morin, & Audet, 1997), participatory democracy (Moote, McClaran, & Chickering,

1997), and cooperative strategies (Lejano & Davos, 1999) being a few of the implementation strategies. Not surprisingly, critics have suggested that some of these implementations are inherently confrontational, and can stall the decision process. But many researchers have pointed out the many long-term partnerships and planning benefits that can accrue by carefully embedding discursive strategies into the participatory decision making process (Healey, 1993; Webler, Tuler, & Krueger, 2001; Wilson & Howarth, 2002).

The Delphi method is a focus-group approach that has been applied in a number of recent studies to structure and incorporate discursive strategies into decision making processes (Gokhale, 2001; Hess & King, 2002). The focus group assembles a small number of individuals in a face-to-face collaborative setting to elaborate the details about a particular issue that is initially chosen for discussion by an investigator who structures or moderates the discussions. The Delphi uses a collaborative approach to create a process of building relationships, awareness, learning, and negotiation. During the Delphi, a neutral facilitator elicits individual, anonymous judgment about an issue from a group by using iterative feedback involving a series of rounds of questioning, in order to explore ideas or achieve a convergence of group opinion (Linstone & Turoff, 1975). There are four phases to the Delphi, with the first phase emphasizing the exploration of ideas through individual comments in a structured, brainstorming session. The second phase captures the collective opinions of the group, focusing on agreements and disagreements. The reasons for the disagreements are explored in the third phase. In the fourth phase, an analysis of the opinion convergence on the issues is presented back to the group for final evaluation. The Delphi allows for improved understanding of the decision problems, goals, and objectives, and is useful when there is limited knowledge and data, strong conflict, and when interpersonal interaction is difficult to organize (Linstone & Turoff, 1975). The Delphi has been integrated within a collaborative GIS design to structure environmental planning and decision-making processes (Balram, Dragicevic, & Meredith, 2003, 2004).

Participation in the collaborative, spatial, decision-making process has been an ongoing issue of concern in environmental and community planning (Brandt, 1998; Ghose, 2001; Harris & Weiner, 1998; Sieber, 2000; Talen, 2000). At the basic level, participation can be interpreted to mean the inclusion of a wide range of stakeholder inputs to all stages of the planning and decision-making process. In order to structure and operationalize the concept of participation, Arnstein's "Ladder of Citizen Participation" (Arnstein, 1969) and Rocha's "Ladder of Empowerment" (Rocha, 1997) are two frameworks of analysis widely used in the planning and decision-making literature. The central arguments of both "ladders" and their adaptations to specific contexts is that through a process of collaboration, participation becomes a knowledge sharing and knowledge producing activity capable of initiating social and political change (Baum, 1999; Healey, 1997). Also, a useful adaptation to Arnstein's ladder is presented in Whitman (1994). Whitman attributes varying levels of expert (those possessing "specialist knowledge" of relevance) involvement in each stage of the Arnstein ladder. At the lower end (individual involvement) of the Arnstein ladder, Whitman attributes a detached expert who is removed from the end user in the decision-making process. At the upper end (community involvement) of the Arnstein ladder, Whitman attributes an absent expert, and action is initiated from collective community initiatives. Collaborative GIS targets a middle ground and works at the "partnership" level of the Arnstein ladder, which has been mapped to the "expert as a team member" in the Whitman ladder. Adopting this position in the "ladder" hierarchy makes the focus one of balancing issues of concern gathered at the individual, expert, and public levels.

COLLABORATIVE GIS: A STRUCTURE OF THE RESEARCH AREA

The intellectual landscape of collaborative GIS can be structured by considering two scales. The first scale can be termed a local interdisciplinary view, where the research agenda of geographic information science situated in the upper hierarchy guides the research directions of collaborative GIS located at a lower level in the hierarchy (Figure 2). The second scale can be termed a global transdisciplinary view, where the adoption of new ideas into group spatial decision support systems (GSDSS) from diverse disciplines, coupled with improvements in Internet and wireless technologies are evolving towards a *distributive turn* to planning, problem solving, and decision making.

GIScience is now fairly well established as a discipline, with a diverse set of themes and subareas complete with research challenges and agendas (McMaster & Usery, 2005). Figure 2 shows the themes of GIScience and the subareas, such as spatial data acquisition and integration, cognition, scale, and so on. Of the subareas, GIS and Society is the most relevant for collaborative GIS (Elmes, Epstein, McMaster, Niemann, Poore, Sheppard et al., 2005). GIS and Society addresses institutional, legal and ethical, intellectual history, critical social theory, and participatory GIS issues. There may be some disagreement on whether participatory GIS or public participation GIS should be higher in the GIScience hierarchy. We suggest that PGIS is a more general concept, and should appear higher in the hierarchy. Both GSDSS (small groups) and PPGIS (large groups) are directly related to group decision making, and are members of the participatory GIS category. Collaborative GIS, geocollaboration, and planning support systems are all GSDSS implementations. However, the presence of fuzzy linguistic terms such as "small," "groups," and "public" will make the structure presented here open to further refinement.

Figure 2. The intellectual structure containing collaborative GIS

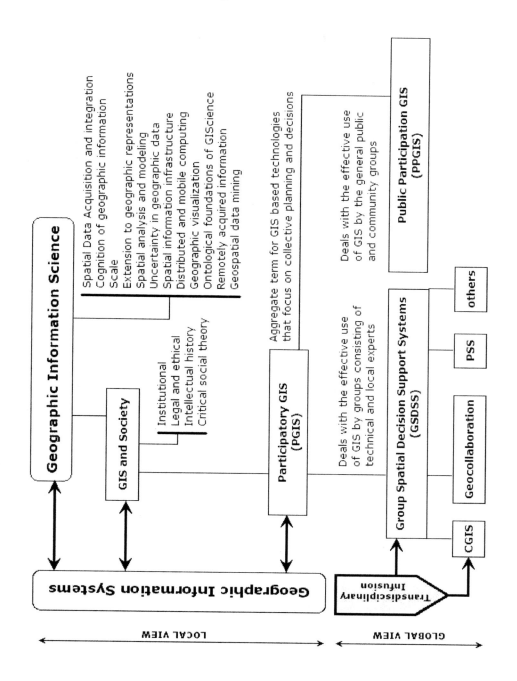

The local interdisciplinary view of collaborative GIS is guided by the geographic information science research agenda. The concepts are interdisciplinary, meaning that the goal is to synthesize two or more disciplines with the intention of creating a coordinated whole. In this view, the research and application focus of collaborative GIS is towards establishing stronger linkages with GIScience. Geographic information systems intervene at all levels of the hierarchy.

The global transdisciplinary view of collaborative GIS is guided mostly by the concepts of theory, experimentation, and simulation as means to explore reality. The concepts are transdisciplinary, meaning that multiple perspectives are integrated and transformed to create new knowledge to solve complex societal problems. It is in this transdisciplinary direction that current collaborative GIS initiatives seem to be focused. The most likely scenario is a *distributive turn* to planning, problem solving and decision making. There are already signals in the research literature (for example: Dymond, Regmi, Lohani, & Dietz, 2004; Schafer, Ganoe, Xiao, Coch, & Carroll, 2005) to suggest that a distributive turn is underway.

CONCLUSION

Progress in collaborative GIS is hinged on an understanding of the historical background of concepts, and the dynamics that are shaping its future. This study has proposed a working definition of collaborative GIS, and presented a philosophical description of the research area. A discussion about the historical background adds justification to the proposed definition. Conceptual linkages between GIScience, public participation GIS, participatory GIS, and CGIS are also presented. An important conclusion is that collaborative GIS is centrally positioned on a participation spectrum that ranges from the individual to the general public, and that argumentation, deliberation, and maps are the common means used to structure and recon-

cile differences between representative interest groups. Collaborative GIS must give consideration to integrating experts and the general public in synchronous and asynchronous space-time interactions. It is suggested that collaborative GIS theory provides a foundation to conceptualize a *distributive turn* to planning, problem solving, and decision making.

ACKNOWLEDGMENT

This work was supported by the Natural Sciences and Engineering Research Council (NSERC) of Canada Discovery Grant awarded to Suzana Dragievi.

REFERENCES

Aberley, D. (1993). *Boundaries of home: Mapping for local empowerment*. Gabriola Island, British Columbia, Canada: New Society Publishers.

Armstrong, M. P. (1994). Requirements for the development of GIS-based group decision support systems. *Journal of the American Society for Information Science, 45*(9), 669-677.

Arnstein, S. R. (1969). A ladder of citizen participation. *Journal of the American Institute of Planners, 35*(4), 216-224.

Badre, A., & Shneiderman, B. (Eds.). (1982). *Directions in human-computer interaction*. Norwood, NJ: Ablex Pub. Corp.

Balram, S. (2005). *Collaborative GIS process modelling using the Delphi method, systems theory and the unified modelling language (UML)*. PhD Thesis, McGill University, Montreal, Canada.

Balram, S., Dragicevic, S., & Meredith, T. (2003). Achieving effectiveness in stakeholder participation using the GIS-based collaborative spatial Delphi methodology. *Journal of Environmental*

Assessment Policy and Management, 5(3), 365-394.

Balram, S., Dragicevic, S., & Meredith, T. (2004). A collaborative GIS method for integrating local and technical knowledge in establishing biodiversity conservation priorities. *Biodiversity and Conservation, 13*(6), 1195-1208.

Bannon, L. J., & Schmidt, K. (1989). CSCW: Four characteristics in search of a context. In J. Bowers & S. Benford (Eds.), *Studies in computer supported cooperative work: Theory, practice and design* (pp. 3-16). Amsterdam: North-Holland.

Baum, H. S. (1999). Community organizations recruiting community participation: Predicaments in planning. *Journal of Planning and Education Research, 18*, 187-199.

Brandt, M. (1998). Public participation GIS - Barriers to implementation. *Cartography and Geographic Information Systems, 25*(2), 105-112.

Bruffee, K. A. (1995). Sharing our toys: Cooperative learning vs. collaborative learning. *Change*, (January/February), 12-18.

Carver, S. (1991). Integrating multicriteria evaluations with geographical information systems. *International Journal of Geographical Information Systems, 5*(3), 321-339.

Craig, W. J., Harris, T. M., & Weiner, D. (Eds.). (2002). *Community participation and geographic information science.* London: Taylor and Francis.

Densham, P. J., & Rushton, G. (1996). Providing spatial decision support for rural public service facilities that require a minimum workload. *Environment and Planning B-Planning & Design, 23*(5), 553-574.

DeSanctis, G., & Gallupe, R. B. (1985). Group decision support systems: A new frontier. *Database, 16*(1), 377-387.

Dewey, J. (1916). *Democracy and education.* New York: The Macmillan Company.

Duckham, M., Goodchild, M. F., & Worboys, M. F. (Eds.). (2003). *Foundations of geographic information science.* New York: Taylor & Francis.

Dymond, R. L., Regmi, B., Lohani, V. K., & Dietz, R. (2004). Interdisciplinary Web-enabled spatial decision support system for watershed management. *Journal of Water Resources Planning and Management, 130*(4), 290-300.

Elmes, G. A., Epstein, E. F., McMaster, R. E., Niemann, B. J., Poore, B., Sheppard, E., et al. (2005). GIS and Society: Interrelation, integration, and transformation. In R. B. McMaster & E. L. Usery (Eds.), *A research agenda for geographic information science* (pp. 287-312). Boca Raton, FL: CRC Press.

Faber, B. G., Watts, R., Hautaluoma, J. E., Knutson, J., Wallace, W. W., & Wallace, L. (1996). A groupware-enabled GIS. In M. Heit, H. D. Parker, & A. Shortreid (Eds.), *GIS applications in natural resources 2* (pp. 3-13). Fort Collins, CO: GIS World Inc.

Fall, A., Daust, D., & Morgan, D. G. (2001). A framework and software tool to support collaborative landscape analysis: Fitting square pegs into square holes. *Transactions in GIS, 5*(1), 67-86.

Faust, N. L. (1995). The virtual reality of GIS. *Environment and Planning B: Planning and Design, 22*, 257-268.

Feick, R. D., & Hall, B. G. (1999). Consensus-building in a multiparticipant spatial decision support system. *URISA Journal, 11*(2), 17-23.

Fischer, F., & Forester, J. (Eds.). (1993). *The argumentative turn in policy analysis and planning.* Durham, NC: Duke University Press.

Forester, J. (1999). *The deliberative practitioner: Encouraging participatory planning processes.* Cambridge, MA: MIT Press.

Fuller, S. (2002). *Social epistemology* (2nd ed.). Bloomington, IN: University Press.

Geertman, S. (2002). Participatory planning and GIS: A PSS to bridge the gap. *Environment and Planning B: Planning and Design, 29*(1), 21-35.

Ghose, R. (2001). Use of information technology for community empowerment: Transforming geographic information systems into community information systems. *Transactions in GIS, 5*(2), 141-163.

Gimblett, H. R. (Ed.). (2002). *Integrating geographic information systems and agent-based modeling techniques for simulating social and ecological processes.* New York: Oxford University Press.

Godschalk, D. R., McMahon, G., Kaplan, A., & Qin, W. (1992). Using GIS for computer-assisted dispute resolution. *Photogrammetric Engineering & Remote Sensing, 58*(8), 1209-1212.

Gokhale, A. A. (2001). Environmental initiative prioritization with a Delphi approach: A case study. *Environmental Management, 28*(2), 187-193.

Gómez-Pérez, A., Fernández-López, M., & Corcho, O. (2004). *Ontological engineering.* London; New York: Springer-Verlag.

Goodchild, M. F. (1992). Geographical information science. *International Journal of Geographical Information Systems, 6*(1), 31-45.

Gorry, G. A., & Scott Morton, M. S. (1971). A framework for management information systems. *Sloan Management Review, 13*(1), 55-70.

Gould, P. R. (1966). *On mental maps.* Ann Arbor: University of Michigan.

Gruber, T. R. (1992). A translation approach to portable ontology specifications. *Knowledge Acquisition, 5*, 199-220.

Habermas, J. (1971). *Knowledge and human interests* (J. J. Shapiro, Trans.). Boston: Beacon Press.

Harris, B. (1960). Plan or projection: An examination of the use of models in planning. *Journal of the American Institute of Planners, 26*, 265-272.

Harris, T., & Weiner, D. (1998). Empowerment, marginalization and community-integrated GIS. *Cartography and Geographic Information Systems, 25*(2), 67-76.

Healey, P. (1992). Planning through debate: The communicative turn in planning theory. *Town Planning Review, 63*(2), 143-162.

Healey, P. (1993). Planning through debate: The communicative turn in planning theory. In F. Fischer & J. Forester (Eds.), *The argumentative turn in policy analysis and planning* (pp. 233-253). Durham, NC: Duke University Press.

Healey, P. (1997). *Collaborative planning: Shaping places in fragmented societies.* Vancouver, BC: UBC Press.

Hess, G. R., & King, T. J. (2002). Planning open spaces for wildlife I: Selecting focal species using a Delphi survey approach. *Landscape and Urban Planning, 58*, 25-40.

Horita, M. (2000). Mapping policy discourse with CRANES: Spatial understanding support systems as a medium for community conflict resolution. *Environment and Planning B: Planning and Design, 27*, 801-814.

Jankowski, P. (1989). Mixed-data multicriteria evaluation for regional planning: A systematic approach to the decision-making process. *Environment and Planning A, 21*, 349-362.

Jankowski, P. (1995). Integrating geographical information systems and multiple criteria decision-making methods. *International Journal of Geographical Information Systems, 9*(3), 251-273.

Jankowski, P., & Nyerges, T. (2001a). *Geographic information systems for group decision making: Towards a participatory geographic information science.* New York: Taylor and Francis.

Jankowski, P., & Nyerges, T. (2001b). GIS-supported collaborative decision making: Results of an experiment. *Annals of the Association of American Geographers, 91*(1), 48-70.

Jankowski, P., Nyerges, T., Smith, A., Moore, T. J., & Horvath, E. (1997). Spatial group choice: An SDSS tool for collaborative spatial decision-making. *International Journal of Geographical Information Science, 11*(6), 577-602.

Jiang, J., & Chen, J. (2002). A GIS-based computer-supported collaborative work (CSCW) system for urban planning and land management. *Photogrammetric Engineering & Remote Sensing, 68*(4), 353-359.

Klosterman, R. E. (1999). The What if? Collaborative planning support system. *Environment and Planning B: Planning and Design, 26*(3), 393-408.

Klosterman, R. E. (2001). Planning support systems: A new perspective on computer-aided planning. In R. K. Brail & R. E. Klosterman (Eds.), *Planning support systems: Integrating geographic systems, models, and visualization tools* (pp. 1-23). Redlands, CA: ESRI Press.

Kyem, P. K. (2000). Embedding GIS applications into resource management and planning activities of local and indigenous communities: A desirable innovation or a destabilizing enterprise? *Journal of Planning Education and Research, 20*, 176-186.

Kyem, P. K. (2004). On intractable conflicts and participatory GIS applications: The search for consensus amidst competing claims and institutional demands. *Annals of the Association of American Geographers, 94*(1), 37-57.

Laurini, R., & Milleret-Raffort, F. (1990, July 23-27). Principles of geomatic hypermaps. In K. Brassel (Ed.), *Proceedings of the 4th International Symposium on Spatial Data Handling* (pp. 642-651). Zurich.

Lejano, R. P., & Davos, C. A. (1999). Cooperative solutions for sustainable resource management. *Environmental Management, 24*(2), 167-175.

Linstone, H. A., & Turoff, M. (1975). *The Delphi method: Techniques and applications.* Reading, MA: Addision-Wesley.

Luscombe, B. W., & Peucker, T. K. (1975). *The Strabo technique.* Burnaby, Canada: Simon Fraser University, Department of Geography Discussion Paper Series.

MacEachren, A., Brewer, I., Cai, G., & Chen, C. (2003, August 10-16). Visually enabled geocollaboration to support data exploration and decision-making. Paper presented at the *Proceedings of the 21st International Cartographic Conference*, Durban, South Africa.

MacEachren, A. M., & Kraak, M. (2001). Research challenges in geovisualization. *Cartography and Geographic Information Science, 28*(1), 3-12.

Malczewski, J. (1996). A GIS-based approach to multiple criteria group decision-making. *International Journal of Geographical Information Systems, 10*(8), 955-971.

McHarg, I. L. (1969). *Design with nature.* Garden City, NY: Natural History Press.

McMaster, R. B., & Usery, E. L. (Eds.). (2005). *A research agenda for geographic information science.* Boca Raton, FL: CRC Press.

Mitcham, C. (1997). Justifying public participation in technical decision making. *IEEE Technology and Society, 16*(1), 40-46.

Moote, M. A., McClaran, M. P., & Chickering, D. K. (1997). Theory in practice: Applying participatory democracy theory to public land planning. *Environmental Management, 21*(6), 877-889.

Nyerges, T. L., & Jankowski, P. (1997). Enhanced adaptive structuration theory: A theory of GIS-supported collaborative decision making. *Geographical Systems, 4*(3), 225-257.

Nyerges, T., Jankowski, P., & Drew, C. (2002). Data-gathering strategies for social-behavioural research about participatory geographical information system use. *International Journal of Geographical Information Science, 16*(1), 1-22.

Oakeshott, M. (1975). *On human conduct.* Oxford, UK: Oxford University Press.

Palo Alto Research Center. (1994). *Xerox PARC map viewer.* Retrieved November 11, 2005, from http://www2.parc.com/istl/projects/mapdocs/

Rescher, N. (2003). *Epistemology: An introduction to the theory of knowledge.* Albany: State University of New York Press.

Rittel, H., & Webber, M. (1973). Dilemmas in a general theory of planning. *Policy Sciences, 4,* 155-169.

Rocha, E. M. (1997). A ladder of empowerment. *Journal of Planning and Education Research, 17*(1), 31-44.

Sager, J. C. (Ed.). (2000). *Essays on definition.* Amsterdam; Philadelphia: J. Benjamins Publishers.

Schafer, W. A., Ganoe, C. H., Xiao, L., Coch, G., & Carroll, J. M. (2005). Designing the next generation of distributed, geocollaborative tools. *Cartography and Geographic Information Science, 32*(2), 81-100.

Shiffer, M. J. (1992). Towards a collaborative planning system. *Environment and Planning B: Planning and Design, 19*(6), 709-722.

Sieber, R. E. (2000). GIS implementation in the grassroots. *URISA Journal, 12*(1), 15-29.

Smith, J. B. (1994). *Collective intelligence in computer-based collaboration.* Hillsdale, NJ: Lawrence Erlbaum and Associates.

Talen, E. (1999). Constructing neighborhoods from the bottom up: The case for resident-generated GIS. *Environment and Planning B: Planning and Design, 26*(4), 533-554.

Talen, E. (2000). Bottom-up GIS: A new tool for individual and group expression in participatory planning. *Journal of the American Planning Association, 66*(3), 279-294.

Tomlinson, R. F. (1967). *An introduction to the geographic information system of the Canada Land Inventory.* Ottawa, Canada: Department of Forestry and Rural Development.

Torres-Fonseca, F., & Egenhofer, M. (2000). Ontology-driven geographic information systems. *Computers, Environment and Urban Systems, 24*(3), 251-271.

Toulmin, S. E. (1958). *The uses of argument.* Cambridge, UK; New York: Cambridge Press.

Vasseur, L., LaFrance, L., Ansseau, C., Renaud, D., Morin, D., & Audet, T. (1997). Advisory committee: A powerful tool for helping decision makers in environmental issues. *Environmental Management, 21*(3), 359-365.

Webler, T., Tuler, S., & Krueger, R. (2001). What is a good public participation process? Five perspectives from the public. *Environmental Management, 27*(3), 435-450.

Whitman, P. (1994). Experts and their interventions: A model of the field of urban improvement. *Environment and Planning B: Planning and Design, 21*(6), 759-768.

Wilson, M. A., & Howarth, R. B. (2002). Discourse-based valuation of ecosystem services: Establishing fair outcomes through group deliberations. *Ecological Economics, 41,* 431-443.

Wright, D. J., Goodchild, M. F., & Proctor, J. D. (1997). Demystifying the persistent ambiguity of GIS as "tool" vs. "science." *Annals of the Association of American Geographers, 87*(2), 346-362.

This work was previously published in E-Learning Methodologies and Computer Applications in Archaeology, edited by D. Politis, pp. 325-341, copyright 2008 by Information Science Reference, formerly known as Idea Group Reference (an imprint of IGI Global).

Glossary of Terms

Academic Integrity:
Accepted academic standards of practice whereby all use of or reference to the work of others should be clearly and openly acknowledged.

Access Management:
Arrangements for granting access to learning resources.

ADEPT:
Alexandria Digital Earth Project

ANGEL:
A collaboratively developed virtual learning environment software suite.

Athens:
Authentication system used to afford access to UK JISC resources, provided by Eduserve.

Atmospheric Correction:
Correction needed to remotely sensed data to allow for atmospheric distortion.

Blackboard VLE:
A proprietary virtual learning environment software suite.

Blended Learning:
Learning through a mixture of face-to-face lectures and online activities.

Bodington VLE:
An open source virtual learning environment software suite.

BRDF:
Bidirectional reflectance distribution function.

CHCC:
Collection of Historical and Contemporary Censuses, a JISC-funded program of learning materials covering UK censuses.

CiteSeer:
Search engine for scientific literature.

CLAD:
Collaborative learning activity design.

CMAP:
Software for designing concept maps

CMS:
Course management system

ConceptVista:
Software that provides an interactive environment for developing concept maps

DEM:
Digital elevation model, a data structure for the representation of a surface, usually the surface of the earth.

DialogPLUS:
A study jointly undertaken by the Universities of Southampton and Leeds in the UK and Pennsylvania State University and The University of California Santa Barbara in the U.S. as part of the JISC/NSF-funded Digital Libraries in the Classroom Programme.

Digital Libraries in the Classroom:
A jointly funded programme of JISC/NSF.

Directional Reflectance:
The dependence of light reflection on the directional orientation of the reflecting surface.

Discussion Board:
See Discussion Room

Discussion Forum:
See Discussion Room

Discussion Room:
A software utility, generally provided within a virtual learning environment, where contributors can view and post comments in a themed discussion.

Distance Learning:
Delivery of education whereby students are remote from the physical location of teachers and learning resources.

DL:
Digital library

DLE:
Digital learning environment

DLESE:
Digital Library for Earth Science Education

Drainage Basin:
The geographical area drained by a river and its tributaries, a fundamental unit of the terrestrial landscape, also catchment.

EIA:
Environmental impact assessment

E-Learning:
See Electronic Learning

Electronic Learning:
Learning that occurs when a computer is used as the means of accessing learning resources and activities.

EML:
Educational modeling language

ENVI:
A software system for processing remotely senses images.

EO:
Earth observation.

EO E-Activities:
Earth observation electronic (interactive) tasks.

EO Tutorials:
Earth observation learning/teaching materials.

EO Online Materials:
Earth observation learning/teaching materials on the Web.

EPC:
Environmental Processes and Change (University of Southampton research theme).

ESRC:
Economic and Social Research Council (UK).

FAQ:
Frequently asked question

Field Trip:
ee Fieldwork

Fieldwork:
Learning activity which involves visit to site of geographical interest.

Fluvial Geomorphology:
The geomorphology of river systems and drainage basins.

Forcing:
In physical geography, an external perturbation to the system under observation.

Formative:
Assessment that does not contribute to the final grade awarded.

Geography Exemplars:
Examples of e-learning materials used in geography teaching.

Geodemographic Classification:
Multivariate small area classification methodology, typically using census-type datasets.

Geomorphology:
The study of the form of the earth's surface.

GIS:
Geographic information system

GPS:
Global positioning system

HEFCE:
Higher Education Funding Council for England

HorizonWimba:
Desktop conferencing tool

Hours of Student Learning:
Calculated time in hours which students are expected to spend on a defined set of learning activities.

HTML:
Hypertext Markup Language

IMAGINE:
A software system for processing remotely senses images.

IMS:
Originally "instructional management systems", now only used as acronym.

Intute:
Online service providing access to Web resources for education and research.

IT Literacy:
Facility in the understanding and use of information technology (IT).

JISC:
Joint Information Systems Committee (UK).

JISCPAS:
JISC Plagiarism Advisory Service.

Jorum:
Repository for e-learning materials provided by JISC.

KB:
Knowledge base

Learning Nugget:

Set of learning materials including learning objective(s), activities, resources and assessments.

Learning Objectives:

Statements of what a student is expected to be able to do, having followed a specified learning activity.

Learning Objects:

Discrete and separable elements of student learning; extensive discussion exists regarding qualifying size and content.

Learning Outcomes:

Statements of what a student is expected to be able to do, having followed a specified learning activity.

Level:

Stages of study, used to describe progression of a student through formal educational system.

LMS:

Learning management system

Marking Criteria:

Specification of the characteristics required of student work in order to achieve a range of marks or grades.

MCQ:

Multiple choice quiz, multiple choice question.

MECA-ODL:

Methodology for analysis of quality of ODL through Internet.

M-learning:

Learning that can take place anytime, anywhere through use of a mobile computer device.

Moodle VLE:

An open source virtual learning environment software suite.

NSF:

National Science Foundation (U.S.).

Nugget:

Set of learning materials including learning objective(s), activities, resources and assessments.

NUTS:
Nomenclature of territorial units for statistics, a European hierarchy of areas for official statistics.

ODL:
Open distance learning

PBL:
Problem-based learning

People Exchanges:
Movement of students to learning resources.

Plagiarism:
The practice of presenting the work of others without acknowledgement as if it were one's own.

Practical:
Classes in which students engage in a practical (including computer-based) activity.

Program:
A scheme of academic study.

Program Specification:
A formal document setting out the rationale, learning objectives, structure and assessment of a formal program of academic study.

QTI:
Question and test interoperability

Reload:
Reusable eLearning object authoring and delivery.

Remote Sensing:
Capture of data about the earth's surface by remote instruments.

Repository:
Archive of online materials and resources.

Repurpose:
The reuse of some or all of a set of existing learning materials in a different context.

RHS:
River Habitat Survey

Shibboleth:
Software for effecting federated access management of user access to learning resources.

SCORM:
Shareable content object reference model.

Semantic Web:
Extension of the Web in which the semantics (meaning) of the information held is defined.

SLS:
Substantive learning stages

SPSS:
Statistical Package for the Social Sciences.

SSM:
Strongly structured model

Summative:
Assessment that contributes to the final grade awarded.

swiki:
Software for project teams to share files and jointly author content.

Tiddly Wiki:
A wiki program for electronic learning diaries.

UCM:
Upland Catchment Management, an undergraduate module taught at the University of Leeds.

UCSB:
University of California Santa Barbara.

UDM:
Universal data map

UKeU:
UK eUniversities, a company set up to deliver UK e-learning programs worldwide

Virtual Globe:
A software tool for interactive display of map and image data related to the surface of the earth, e.g., Google Earth

VLE:
Virtual learning environment, a suite of online software for the delivery of learning materials to students

Web 2.0:
Web technology used to enhance creativity, information sharing, and, collaboration among users.

Wikipedia:
Open access community encyclopedia project.

WILSIM:
Web-based Interactive Landform Simulator.

World Campus:
The "online campus" of The Pennsylvania State University.

XML:
Extensible Markup Language

Compilation of References

ADEPT. (2003). *Virtual learning environment.* Retrieved February 15, 2007, from http://www.alexandria.ucsb. edu/research/learning/index.htm

Adobe. (2008). *Adobe Flash CS3 Professional.* Retrieved March 27, 2008, from http://www.adobe.com/products/flash/

Agosti, M., Ferro, N., Frommholz, I., & Thiel, U. (2004). Annotations in digital libraries and collaboratories: Facets, models and usage. In R. Heery & L. Lyon (Eds.), *Proceedings of the Eighth European Conference on Digital Libraries* (Lecture Notes in Computer Science 3232, pp. 244-255). Berlin: Springer-Verlag.

Åkerlind, G. S., & Trevitt, A. (1999). Enhancing self-directed learning through educational technology: When students resist the change. *Innovations in Education and Teaching International, 36*(2), 96-105.

Allen, B.S., Hoffman, R.P., Kompella, J., & Sticht, T.G. (1993). Computer-based mapping for curriculum development. In *Proceedings of selected Research and Development Presentations Technology sponsored by the Research and Theory Division,* New Orleans, LA. (Eric Document Reproduction Services No. ED 362 145).

Althausen, J. D., & Mieczkowski, T. M. (2001). The merging of criminology and geography into a course on spatial crime analysis. Journal of Criminal Justice Education, 12(2), 367-383.

Ancona, D., & Smith, T. (2002). *Visual explorations for the Alexandria Digital Earth Prototype.* Paper presented at the Second International Workshop on Visual Interfaces to Digital Libraries, The ACM+IEEE Joint Conference on Digital Libraries (JCDL), Portland, OR. Retrieved February 15, 2007, from http://vw.indiana. edu/visual02/Ancona.pdf

Ancona, D., Frew, J., Janée, G., & Valentine, D. (2005). Accessing the Alexandria digital library from geographic information systems. In T. Sumner & F. Shipman (Eds.), *Proceedings of the Fifth ACM/IEEE-CS Joint Conference on Digital Llibraries* (pp. 74-75). New York: ACM Press.

Anderson, L.W., & Krathwohl, D. (Eds.). (2001). *A taxonomy for learning, teaching, and assessing: A revision of Bloom's Taxonomy of educational objectives.* New York: Longman.

Andrienko, G., & Andrienko, N. (2006). *Exploratory analysis of spatial and temporal data: A systematic approach.* London: Springer.

Angel Learning. (2008). *Angel learning: Recognized innovator of enterprise eLearning software.* Retrieved March 20, 2008, from http://www.angellearning.com/

Anselin, L. (1995). *Local Indicators of Spatial Association—LISA.* Geographical Analysis, 27(2), 93-115.

Anselin, L. et al. (2006b). *GeoDa: An introduction to spatial analysis.* Champaign-Urbana: Spatial Analysis Lab, University of Illinois. Retrieved February 20, 2008, from https://www.geoda.uiuc.edu/

Anselin, L., Syabri, I., & Kho, Y. (2006a). GeoDa: An introduction to spatial analysis. *Geographical Analysis, 38*(1), 5-22.

Apple Inc. (2008). *Keynote: Cinema-quality presentations for everyone.* Retrieved March 27, 2008, from http://www.apple.com/iwork/keynote/

Arias, E., Eden, H., Fischer, G., Gorman, A., & Scharff, E. (2000). Transcending the individual human mind - creating shared understanding through collaborative design. ACM Transactions on Computer-Human Interaction, 7(1), 84-113

Arnell, N. W. (2003). Effects of IPCC SRES emissions scenarios on river runoff: A global perspective. *Hydrology and Earth System Sciences, 7,* 619-641.

Ashworth, P., Bannister, P., Thorne, P., & Unit Students on the Qualitative Research Methods Course. (1997). Guilty in whose eyes? University students' perceptions of cheating and plagiarism in academic work and assessment. *Studies in Higher Education, 22*(2), 187-203.

Austin, J., & Brown, L. (1999). Internet plagiarism: Developing strategies to curb student academic dishonesty. *The Internet and Higher Education, 2*(1), 21-23.

Avancini, H., & Straccia, U. (2004). Personalization, collaboration, and recommendation in the digital library environment CYCLADES. In *Proceedings of IADIS conference on applied computing 2004* (pp. 67-74). Lisbon, Portugal: IADIS Press.

C., Zalfan M., Davis H., Fill K. & Conole G. (2006). Panning for gold: Designing pedagogically-inspired learning nuggets. *Educational Technology & Society, 9*(1), 113-122.

Barker, P. (2005). *What is IEEE learning object metadata/IMS learning resource metadata?* CETIS Standards briefing series. Retrieved February 18, 2008, from http://wiki.cetis.ac.uk/What_is_IEEE_LOM/IMS_LRM

Barrett, R., & Cox, A. L. (2005). At least they're learning something: The hazy line between collaboration and collusion. *Assessment and Evaluation in Higher Education, 30*(2), 107-122.

Barrow, C. J. (1999). *Environmental management: Principles and practice.* London: Routledge.

BBC News. (2008). *Teachers voice plagiarism fears.* Retrieved February 8, 2008, from http://news.bbc.co.uk/1/hi/education/7194772.stm

Beetham, H. (2004). *Review: Developing e-learning models for the JISC practitioner communities.* Report for the Pedagogies for e-Learning Programme, Version 2.1. Retrieved February 22, 2007, from http://www.jisc.ac.uk/uploaded_documents /Review 20models.doc

Beetham, H., & Sharpe, R. (Eds.) (2007). *Rethinking pedagogy for a digital age. Oxford.* UK: Routledge-Falmer.

Bennett, S., & Marsh, D. (2001). Are we expecting online tutors to run before they can walk? *Innovations in Education and Teaching International, 39*(1), 14-20.

Berners-Lee, T., Hendler, J., & Lassila, O. (2001, May 17). The Semantic Web. *Scientific American.* Retrieved March 27, 2008, from http://www.sciam.com/article.cfm?id=the-semantic-web

Biggs, J. (2003). *Teaching for quality learning at university* (2nd ed.). Bury St Edmunds, UK: The Society for Research into Higher Education and the Open University Press.

Binwal, J. C., & Lalhmachhuana. (2001). Knowledge representation: Concept, techniques, and the analytico-synthetic paradigm. *Knowledge Organization, 28*(1), 5-16.

Bloom, B. S. (Ed.) (1956). *Taxonomy of educational objectives, the classification of educational goals—Handbook I: Cognitive domain.* New York: McKay.

Bond, S., Wodehouse, A., Leung, S., & Wallace, I. (2007). It takes a bit of imagination.... In *Institutional Transformation: The Proceedings of the JISC Innovating E-Learning 2007 Online Conference* (pp. 18-21). Retrieved February 18, 2008, from http://www.jisc.org.uk/media/documents/programmes/elearningpedagogy/ebookone2007.pdf

Borgman, C. L. (2006). What can studies of e-learning teach us about collaboration in e-research? Some findings from digital library studies. *Computer Supported Cooperative Work, 15,* 359-383.

Borgman, C. L., Gilliland-Swetland, A. J., Leazer, G. H., Mayer, R., Gwynn, D., Gazan, R., & Mautone, P. (2000). Evaluating digital libraries for teaching and learning in undergraduate education: A case study of the Alexandria Digital Earth Prototype (ADEPT). *Library Trends, 49*(2), 228-250.

Borgman, C. L., Leazer, G. H., Swetland, A., Millwood, K., Champeny, L., Finley, J., et al. (2004). How geography professors select materials for classroom lectures: implications for the design of digital libraries. In E. P. Lim & M. Christel (Eds.), *Proceedings of the Fourth ACM/IEEE-CS Joint Conference on Digital Libraries* (pp. 179-185). New York: ACM Press.

Borgman, C. L., Smart, L. J., Millwood, K. A., Finley, J. R., Champeny, L., Gilliland, A. J., & Leazer, G. H. (2005). Comparing faculty information seeking in teaching and research: Implications for the design of digital libraries. *Journal of the American Society for Information Science and Technology, 56*(6), 636-657.

Boshier, R., Mohapi, M., Moulton, G., Qayyum, A., Sadownik, L., & Wilson, M. (1997). Best and worst dressed Web courses: Strutting into the 21st Century in comfort and style. *Distance Learning Education: An International Journal, 18*(2), 327-348.

Boyle, A., Conchie, S., Maguire, S., Martin, A., Milsom, C., Nash, R., Rawlinson, S., Turner, A., & Wurthmann, S. (2003). Fieldwork is good? The student experience of field courses. *Planet Special Edition 5: Linking Teaching and Research in Geography, Earth and Environmental Sciences,* (11), 48-51.

Boyle, T., & Cook, J. (2001). Towards a pedagogically sound basis for learning object portability and re-use. In G. Kennedy, M. Keppell, C. McNaught, & T. Petrovic (Eds.), *Meeting at the crossroads*: Proceedings of the 18th Annual Conference of the Australasian Society for Computers in Learning in Tertiary Education (AS-CILITE 2001) (pp. 101-109). Melbourne: The University of Melbourne. Retrieved October 05, 2007, from http://www.ascilite.org.au/conferences/melbourne01/pdf/papers/boylet.pdf

Bricheno, P., Higgison, C., & Weedon, E. (2004). *The impact of networked learning on education institutions.* Bristol, UK: Joint Information Systems Committee (JISC). Retrieved March 17, 2008, from http://www.sfeuprojects.org.uk/inlei/

Brooks, Jr, F. P. (1975). The mythical man-month: Essays on software engineering. Reading, MA: Addison-Wesley Pub. Co.

Brown, S., Race, P., & Bull, J. (1999). *Computer-assisted assessment in higher education.* London, UK: SEDA and Kogan Page.

Browne, T., & Jenkins, M. (2003). *VLE surveys: A longitudinal perspective between March 2001 and March 2003 for higher education in the United Kingdom.* UCISA. Retrieved on March 27, 2008, from http://www.immagic.com/eLibrary/UNPROCESSED/Unprocessed%20eLibrary/eLibrary_uprocessed/JISC_Briefing_Papers/vle2003.pdf

Bruer, J. T. (1993). *Schools for thought: A science of leaning in the classroom.* Cambridge, MA: MIT Press.

Brutoco, D., & Maurissa, G. (1997). *Making the grade: Cheating at Santa Clara University.* Unpublished report. Santa Clara, CA: Santa Clara University.

Bryant, R.L., & Wilson, A. (1998). Rethinking environmental management. *Progress in Human Geography, 22*(3), 321-343.

Buendia-Garcia, F., & Diaz, P. (2003). A framework for the specification of the semantics and the dynamics of instructional applications. *Journal of Educational Multimedia and Hypermedia, 12*(4), 399-424.

Buzan, T., & Buzan, B. (1996). *The mind map book: How to use radiant thinking to maximize your brain's untapped potential.* New York: Plume.

Carter, J. (2002). A framework for the development of multimedia systems for use in engineering education. *Computers & Education, 39,* 111-128.

CAS. (1999). *1999 CA index guide.* Columbus, OH: Chemical Abstracts Services.

CAS. (2008). *The CAS registry.* Retrieved March 5, 2008, from http://www.cas.org/EO/regsys.html

Casey, J. (2006). *Intellectual property rights (IPR) in networked e-learning: A beginner's guide for content developers.* Glasgow: JISC Legal. Retrieved March 28, 2008, from http://www.jisclegal.ac.uk/pdfs/johncasey.pdf

CeLSIUS. (2008). *Welcome to CeLSIUS.* London: Centre for Longitudinal Study Information and User Support, London School of Hygiene and Tropical Medicine. Retrieved April 27, 2008, from http://www.celsius.lshtm.ac.uk/

Census Dissemination Unit (CDU). (2008). *Accessing the census data.* Manchester: Census Dissemination Unit, University of Manchester. Retrieved April 27, 2008, from http://www.census.ac.uk/cdu/

Census Dissemination Unit. (2008) *GeoConvert: A resource for geographical lookups and conversion.* Manchester: University of Manchester. Retrieved February 20, 2008, from http://geoconvert.mimas.ac.uk/

Census Dissemination Unit. (2008). *CASWEB: Web interface to census aggregate outputs and digital boundary data.* Retrieved February 20, 2008, from http://casweb.mimas.ac.uk/

Centre for Census and Survey Research (CCSR). (2008). *The samples of anonymised records.* Manchester: Cathie Marsh Centre for Census and Survey Research, University of Manchester. Retrieved April 27, 2008, from http://www.ccsr.ac.uk/sars/

Centre for Interaction Data Estimation and Research (CIDER). (2008). *Welcome to CIDER.* Leeds: Centre for Interaction Data Estimation and Research, University of Leeds. Retrieved April 27, 2008, from http://www.census.ac.uk/cids/

CETIS. (2005). CETIS briefing on e-learning standards. Retrieved August 26, 2008, from http://zope.cetis.ac.uk/static/briefings.html

Champeny, L., Borgman, C. L., Leazer, G. H., Gilliland-Swetland, A. J., Millwood, K.A., D'Avolio, L., Finley, J.

R., Smart, L. J., Mautone, P. D., Mayer, R. E., & Johnson, R. A. (2004). Developing a digital learning environment: An evaluation of design and implementation processes. In *ACM/IEEE-CS Joint Conference on Digital Libraries* (JCDL 2004), Tucson, AZ. New York: ACM Press.

CHCC. (2003). *Developing the collection of historical and contemporary census data and related materials (CHCC) into a major learning and teaching resource.* Retrieved February 14, 2008, from http://www.chcc.ac.uk

Childs, M. (2004). Is there an e-pedagogy of resource-and-problem based learning? Using digital resources in three case studies. *Interactions, 8*(2). Retrieved March 30, 2005, from www2.warwick.ac.uk/services/cap/resources/interactions/archive/issue24/childs/

Church, M. (2005). Continental drift. *Earth Surface Processes and Landforms, 30*(1), 129-130.

Cierniewski, J. (1987). A model for soil surface roughness influence on the spectral response of bare soils in the visible and near infrared range. *Remote Sensing of Environment, 23,* 97-115.

Cisco Systems. (2003). *Reusable learning object strategy: Designing and developing learning objects for multiple learning approaches* (White paper). Cisco Systems Inc.

Clark, E. A. (2005). *SURF X4L - Reuse and repurposing of resources for content exchange, including technical considerations on interoperability.* A report from the SURF X4L Project, a JISC Exchange for Learning Project. Retrieved April 25, 2008, from http://www.staffs.ac.uk/COSE/X4L/X4Ltechnical.pdf

Clark, G. (1998). Maximising the benefits from work-based learning: The effectiveness of environmental audits. *Journal of Geography in Higher Education, 22*(3), 325-334.

Clark, M.J., Ball, J.H., & Sadler, J.D. (1995). Multimedia delivery of coastal zone management training. *Innovations in Education and Training International, 32*(3), 229-238.

College of Information Sciences and Technology. (no date). *CiteSeer.IST: Scientific Literature Digital Library.* University Park, PA: The Pennsylvania State University. Retrieved January 26, 2008, from http://citeseer.ist.psu. edu/

Columbia University. (2008). *Centre for International Earth Science Information Network (CIESIN).* New York: Earth Institute, Columbia University. Retrieved March 23, 2008, from http://www.ciesin.org/

Computing. (2004). *The failure of UKeU: Computing's high-profile investigation into the government's disastrous £62m e-learning scheme.* Retrieved March 30, 2008, from http://www.computing.co.uk/computing/ specials/2071853/failure-ukeu

Conole, G. (2008). Capturing practice: The role of mediating artefacts in learning design. In L. Lockyer, S. Bennett, S. Agostinho, & B. Harper (Eds.), *Handbook of research on learning design and learning objects: Issues, applications and technologies* (forthcoming). Hershey, PA: Information Science Publishing.

Conole, G., & Oliver, M. (1998). A pedagogical framework for embedding C and IT into the curriculum. *Association of Learning Technology Journal, 6*(2), 4-16.

Conole, G., & Oliver, M. (2007). Introduction. In G. Conole & M. Oliver (Eds.), *Contemporary perspectives in e-learning research: themes, methods and impact on practice.* London: Routledge.

Conole, G., & Weller, M. (2008). Using learning design as a framework for supporting the design and reuse of OER. *Journal of Interactive Media in Education, 2008(5), 1-12. Retrieved from http://jime.open.ac.uk/2008/05/jime-2008-05.pdf*

Conole, G., Carusi, A., de Laat, M., Wilcox, P., & Darby, J. (2006). Managing differences in stakeholder relationships and organizational cultures in e-learning development: Lessons from the UK eUniversity experience. *Studies of Continuing Education, 28*(2), 135-150.

Conole, G., Crewe, E., Oliver, M., & Harvey, J. (2001). A toolkit for supporting evaluation. *Association of Learning Technology Journal, 9*(1), 38-49.

Conole, G., De Laat, M., Dillon, T., & Darby, J. (2008). Disruptive technologies,' 'pedagogical innovation:' What's new? Findings from an in-depth study of students' use and perception of technology. *Computers and Education, 50*(2), 511-524.

Conole, G., Littlejohn, A., Falconer, I., & Jeffrey, A. (2005). *Pedagogical review of learning activities and use cases.* LADIE Project report. Retrieved December 22, 2006, from http://www.elframework.org/refmodels/ ladie/ouputs/LADIE lit review v15.doc

Conole, G., Scanlon, E., Kerawalla, C., Mullholland, P., Anastopulou, S., & Blake, C. (2008). *From design to narrative: The development of inquiry-based learning models.* Ed-Media Conference, July 2008, Vienna.

Conole, G., Thorpe, M., Weller, M., Wilson, P., Nixon, S., & Grace, P. (2007). *Capturing practice and scaffolding learning design.* Paper presented at European Distance and E-Learning Network Conference, Naples, Italy.

Culwin, F., & Lancaster, T. (2001) *Plagiarism, prevention, deterrence and detection.* Retrieved June 30, 2007, from http://www.heacademy.ac.uk/resources. asp?process=full_record&section=generic&id=426

Currier, S., & Campbell, L. M. (2005). Evaluating 5/99 content for reusability as learning objects. *VINE: The Journal of Information and Knowledge Management Systems, 35*(1/2), 85-96.

Currier, S., Campbell, L., & Beetham, H. (2006). *Pedagogical vocabularies review.* JISC Pedagogical vocabularies project report 1. Retrieved February 28, 2007, from http://www.jisc.ac.uk/uploaded_documents/ PedVocab_VocabsReport_v0p11.doc

D'Avolio, L. W., Borgman, C. L., Champeny, L., Leazer, G. H., Gilliland, A. J., & Millwood, K. A. (2005). From prototype to deployable system: Framing the adoption of digital library services. In A. Grove (Ed.), *Proceedings 68th Annual Meeting of the American Society for Information Science and Technology (ASIST) 42,* Charlotte, NC, US. Retrieved from http://eprints.rclis. org/archive/00005052/

Dalziel, J. (2003). Implementing learning design: The learning activity management system (LAMS). In G. Crisp, D. Thiele, I. Scholten, S. Barker, & J. Baron (Eds.), Interact, Integrate, Impact: Proceedings of the 20th Annual Conference of the Australasian Society for Computers in Learning in Tertiary Education, Adelaide.

Dana, P. H. (1997). Global positioning system overview. NCGIA Core Curriculum in GIScience. Retrieved April 11, 2008, from http://www.ncgia.ucsb.edu/giscc/units/u017/u017.html,

Dana, P. H. (1999). Global positioning system overview. The Geographer's Craft Project, Department of Geography, The University of Colorado at Boulder. Retrieved March 1, 2008, from http://www.colorado.edu/geography/gcraft/notes/gps/gps_f.html.

Davis, H. C., & Fill, K. (2007). Embedding blended learning in a university's teaching culture: Experiences and reflections. *British Journal of Education Technology, 38*(5), 817-828.

Devlin, M. (2002). *Plagiarism detection software: How effective is it?* Retrieved June 30, 2007, from http://www.cshe.unimelb.edu.au/assessinglearning/03/Plag2.html

DialogPLUS (2004). DialogPlus: Digital libraries in support of innovative approaches to learning and teaching in geography. *Proceedings of the Fourth ACM/IEEE-CS Joint Conference on Digital Libraries (JCDL 2004).*

DialogPLUS. (2006). *Nugget developer guidance toolkit.* Retrieved March 20, 2008, from http://www.nettle.soton.ac.uk/toolkit/

DiBiase, D. (2000). Is distance education a Faustian bargain? *Journal of Geography in Higher Education, 24*(1), 130-135.

DiBiase, D. (2004). The impact of increasing enrollment on faculty workload and student satisfaction over time. Journal of Asynchronous Learning Networks, 8(1), 45-60. Retrieved January 26, 2008, from http://www.aln.org/publications/jaln/v8n2/pdf/v8n2_dibiase.pdf

DiBiase, D. (2005). Using concept mapping to design reusable learning objects for e-education in cartography

and GIS. In Proceedings of the XXII International Cartographic Association (ICC2005), A Coruña, Spain.

DiBiase, D. (2008). *Nature of geographic information: An open geospatial textbook. University Park, PA:* The Pennsylvania State University. Retrieved January 26, 2008, from http://natureofgeoinfo.com

DiBiase, D., & Kidwai, K. (2007). Wasted on the young? Comparing the efficacy of instructor-led online education in GIScience for post-adolescent undergraduates and adult professionals. In *Proceedings of the Association of American Geographers 2007 Annual Meeting*, San Francisco, USA.

DiBiase, D., & Rademacher, H. J. (2005). Scaling up: How increasing enrollments affect faculty and students in an asynchronous online course in geographic information science. *Journal of Geography in Higher Education, 29*(1), 141-160.

Diduck, A. (1999). Critical education in resource and environmental management: Learning and empowerment for a sustainable future. *Journal of Environmental Management, 57*(2), 85-97.

DLESE. (no date). *Digital library for earth systems education.* Retrieved January 26, 2008, from http://www.dlese.org/library/index.jsp

Dordoy, A. (2002). Cheating and plagiarism: Student and staff perceptions at Northumbria. In *Proceedings of the Northumbria Conference: Educating for the Future, Newcastle, UK.* Retrieved February 18, 2008, from http://www.jiscpas.ac.uk/images/bin/AD.doc

Dorling, D. (1995). Visualising changing social-structures from a census. *Environment and Planning A, 27,* 353-78.

Downes, S. (2000). *Learning objects* (Essay). Retrieved October 23, 2007, from http://www.downes.ca/files/Learning_Objects.htm

Drennon, C. (2005). Teaching geographic information systems in a problem-based learning environment. *Journal of Geography in Higher Education, 29*(3), 385-402.

Duguid, P., & Atkins, D. E. (Eds). (1997). *Report of the Santa Fe Planning Workshop on Distributed Knowledge Work Environments.* Retrieved from http://www.si.umich.edu/SantaFe/

Duke-Williams, O., & Rees, P. H. (1998). Can Census Offices publish statistics for more than one small area geography? An analysis of the differencing problem in statistical disclosure. *International Journal of Geographical Information Systems, 12*(6), 579-605.

Duncan, C. (2003). Granularization. In A. Littlejohn (Ed.), *Reusing online resources: A sustainable approach to e-learning* (pp. 12-19). London: Kogan.

Durham, H., & Arrell, K. (2006). *Introducing new cultural and technological approaches into institutional practice: An experience from geography.* Paper presented at European Conference on Digital Libraries, Alicante, Spain. Retrieved March 21, 2008, from http://www.csfic.ecs.soton.ac.uk/Durham.doc

Durham, H., & Arrell, K. (2007). Introducing new cultural and technological approaches into institutional practice: An experience from geography. British Journal of Educational Technology, 38(5), 795-804.

Durham, H., & Rees, P. (2006) *Census analysis and GIS - unit 4.* Retrieved February 6, 2008, from http://repository.jorum.ac.uk/intralibrary/IntraLibrary?command=preview&learning_object_id=3645

Durham, H., Dorling, D., & Rees, P. (2003). *Online census atlas.* Retrieved February 16, 2008, from http://www.ccg.leeds.ac.uk/teaching/chcc/; http://devchcc.mimas.ac.uk/cgi-bin/CAS/atlas/showdata.cgi

Durham, H., Dorling, D., & Rees, P. (2006). An online census atlas for everyone. *Area, 38,* 336-341.

Duval, E., & Hodgins, W. (2003). A LOM research agenda. In *Proceedings of the 12th International World Wide Web Conference*, Budapest, Hungary. Retrieved October 23, 2007, from http://www2003.org/cdrom/papers/alternate/P659/p659-duval.html.html.

Duval, E., Hodgins, W., Sutton, S., & Weibel, S. L. (2002). Metadata principles and practicalities. *D-Lib Magazine,*

8(4). Retrieved March 27, 2008, from http://www.dlib.org/dlib/april02/weibel/04weibel.html.

Dye, A., Jones, B., & Kismihok, G. (2003). Mobile learning: The next generation of learning exploring online services in a mobile environment. Retrieved March 29, 2008, from http://www.dye.no/articles/mlearning/exploring_online_services_in_a_mobile_environmnet.pdf

Dyke, M., Conole, G., Ravenscroft, A., & de Freitas, S. (2007). Learning theories and their application to e-learning. In G. Conole & M. Oliver (Eds.), *Contemporary perspectives in e-learning research: themes, methods and impact on practice.* London: Routledge.

EDINA. (2008). *Welcome to UKBORDERS.* Edinburgh: Edinburgh Data Library, University of Edinburgh. Retrieved April 27, 2008, from http://www.edina.ac.uk/ukborders/

EDIT4L. (2007). *Evaluation of design and implementation tools for learning.* Project Web site. Retrieved March 21, 2008, from http://www.edit4l.soton.ac.uk:8081/

Edmondson, K. M. (1993). *Concept mapping for the development of medical curricula.* Paper presented at the Annual Conference of the American Educational Research Association, Atlanta, GA. (Eric Document Reproduction Services No. ED 360 322).

E-Learning Centre. (2008). *What is e-learning?* Sheffield: Learning Light Ltd. Retrieved March 25, 2008, from http://www.e-learningcentre.co.uk/eclipse/Resources/whatise.htm

Elearningeuropa.info. (2008). *Directory.* Retrieved March 29, 2008, from: http://www.elearningeuropa.info/directory/index.php?page=home

Englebart, D. C. (1995). Towards augmenting the human intellect and boosting our collective IQ. Communications of the ACM, 38(8), 30-32.

Environment Agency. (1998). *River Habitat Quality: The physical characteristics of rivers and streams in the UK and the Isle of Man.* (RHS Report No. 2). Bristol, UK: Environment Agency.

Ercegovac, Z., & Richardson, J. V. (2004). Academic dishonesty, plagiarism included, in the digital age: A literature review. *College and Research Libraries, 65*(4), 301-318.

ESPON. (2006, October). *ESPON ATLAS: Mapping the structure of the European territory. The European Spatial Planning Observation Network and the partners of the ESPON programme.* Retrieved March 8, 2008, from http://www.espon.eu/mmp/online/website/content/publications/98/1235/file_2489/final-atlas_web.pdf

ESRC Census Programme. (2008). *The ESRC Census Programme.* Retrieved April 27, 2008, from http://census.ac.uk/censusprogramme/Default.aspx

ESRC. (2006). *Moving you closer to the data.* Economic and Social Research Council. Retrieved March 26, 2008, from http://census.ac.uk/

Eurostat. (2008). *Regions: Databases and publications.* Luxembourg: Statistical Office of the European Communities. Retrieved March 23, 2008, from http://epp.eurostat.ec.europa.eu/portal/page?_pageid=1335,47078146&_dad=portal&_schema=PORTAL

Eurostat. (2008). *Nomenclature of territorial units for statistics—NUTS statistical regions of Europe.* Retrieved March 25, 2008, from http://ec.europa.eu/comm/eurostat/ramon/nuts/home_regions_en.html

Evans, A. (2008, April 24). Happier, more productive...(E-mail to the School of Geography, University of Leeds).

Facebook. (2008) *Facebook is a social utility that connects you with the people around you.* Retrieved February 14, 2008, from http://www.facebook.com/

Falconer, I., & Conole, G. (2006). *LADIE gap analysis. Report for the JISC-funded LADIE project.* Retrieved February 22, 2007, from http://www.elframework.org/refmodels/ladie/guides/LADiE 20Gap 20Analysis.doc

Falconer, I., & Littlejohn, A. (2006). *Mod4L Report: Case Studies, Exemplars and Learning Designs.* Glasgow, UK: Glasgow Caledonian University. Retrieved February 22, 2007, from http://mod4l.com/tiki-download_file.php?fileId=2

Farrell, J., & Shapiro, C. (1989). Optimal contracts with lock-in. *American Economic Review, 79*(1), 51-68.

Fernandez-Young, A., Ennew, C., Owen, N., DeHaan, C., & Schoefer, K. (2006). Developing material for online management education - a UK eUniversity experience. *International Journal of Management Education, 5*(1), 45-55.

Ferreira, H. S., Florenzano, T. G., Dias, N. W., Mello, E. M. K., Moreira, J. C., & Moraes, E. C. (2005). Distance learning courses for disseminating remote sensing technology and enhancing undergraduate education. In G. König, H. Lehmann, & R. Köhring (Eds.), *ISPRS Workshop on Tools and Techniques for E-Learning* (pp. 110-113). Potsdam, Germany: Institute of Geodesy and Geoinformation Science.

Fill, K. (2005). *Student-focused evaluation of eLearning activities.* Paper presented at European Conference on Educational Research, Dublin, Ireland. Retrieved March 27, 2008, from http://www.leeds.ac.uk/educol/documents/143724.htm

Fill, K., Leung, S., DiBiase, D., & Nelson, A. (2006). Repurposing a learning activity on academic integrity: The experience of three universities. *Journal of Interactive Media in Education, 2006*(1). Retrieved June 30, 2007, from http://jime.open.ac.uk/2006/01/

Fischer, G., Grudin, J., Lemke, A., McCall, R., Ostwald, J., Reeves, B., & Shipman, F. (1992). Supporting indirect collaborative design with integrated knowledge-based design environments. Human-Computer Interaction, 7, 281-314.

Fletcher, S., France, D., Moore, K., & Robinson, G. (2003). Technology before pedagogy? A GEES C&IT perspective. *Planet Special Edition 5 - Part B Pedagogic Research in Geography, Earth and Environmental Sciences,* pp. 52-55.

Fletcher, S., France, D., Moore, K., & Robinson, G. (2007). Practitioner perspectives on the use of technology in fieldwork teaching. *Journal of Geography in Higher Education, 31*(2), 319-330.

Ford, C. E. (1998). Supporting fieldwork using the Internet. *Computers and Geosciences, 24*(7), 649-651.

Fox S., & MacKeoch K. (2003). Can eLearning promote higher-order learning without tutor overload? *Open Learning, 18*, 121-134.

France, D., Fletcher, S., Moore, K., & Robinson, G. (2002). Fieldwork education and technology. The GEES Subject Centre. Retrieved May 6, 2008, from http://www.gees.ac.uk/pedresfw/pedrcit.htm.

Friesen, N. (2003). Three objections to learning objects and e-learning standards. In R. McGreal (Ed.), *Online education using learning objects* (pp. 59-70). London: Routledge.

Gahegan, M., Agrawal, R., & DiBiase, D. (2007). Building rich, semantic descriptions of learning activities to facilitate reuse in digital libraries. *International Journal on Digital Libraries, 7*(1-2), 81-97.

Galster, G. (2001). On the nature of neighborhood. *Urban Studies, 38*(12), 2111-2124.

Gärdenfors, P. (2000). *Conceptual spaces: The geometry of thought.* Cambridge, MA: MIT Press.

Gardiner, V., & Unwin, D. (1986). Computers and the field class. *Journal of Geography in Higher Education, 10*, 169-179.

Garrett, J. (2005). *Ajax: A new approach to Web applications.* Retrieved, March 01, 2007, from http://www.adaptivepath.com/publications/essays/archives/000385.php

Garrison, D. R. (1989) *Understanding distance education.* London/New York: Routledge.

Garrison, D.R. (1993). Quality and access in distance education: Theoretical considerations. In D. Keegan (Ed.), *Theoretical principles of distance education.* London/New York: Routledge.

General Register Office for Scotland (GROS). (2008). *Welcome to SCROL: Scotland's Census Results Online.* Edinburgh: General Register Office for Scotland. Retrieved April 27, 2008, from: http://www.gro-scotland.gov.uk/

GeoVISTA Center. (2005). *ConceptVista: Ontology management.* University Park, PA: The Pennsylvania State University. Retrieved January 26, 2008, from http://www.geovista.psu.edu/ConceptVISTA/index.jsp

Goh, D., Fu, L., & Foo, S. (2002). A work environment for a digital library of historical resources. In E. P. Lim, S. Foo, C. S. G. Khoo, H. Chen, E. A. Fox, S. R. Urs, et al. (Eds.), *Proceedings of the 5th International Conference on Asian Digital Libraries* (LNCS 2555, pp. 260-261). Berlin: Springer-Verlag.

Gold, J. R., Jenkins, A., Lee, R., Monk, J., Riley, J., Shepherd, I., & Unwin, D. (1991). *Teaching geography in higher education.* Oxford, UK: Blackwell.

Gould, M., Stillwell, J., & Vanderbeck, R. (2007-2008). *Research methods in human geography.* University of Leeds Module GEOG2680. Module outline and materials. Retrieved February 16, 2008, from http://webprod1.leeds.ac.uk/catalogue/dynmodules.asp?Y=200708&M=GEOG-2680; http://vle.leeds.ac.uk/site/nbodington/geography/geoglev2/geog2680/sem1/

Graham, C. R. (2004). Blended learning systems: Definition, current trends, and future directions. In C. J. Bonk & C. R. Graham (Eds.), *Handbook of blending learning: Global perspectives, local designs* (pp. 3-11). San Francisco, CA: Pfeiffer Publishing.

Graham, L. (1999). *The principles of interactive design.* Devon: Delmar Publishing.

Groark, M., Oblinger, D., & Choa, M. (2001). Terms paper mills, anti-plagiarism tools, and academic integrity. *Educause Review, 36*(5), 40-48.

GROS. (2008). *Welcome to SCROL: Scottish Census Results Online.* Edinburgh: General Register Office. Retrieved February 20, 2008, from http://www.scrol.gov.uk/scrol/common/home.jsp

GROS. (2008). *Scottish neighbourhood statistics.* Edinburgh: General Register Office. Retrieved February 20, 2008, from http://www.sns.gov.uk/

Grumbine, R. E. (1994). What is ecosystem management? *Conservation Biology, 8*(1), 27-38.

Gunn, C., Woodgate, S., & O'Grady, W. (2007). Repurposing learning objects: A sustainable alternative? *Association for Learning Technology Journal, 13*(3), 189-200.

Hamal, C. J., & Ryan-Jones, D. (2002, November). Designing instruction with learning objects. *International Journal of Educational Technology, 3*(1). Retrieved October 23, 2007, from http://www.ed.uiuc.edu/ijet/v3n1/hamel/index.html.

Hansen, B. (2003). Combating plagiarism: Is the Internet causing more students to copy? *CQ Researcher, 13*(32), 773-796.

Harris, J., Hayes, J., & Cole, K. (2002). Disseminating census area statistics over the Web. In P. Rees, D. Martin, & P. Williamson (Eds.), *The census data system* (pp.113-122). London: Wiley.

Hart, M., & Friesner, T. (2004). Plagiarism and poor academic practice - a threat to the extension of e-learning in higher education? *Electronic Journal on e-Learning, 2*(1), 89-96.

Harvey, J. (Ed.) (1998). *Evaluation cookbook.* Edinburgh, UK: Heriot-Watt University.

Harwood, I.A. (2005). When summative computer-aided assessments go wrong: Disaster recovery after a major failure. *British Journal of Educational Technology, 36, Special issue on Thwarted Innovation in e-Learning.*

Henze, N., Dolog, P., & Nejdl, W. (2004). Reasoning and ontologies for personalized e-learning in the Semantic Web. *Educational Technology & Society, 7*(4), 82-97.

Hernández-Leo, D., Harrer, A., Dodero, J. M., Asensio-Pérez, J. I., & Burgos, D. (2006). Creating by reusing learning design solutions. In *Proceedings of 8th Simposo Internacional de Informática Educativa, León, Spain: IEEE Technical Committee on Learning Technology.* Retrieved March 11, 2008, from http://dspace.ou.nl/handle/1820/788

Higher Education Funding Council for England (HEFEC). (1996). *Evaluation of the Teaching and Learning Technology Programme.* Bristol, UK: Higher Education Funding Council for England (HEFEC).

Hill, L., Buchel O., Janee, G., & Zeng, M. L. (2002). Integration of knowledge organization systems into digital library architectures: Position paper. In J.-E. Mai, C. Beghtol, J. Furner, & B. Kwasnik (Eds.), *Advances in classification research. Proceedings of the 13th ASIS&T SIG/CR Workshop* (Vol. 13, pp. 62-68). Medford, NJ: Information Today. Retrieved February 15, 2007, from http://alexandria.sdc.ucsb.edu/~lhill/paper_drafts/KOSpaper7-2-final.doc

Hinman, L. M. (2002). Academic integrity and the World Wide Web. *Computers and Society, 32,* 33-42.

Hoffmann-Wellenhof, B., Lichtenegger, H., & Collins, J. (2001). Global positioning system: Theory and practice (5th rev. ed.). Austria: Springer-Verlag.

Hogan, K. (2002). Small groups' ecological reasoning while making an environmental decision. *Journal of Research in Science Teaching, 39*(4), 341-368.

Höhle, J., (2004). Designing of course material for e-learning in photogrammetry. *International Archives of Photogrammetry, Remote Sensing and Spatial Information Sciences, 35*(B6), Commission VI, WG VI/2, 89-94.

Holmberg, B. (1960). *On the methods of teaching by correspondence.* Lund: Gleerup.

Holmberg, B. (2001) A theory of distance education based on empathy. In M. Moore & W. G. Anderson (Eds.) *Handbook of distance education* (pp.79-86). Mahwah, NJ: Lawrence Erlbaum Associates.

Honebein, P. (1996). Seven goals for the design of constructivist learning environments. In B. Wilson (Ed.), *Constructivist learning environments* (pp. 17-24). New Jersey: Educational Technology Publications.

Honey, P., & Mumford, A. (1992). *The manual of learning styles.* Maidenhead, UK: Peter Honey.

Hoon, E., & Van der Graaf, M. (2006). Copyright issues in open access research journal: The authors' perspective. *D-Lib Magazine, 12*(2). Retrieved February 18, 2008, from http://dlib.org/dlib/february06/vandergraaf/02vandergraaf.html.

House of Commons Education & Skills Committee. (2005). *UK e-University: Third report of session 2004-5.* London: The Stationery Office. Retrieved March 30, 2008, from http://www.publications.parliament.uk/pa/cm200405/cmselect/cmeduski/205/205.pdf

Hurley, J. M., Proctor, J. D., & Ford, R. E. (1999). Collaborative inquiry at a distance: Using the Internet in geography education. *Journal of Geography, 98*(3), 128-140.

Iceland, J., Weinberg, D. H., & Steinmetz, E. (2002). *Racial and ethnic residential segregation in the United States: 1980-2000.* U.S. Census Bureau, Series CENSR-3. Washington DC: U.S. Government Printing Office.

IEEE Learning Technology Standards Committee. (2006). *Standard for information technology, education and training systems, learning objects and meta-data.* Retrieved November 10, 2006, from http://ltsc.ieee.org/wg12/

IMS Global Learning Consortium. (2003). *Learning design specification version 1.* Retrieved May 25, 2007, from http://www.imsglobal.org/learningdesign/

IMS. (2004). *Content packaging specification.* Retrieved March 21, 2008, from http://www.imsglobal.org/content/packaging/

INSEE. (2008). *Portail INSEE.* Paris: Institut National de la Statistique er des Etudes Economiques. Retrieved March 23, 2008, from http://www.insee.fr/en/home/home_page.asp

Intergovernmental Panel on Climate Change (IPCC). (2000). *Special report on emissions scenarios.* Cambridge, UK: Cambridge University Press.

Internet Systems Consortium. (2008, January). *ISC Internet Domain Survey.* Retrieved on February 19, 2008, from http://www.isc.org/index.pl?/ops/ds/.

Iredale, A. (2006). Successful learning or failing premise? A situated evaluation of a virtual learning environment. In D. Whitelock & S. Wheeler (Ed.), *Research proceedings: ALT-C 2006 The next generation*, Edinburgh, UK (pp. 1-10). .

Janee, G., & Frew, J. (2002). The ADEPT digital library architecture. In *Proceedings of the Second ACM/IEEE-CS Joint Conference on Digital Libraries (JCDL)* (pp. 342-350). New York: ACM Press.

Jenkins, M., Browne, T., & Walker, R. (2005). *VLE surveys: A longitudinal perspective between March 2001, March 2003 and March 2005 for higher education in the United Kingdom.* Oxford: Universities and Colleges Information Systems Association.

Jimenez-Aleixandre, M. P. (2002). Knowledge producers or knowledge consumers? Argumentation and decision-making about environmental management. *International Journal of Science Education, 24*(11), 1171-1190.

JISC. (2008). *Design for learning.* Retrieved March 25, 2008, from http://www.jisc.ac.uk/whatwedo/programmes/elearning_pedagogy/elp_designlearn.aspx

JISC. (2008). *Jorum to move to open access.* Bristol: Joint Information Systems Committee of the Higher Education Funding Councils of the UK. Retrieved April 25, 2008, from http://www.jisc.ac.uk/Home/news/stories/2008/04/jorumopen.aspx

Jocoy, C. L., & DiBiase, D. (2006). Plagiarism by adult learners online: A case study in detection and remediation. *International Review of Research in Open and Distance Learning, 7*(1). Retrieved June 30, 2007, from http://www.irrodl.org/index.php/irrodl/article/view/242/466

Jorum. (2007). *Jorum user terms of use.* Retrieved May 31, 2007, from http://www.jorum.ac.uk/user/termsofuse/index.html

Jorum. (2008). *Jorum: Helping to build a community for sharing.* Retrieved March 29, 2008, from http://www.jorum.ac.uk/

Keegan, D. (1988). Problems in defining the field of distance education. *The American Journal of Distance Education, 2*(2), 4-11.

Keegan, D. (1988). Theories of distance education. In D. Sewart, D. Keegan, & B. Holmberg (Ed.), *Distance education: International perspectives* (pp 63-67). London: Routledge.

Keegan, D. (Ed.) (1994). *Otto Peters on distance education: The industrialization of teaching and learning.* London: Routledge.

Kerres, M., & De Witt, C. (2003). A didactical framework for the design of blended learning arrangements. *Journal of Education Media, 28*, 101-113.

Keylock, C. J., & Dorling, D. (2004). What kind of quantitative methods for what kind of geography? *Area, 36*(4), 358-366.

Kolb, D. A. (1984). *Experiential learning experience as a source of learning and development.* Englewood Cliffs, NJ: Prentice Hall.

König, G., Jaeger, M., Reigber, A., & Weser, T. (2005). An e-learning tutorial for RADAR remote sensing with RAT. In G. König, H. Lehmann, & R. Köhring (Eds.), *ISPRS Workshop on Tools and Techniques for E-Learning* (pp. 28-32). Potsdam, Germany: Institute of Geodesy and Geoinformation Science.

Koper, R., & Manderveld, J. (2004). Educational modelling language: Modelling reusable, interoperable, rich and personalised units of learning. *British Journal of Educational Technology, 35*(5) 537-551.

Koper, R., Pannekeet, K., Hendriks, M., & Hummel, H. (2004). Building communities for the exchange of learning objects: Theoretical foundations and requirements. *ALT-J Research in Learning Technology, 12*(1), 21-35.

Krauss, F., & Ally, M. (2005). A study of the design and evaluation of a learning object and implication for content development. *Interdisciplinary Journal of Knowledge and Learning Objects, 1*, 1-22.

Krüger, A., & Brinkhoff, T. (2005). Development of e-learning modules in spatial data management. In G. König, H. Lehmann, & R. Köhring (Eds.), *ISPRS Workshop on Tools and Techniques for E-Learning* (pp. 18-22). Potsdam, Germany: Institute of Geodesy and Geoinformation Science.

Kvan, T. (2000). Collaborative design: What is it? Automation in Construction, 9(2000), 409-415.

L'Allier, J. J. (1997). *Frame of reference: NETg's map to its products, their structures and core beliefs.* Intermedia. Retrieved July 18, 2006, from http://www.im.com.tr/framerefer.htm

Lagoze, C., & Van de Sompel, H. (2001). The open archives initiative: Building a low-barrier interoperability framework. In E. A. Fox & C. L. Borgman (Eds.), *Proceedings of the First ACM/IEEE-CS Joint Conference on Digital Libraries* (pp. 54-62). New York: ACM Press.

LAMS International. (2007). *Learning activity management system.* Retrieved March 25, 2008, from http://www.lamsinternational.com/

Larkham, P. J., & Manns, S. (2002). Plagiarism and its treatment in higher education. *Journal of Further and Higher Education, 26*(4), 339-349.

Laurillard, D. (1997). Learning formal representations through multimedia. In F. Marton, D. J. Hounsell, & N. Entwistle (Eds.), *The experience of learning* (2nd ed.) (pp. 172-183). Scottish Academic Press.

Laurillard, D. (2002) *Rethinking university teaching: A conversational framework for the effective use of Learning technologies* (2nd ed.). London: RoutledgeFalmer.

Laurillard, D. (2002). Design tools for e-learning. Keynote presentation for ASCILITE 2002. Winds of change in the sea of learning: Charting the course of Digital education, Auckland, New Zealand, 8-11 December 2002. Retrieved May 6, 2008, from http://www.ascilite.org.au/conferences/auckland02/proceedings/papers/key_laurillard.pdf.

Learning Circuits. (2000). *Glossary compiled by Eva Kaplan-Leiserson.* Alexandria, VA: American Society for Training and Developments. Retrieved March 29, 2008, from http://www.learningcircuits.org/glossary

Learning Development Unit. (2007). *Plagiarism - University of Leeds Guide.* Retrieved March 20, 2008, from http://www.ldu.leeds.ac.uk/plagiarism/index.php

Leung, S., Fill, K., DiBiase, D., & Nelson, A. (2005). *Sharing academic integrity guidance: Working towards a digital library infrastructure* (LNCS 3652, pp. 533-534).

Retrieved June 30, 2007, from http://www.springerlink.com/content/2x2x1wuc6dw3n1ev/

Leung, S., Harding, I., Wang, S., & Moloney, J. (2008). *Encouraging academic integrity to discourage plagiarism.* Paper presented at the 3rd International Plagiarism Conference (June 23-25), Newcastle-upon-Tyne, UK.

Lewis Mumford Center. (2008). *Lewis Mumford Center for Comparative Urban and Regional Research.* Retrieved February 20, 2008, from http://www.albany.edu/mumford/

Lim, E. P., Goh, D. H., Liu, Z. H., Ng, W. K., Khoo, C., & Higgins, S. E. (2002). G-Portal: A map-based digital library for distributed geospatial and georeferenced resources. In G. Marchionini (Ed.), *Proceedings of the Second ACM/IEEE-CS Joint Conference on Digital Libraries* (pp. 351-358). New York: ACM Press.

Lim, E. P., Liu, Z., Goh, D. H., Theng, Y. L., & Ng, W. K. (2005). On organizing and accessing geospatial and georeferenced Web resources using the G-Portal system. *Information Processing and Management, 41,* 1277-1297.

Lindström, G., & Gasparini, G. (2003). The Galileo satellite system and its security implications. Occasional Paper No. 44. Paris: European Union Institute for Security Studies. Retrieved March 1, 2008, from http://aei.pitt.edu/682/01/occ44.pdf.

Lisewski, B. & Settle, C. (1996). Integrating multimedia resource-based learning into the curriculum. In S. Brown & B. Smith (Eds.), *Resource-based learning* (pp.109-119). London: Kogan Page.

Liu, Z., Yu, H., Lim, E. P., Ming, Y., Goh, D. H., Theng, Y. L., et al. (2004). A Java-based digital library portal for geography education. *Science of Computer Programming, 53,* 87-105.

Longitudinal Studies Centre - Scotland (LSCS). (2008). *Linking lives through time.* St.Andrews: Longitudinal Studies Centre - Scotland, University of St. Andrews. Retrieved April 27, 2008, from http://www.lscs.ac.uk/

Luo, W., Duffin, K. L., Peronja, E., Stravers, J. A., & Henry, G. M. (2004). A Web-based interactive landform simulation model (WILSIM). *Computers and Geosciences, 30,* 215-220.

Luo, W., Peronja, E., Duffin, K., & Stravers, J. A. (2006). Incorporating non-linear rules in a Web-based interactive landform simulation model (WILSIM). *Computers and Geosciences, 32,* 1512-1518.

Luo, W., Stravers, J. A., & Duffin, K. L. (2005). Lessons learned from using a Web-based interactive landform simulation model (WILSIM) in a general education physical geography course. *Journal of Geoscience Education, 53*(5), 489-493.

Lynch, C. A. (1997, April). The Z39.50 information retrieval standard. *DLib Magazine.* Retrieved November 7, 2005, from http://www.dlib.org/dlib/april97/04lynch.html

MacDonald Ross, G. (2005). Plagiarism the Leeds approach. *Learning & Teaching Bulletin, (8).* Leeds, UK: University of Leeds. Retrieved August 7, 2007, from http://www.ldu.leeds.ac.uk/l&tbulletin/issue8/ross.htm

Madge, C., & O'Connor, H. (2004). Online methods in geography educational research. *Journal of Geography in Higher Education, 28*(1), 143-152.

Marshall, C. C. (1997). Annotation: From paper books to the digital library. In R. B. Allen & E. Rasmussen (Eds.), *Proceedings of the 2nd ACM International Conference on Digital Libraries* (pp. 131-140). New York: ACM Press.

Martin, D. J. (1994). Concept mapping as an aid to lesson planning: A longitudinal study. *Journal of Elementary Science Education, 6*(2), 11-30.

Martin, D., & Treves, R. (2007). DialogPLUS: Embedding eLearning in geographical practice. *British Journal of Education Technology, 38*(5), 773-783.

Martin, D., Harris, J., Sadler, J., & Tate, N. J. (1998). Putting the census on the Web: Lessons from two case studies. *Area, 30*(4), 311-320.

Massey, D. S., & Denton, N. A. (1988). The dimensions of residential segregation. *Social Forces, 67*, 281-315.

MatML. (2001). *MatML overview.* National Institute of Standards and Technology. Retrieved March 28, 2008, from http://www.matml.org/

MatML. (2004). *MatML schema, version 3.1.* MatML Schema Development Working Group. Retrieved March 28, 2008, from http://www.matml.org/schema.htm

Mattesssich, P. W., & Monsey, B. R. (1992). Collaboration: What makes it work? St. Paul, MN: Amherst H. Wilder Foundation

Mayer, R. E. (1991). *The promise of educational psychology: Learning in the content areas.* Upper Saddle River, NJ: Merrill Prentice Hall.

Mayer, R. E. (2001). *Teaching for meaningful learning.* Upper Saddle River, NJ: Merrill Prentice Hall.

Mayer, R. E., & Moreno, R. (2002). Animation as an aid to multimedia learning. *Educational Psychology Review, 14*(1), 87-100.

Mayer, R. E., & Moreno, R. (2003). Nine ways to reduce cognitive load in multimedia learning. *Educational Psychologist, 38*, 43-52.

Mayes, T., & de Freitas, S. (2004). *Review of e-learning frameworks, models and theories.* JISC e-learning models desk study. Retrieved, December 22, 2006, from http://www.jisc.ac.uk/uploaded_documents/Stage 2 Learning Models (Version 1).pdf

McCabe, D. (2000). New research on academic integrity: The success of "modified" honor codes. *Synfax Weekly Report*, (17), 975.

McClellan, J. H., Harvel, L. D., Velmurugan, R., Borkar, M., & Scheibe, C. (2004). CNT: Concept-map based navigation and discovery in a repository of learning content. In *34th Annual ASEE/ISEE Frontiers in Education Conference,* Savannah, GA (Session F1F, pp. 13-18). Retrieved May 5, 2008, from http://ieeexplore.ieee.org/iel5/9652/30543/01408581.pdf?tp=&isnumber=&arnumber=1408581

McConkie, G. W., & Currie, C. B. (1996). Visual stability across saccades while viewing complex pictures. *Journal of Experimental Psychology: Human Perception and Performance, 22*(3), 563-581.

MECA_ODL. (2003). *MECA-ODL: Methodology for the analysis of quality in ODL through the Internet.* Retrieved March 27, 2008, from http://www.adeit.uv.es/mecaodl/

Metros, S. E., & Bennett, K. A. (2004). Learning objects in higher education: The sequel. *ECAR Research Bulletin, 2004*(11), 1-13.

Miller, G. A. (1956). The magical number seven plus or minus two: Some limitations on our capacity for processing information. *Psychological Review, 63,* 81-97.

Milton, E. J. (1994). A new aid for teaching the physical basis of remote sensing. *International Journal of Remote Sensing, 15*, 1141-1147.

Milton, E. J. (1994). Teaching atmospheric correction using a spreadsheet. *Photogrammetric Engineering and Remote Sensing, 60*, 751-754.

Milton, R. (2007). *Dan Vickers' output area classification.* London: Centre for Advanced Spatial Analysis, University College London. Retrieved February 20, 2008, from http://www.casa.ucl.ac.uk/googlemaps/OAC-super-EngScotWales.html

Mitchell, T. (2004). *How do you bridge the CLI vs. GUI gap in app. design?* O'Reilly Digital Media. July 17, 2004. Retrieved August 21, 2007, from http://www.oreillynet.com/digitalmedia/blog/2004/07/how_do_you_bridge_the_cli_vs_g.html

Moodle. (2008). *Moodle - a free, open-source content management system for online learning.* Retrieved March 26, 2008, from http://moodle.org/

Moore, M., & Anderson, W.G. (Eds.) (2007). *Handbook of distance education.* London: Lawrence Erlbaum Associates.

Mowshowitz, A. (1997). Lessons from a cautionary tale. *Communications of the ACM 40*(5), 23-25. Retrieved March 27, 2008, from http://portal.acm.org/citation.cfm?id=253777&coll=portal&dl=ACM

Munowenyu, E. M. (2002). Fieldwork in geography: A review and critique of the relevant literature on the use of objectives. *Educate, 2,* 16-31.

Mutch, A. (2003). Exploring the practice of feedback to students. *Active Learning in Higher Education, 4,* 24-38.

National Research Council (NRC). (1996). *National science education standard.* Washington, DC: National Academy Press.

National Statistics. (2007). *Neighbourhood Statistics home page.* London, UK: Office for National Statistics. Retrieved February 20, 2008, from http://neighbourhood.statistics.gov.uk/

National Statistics. (2007). *UK census based classification of output areas.* London: Office for National Statistics. Retrieved February 20, 2008, from http://www.statistics.gov.uk/census2001/cn_139.asp

National Statistics. (2007). *Area classification for output areas.* London: Office for National Statistics. Retrieved February 20, 2008, from http://www.statistics.gov.uk/about/methodology_by_theme/area_classification/oa/default.asp

National Statistics. (2007). *Understanding the 2001 census area classification for output areas.* London: Office for National Statistics. Retrieved February 20, 2008, from http://neighbourhood.statistics.gov.uk/dissemination/Info.do;jsessionid=ac1f930dce62afd8c80e527429aaefdb6cfe17d733c.e38OaNuRbNuSbi0LbhyNb3eOb3uLe6fznA5Pp7ftolbGmkTy?page=CaseStudies_Classification.htm&bhcp=1

National Statistics. (2008) *The census in England and Wales.* London, Newport, Southport, Titchfield: Office for National Statistics. Retrieved April 27, 2008, from http://www.statistics.gov.uk/census/

Navarro, L. I., Such, M. M., Martin, D. M., Sancho, C. P., & Peco, P. P. (2005). Concept maps and learning objects. In *Proceedings of the Fifth IEEE International Conference on Advanced Learning Technologies* (ICALT'05) (pp. 263-265).

Nesbit, J., Belfer, K., & Leacock, T. (2003). Learning Object Review Instrument (LORI) user manual. *E-Learning Research and Assessment Network.* Retrieved October 23, 2007, from http://www.elera.net/eLera/Home/Articles/LORI 1.5.pdf.

Neven, F., & Duval, E. (2002). Reusable learning objects: A survey of LOM-based repositories. In *Proceedings of the 10th ACM International Conference on Multimedia,* Juan-les-Pins, France (pp. 291-294). New York: ACM.

New, M., Hulme, M., & Jones, P. D. (1999). Representing twentieth century space-time climatic variability. Part 1: Development of a 1961-1990 mean monthly terrestrial climatology. *Journal of Climate, 12,* 829-856.

Nicholson, D. (2002). Optimal use of MS PowerPoint for teaching in the GEES disciplines. *Planet, 4,* 7-9.

NISRA. (2008). *NICA: Northern Ireland census access.* Belfast: Northern Ireland Statistics and Research Agency. Retrieved February 20, 2008, from http://www.nicensus2001.gov.uk/nica/public/index.html

NISRA. (2008). *NINIS: Northern Ireland neighbourhood information service.* Belfast: Northern Ireland Statistics and Research Agency. Retrieved February 20, 2008, from http://www.ninis.nisra.gov.uk/mapxtreme/default.asp

Noble, D. F. (1998). Digital diploma mills: The automation of higher education. *First Monday, 3*(1). Retrieved March 20, 2008, from http://www.firstmonday.dk/issues/issue3_1/noble/

NOMIS. (2008). *Official labour market statistics.* Census 2001 on Nomis. Retrieved February 18, 2008, from https://www.nomisweb.co.uk/home/census2001.asp

Northern Ireland Statistics and Research Agency (NISRA). (2008). *Welcome.* Belfast: Northern Ireland Statistics and Research Agency. Retrieved April 27, 2008, from http://www.nisra.gov.uk/

Novak, J. D. (1990). Concept mapping: A useful tool for science education. *Journal of Research in Science Teaching, 10,* 923-949.

NSF. (2003). *National science, technology, engineering, and mathematics education digital library (NSDL)*

- *program solicitation.* NSF 03-530. Retrieved February 15, 2007, from http://www.nsf.gov/pubs/2003/nsf03530/nsf03530.htm

Oliver, M. (2004). *Against the term blended learning. OU Knowledge Network.* Retrieved March 30, 2005, from http://kn.open.ac.uk/public/document.cfm?docid=5053

Open University (2008). T552 diagramming. Retrieved March 27, 2008, from http://systems.open.ac.uk/materials/t552/index.htm

OpenOffice. (2008). *OpenOffice.org: the free and open productivity suite.* Retrieved March 27, 2008, from http://www.openoffice.org/

Ordnance Survey. (2007). Beginners guide to GPS - accuracies (single GPS receiver) Retrieved July 23, 2007, from http://www.ordnancesurvey.co.uk/oswebsite/gps/information/gpsbackground/beginnersguidetogps/whatisgps_08.html

Ordnance Survey. (2007). GPS background - emerging satellite navigation systems Retrieved July 23, 2007, from http://www.ordnancesurvey.co.uk/oswebsite/gps/information/gpsbackground/satnavsystems.html

Paivio, A. (1986). *Mental representations: A dual coding approach.* Oxford, UK: Oxford University Press.

Park, C. (2003). In other (people's) words: plagiarism by university students - literature and lessons. *Assessment and Evaluation in Higher Education, 28*(5), 471-488.

Pateraki, M., & Baltsavias, E. (2005). Eye Learn - An interactive WEB based e-learning environment in photogrammetry and remote sensing. In G. König, H. Lehmann, & R. Köhring (Eds.), *ISPRS Workshop on Tools and Techniques for E-Learning* (pp. 23-27). Potsdam, Germany: Institute of Geodesy and Geoinformation Science.

Pattinson, S. (2004). *The use of CAA for formative and summative assessment - Student views and outcomes.* Paper presented at the CAA Conference. Retrieved March 22, 2005, from www.caaconference.com/

Patton, M. Q. (1986). *Utilization-focused evaluation.* Thousand Oaks, CA: Sage Publications Inc.

Paulsen, M. F. (2003). E-learning - the state of the art. *NKI Distance Education, March 2003 Work Package One: The Delphi Project.* Retrieved December 12, 2007, from http://home.nettskolen.nki.no/~morten/E-learning/Teaching%20and%20learning%20philosophy.htm

Paulsen, M.F. (2003). *WEB-EDU, Web education systems: A study of learning management systems for online education.* Powerpoint™ presentation. Retrieved March 29, 2008, from http://home.nettskolen.com/~morten/pp/Web-edu.ppt

Peltier, L. C. (1950). The geographical cycle in the periglacial region as it is related to climatic geomorphology. *Annals of the Association of American Geographers, 40*(3), 214-236.

Penn State Population Research Institute. (2005-2007). *Measuring spatial segregation.* State College, PA: Population Research Institute, The Pennsylvania State University. Retrieved February 16, 2008, from http://www.pop.psu.edu/mss/

Perkin, M. (1999). Validating formative and summative assessment. In S. Brown, J. Bull, & P. Race (Eds.), *Computer-assisted assessment in higher education* (pp. 29-37). London: Kogan Page.

Phillips, W. A. (1974). On the distinction between sensory storage and short-term visual memory. *Perception and Psychophysics, 16,* 283-290.

Plowman, L. (1996). Narrative, linearity and interactivity: Making sense of interactive multimedia. *British Journal of Educational Technology, 27*(2), 92-105.

Polsani, P. R. (2003). Use and abuse of reusable learning objects. *Journal of Digital Information, 3*(4). Retrieved July 18, 2006, from http://jodi.tamu.edu/Articles/v03/i04/Polsani/

Prasad, P., & Elmes M. (2005). In the name of the practical: Unearthing the hegemony of pragmatics in the discourse of environmental management. *Journal of Management Studies, 42*(4), 845-867.

Prensky, M. (2005-2006). Listen to the natives. *Educational Leadership, 63*(4), 8-13. Retrieved May 12, 2008, from http://www.ascd.org/authors/ed_lead/el200512_prensky.html

PresentationPro. (2008). *Convert your PowerPoint to Flash with PowerCONVERTER.* Retrieved March 26, 2008, from http://www.presentationpro.com/products/powerconverter.asp

Priest, S., & Fill, K. (2006). Online learning activities in second year environmental geography. In J. O'Donoghue (Ed.), *Technology supported learning and teaching: A staff perspective* (pp. 243-260). Hershey, PA: Information Science Publishing.

Purves, R. S., Medyckyi-Scott, D. J., & MacKaness, W. A. (2005). The e-MapScholar project - an example of interoperability in GIScience education. *Computers & Geosciences, 31*(2), 189-198.

Qin, J., & Finneran, C. (2002). Ontological representation for learning objects. In *Proceedings of the Workshop on Document Search Interface Design and Intelligent Access in Large-Scale Collections,* JCDL'02, Portland, OR, USA.

Quality Assurance Agency for Higher Education (QAA). (2000). *Geography: Subject benchmark statements.* Cheltenham, UK: Quality Assurance Agency for Higher Education. Retrieved January 31, 2005, from http://www.chelt.ac.uk/gdn/qaa/geography.pdf

Race, P. (1996). Helping students to learn from resources. In S. Brown & B. Smith (Eds.), *Resource-based learning* (pp. 22-37). London: Kogan.

Reardon, S. F., Matthews, S. A., O'Sullivan, D., Lee, B. A., Firebaugh, G., & Farrell, C. R. (2006). *The segregation profile: Investigating how metropolitan racial segregation varies by spatial scale.* PRI Working Paper 06-01. State College, PA: Population Research Institute, The Pennsylvania State University. Retrieved February 14, 2008, from http://www.pop.psu.edu/general/pubs/working_papers/psu-pri/wp0601.pdf

Reardon, S. F., Matthews, S. A., O'Sullivan, D., Lee, B. A., Firebaugh, G., & Farrell, C. R. (2006). *Measuring spatial segregation.* Retrieved February 20, 2008, from http://www.pop.psu.edu/mss/

Rees, P. (2003-2008). *GEOG5101 census analysis and GIS.* Distance Learning Masters Module. University of Leeds, Leeds, UK. Retrieved February 6, 2008, from http://vle.leeds.ac.uk/site/nbodington/geography/geog-levm/geogodl/geog5101/. (registered users only)

Rees, P., & Butt, F. (2004). Ethnic change and diversity in England, 1981-2001. *Area, 36*(2), 174-186. Retrieved February 6, 2008, from http://0-www.blackwell-synergy.com.wam.leeds.ac.uk/doi/full/10.1111/j.0004-0894.2004.00214.x

Rees, P., Martin, D., & Williamson, P. (Eds.). (2002). *The census data system.* Chichester, UK: Wiley.

Rehak, D. R., & Mason, R. (2003). Keeping the learning in learning objects. In A. Littlejohn (Ed.), *Reusing online resources: A sustainable approach to e-learning* (pp. 20-34). London: Kogan.

Reload. (2006). *Reusable eLearning object authoring & delivery.* Project Web site. Retrieved March 21, 2008, from http://www.reload.ac.uk/

Richardson, V. (1997). Constructivist teaching and teacher education: Theory and practice. In V. Richardson (Ed.), *Constructivist teacher education: Building new understandings* (pp. 3-14). Washington, DC: Falmer Press.

Riddy, P., & Fill, K. (2003). *Evaluating the quality of elearning resources.* Paper presented at the British Educational Research Association Annual Conference, Edinburgh, UK. Retrieved March 27, 2008, from http://www.leeds.ac.uk/educol/documents/00003331.htm

Riddy, P., & Fill, K. (2004). *Evaluating eLearning resources.* Paper presented at In Networked Learning 2004. Retrieved March 27, 2008, from http://www.networkedlearningconference.org.uk/past/nlc2004/proceedings/individual_papers/riddy_fill.htm

Riddy, P., & Fill, K. (2004). *Evaluating eLearning resources: Moving beyond the MECA ODL methodology.* Paper presented at ALT-C 11th International Conference, Exeter, UK.

Rieh, S. Y., Markey, K., Yakel, E., St. Jean, B., & Kim, J. (2007). Perceived values and benefits of institutional repositories: A perspective of digital curation. In *An International Symposium on Digital Curation* (DigCCurr 2007), Chapel Hill, NC. Retrieved April 28, 2008, from http://www.ils.unc.edu/digccurr2007/papers/rieh_paper_6-2.pdf

Rodríguez-Artacho, M., & Verdejo Maíllo, M. F. (2004). Modeling educational content: The cognitive approach of the PALO language. *Educational Technology & Society, 7*(3), 124-137.

Russian Space Agency. (2006). Information-analytical centre. Retrieved March 1, 2008, from http://www.glonass-ianc.rsa.ru/pls/htmldb/f?p=202:1:2558501722024360015::NO

Rust, C. (2001). *A briefing on assessment of large groups.* Assessment Series No. 12. York, UK: LTSN Generic Centre.

Saad, M., & Maher, M. L. (1996). Shared understanding in computer-supported collaborative design. Computer-Aided Design, 28(3), 183-192

Salmon, G. (2000). *E-moderating: The key to teaching and learning online.* London: Routledge.

SASI. (2007). *The National classification of census output areas.* Sheffield: Sheffield Social and Spatial Inequalities Research Group, Department of Geography, University of Sheffield. Retrieved February 20, 2008, from http://www.sasi.group.shef.ac.uk/area_classification/index.html

SASI. (2008). *Social and Spatial Inequalities Research Group.* Sheffield: Geography Department, University of Sheffield. Retrieved February 6, 2008, from http://www.sasi.group.shef.ac.uk/

Schon, D. (1987). *Educating the reflective practitioner: Toward a new design for teaching and learning in the professions.* San Francisco, CA: Jossey Bass.

Schulman, M. (1998). Cheating themselves. *Issues in Ethics, 9*(1). Retrieved June 30, 2007, from http://www.scu.edu/ethics/publications/iie/v9n1/cheating.html

Schweizer, H. (1999). *Designing and teaching an on-line course, spinning your Web classroom.* Boston: Allyn and Bacon.

Sclater, N. (2008, June 10). Large scale open source e-learning systems at the Open University UK. EDUCAUSE Center for Applied Research. *Research Bulletin,* (12).

Seale, J. K., & Cann, A. J. (2000). Reflection on-line or off-line: The role of learning technologies in encouraging students to reflect. *Computers and Education, 34,* 309-320.

See, L., Gould, M. I., Carter, J., Durham, H., Brown, M., Russell, L., & Wathan, J. (2004). Learning and teaching online with the UK census. *Journal of Geography in Higher Education, 28*(2), 229-245.

Senior, M. (2002). Deprivation indicators. In P. Rees, D. Martin, & P. Williamson (Eds.), *The census data system* (pp. 123-138). Chichester, UK: Wiley.

Sharpe, R., Benfield, G., Roberts, G., & Francis, R. (2006). *The undergraduate experience of blended e-learning: A review of UK literature and practice.* Oxford: The Higher Education Academy. Retrieved March 27, 2008, from http://www.heacademy.ac.uk/resources/detail/ourwork/research/Undergraduate_Experience

Smith, T. R., Ancona, D., Buchel, O., Freeston, M., Heller, W., Nottrott, R., Tierney, T., & Ushakov, A. (2003). The ADEPT concept-based digital learning environment. In *Proceedings of the 7th European Conference on Research and Advanced Technology for Digital Libraries* (ECDL 2003), Trondheim, Norway (pp. 300-312). Berlin: Springer.

Smith, T. R., Zeng, M. L., & ADEPT Knowledge Team. (2002). *Structured models of scientific concepts as a basis for organizing, accessing, and using learning materials* (Technical Report 2002-04). Santa Barbara, CA: University of California Santa Barbara, Department of Computer Science.

Smith, T. R., Zeng, M. L., & ADEPT Knowledge Team. (2002). Structured models of scientific concepts for organizing learning materials. In M. J. Lopez-Huertas & F. J. Munoz-Fernandez (Eds.), *Challenges in knowledge representation and organization for the 21st century: Integration of knowledge across boundaries: Proceedings of the seventh international ISKO conference*, Granada, Spain (pp. 232-239).

Song, L., Singleton, E. S., Hill, J. R., & Koh, M. H. (2004). Improving online learning: Student perceptions of useful and challenging characteristics. *The Internet and Higher Education, 7*(1), 59-70.

Spellman, G. (2000). Evaluation of CAL in higher education geography. *Journal of Computer Assisted Learning, 16,* 72-82.

Spicer, J. J., & Stratford, J. (2001). Student perceptions of a virtual field trip to replace a real field trip. *Journal of Computer Assisted Learning, 17,* 345-354.

SPSS. (2008). *Data analysis with comprehensive statistics software.* Retrieved March 25, 2008, from http://www.spss.com/spss/

Statistics Canada. (2005). *2006 Census.* It's not too late. About the online questionnaire. Ottawa: Statistics Canada/Statistique Canada. Retrieved February 17, 2008, from http://www12.statcan.ca/IRC/english/about_e.htm

Statistics Canada. (2008). *2006 Census, 2001 Census and 1996 Census.* Ottawa: Statistics Canada/Statistique Canada. Retrieved March 23, 2008, from http://www12.statcan.ca/english/census/index.cfm

Statistics South Africa (2008). *Census 2001: Digital census atlas.* Retrieved March 23, 2008, from http://www.statssa.gov.za/census2001/digiAtlas/index.html

Statistics South Africa. (2008). *Census 2001: Interactive & electronic products.* Retrieved March 23, 2008, from http://www.statssa.gov.za/census01/html/C2001Interactive.asp

Stefani, L. A., Clarke, J., & Littlejohn, A. H. (2000). Developing a student-centred approach to reflective learning. *Innovations in Education and Teaching International, 37*(2), 163-171.

Stein, M., Edwards, T., Norman, J., Roberts, S., Sales, J., & Alec, R. (1994). *A constructivist vision for teaching, learning, and staff development.* Detroit, MI: Detroit Public

Schools Urban Systemic Initiative. Retrieved March, 18, 2008, from http://www.eric.ed.gov/ERICWebPortal/contentdelivery/servlet/ERICServlet?accno=ED383557

Stevenson, A. (2005). *Jorum application profile v1.0 draft.* Retrieved February 18, 2008, from http://www.jorum.ac.uk/docs/pdf/japv1p0.pdf

Stiles, M. (2007). Death of the VLE?: A challenge to a new orthodoxy. *Serials, 20*(1), 31-36.

Stoltman J. P., & Fraser, R. (2000). Geography fieldwork: Tradition and technology meet. In R. Gerber, & G. K. Chuan (Eds.), Fieldwork in geography: Reflections, perspectives, and actions. Kluwer Academic Publishers

Sumner, T., & Marlino, M. (2004). Digital libraries and educational practice: A case for new models, In E. P. Lim & M. Christel (Eds.), *Proceedings of the 4th ACM/IEEE-CS Joint Conference on Digital Libraries* (pp. 170-178). New York: ACM Press.

Swales, J., & Feak, C. (1994). *Academic writing for graduate students.* Ann Arbor, MI: University of Michigan Press.

Sweller, J. (1988). Cognitive load during problem solving. *Cognitive Science, 12,* 257-285.

Szabo, A., & Underwood, J. (2004). Cybercheats: Is information and communication technology fuelling academic dishonesty? *Active Learning in Higher Education, 5*(2), 180-199.

TechSmith. (2008). *Camtasia screen recorder.* Retrieved March 26, 2008, from http://www.techsmith.com/camtasia.asp

Tempfli, K. (2003). LIDAR and INSAR embedded in the OEEPE's distance learning initiative. In A. Grün & H. Kahmen (Eds.), *Optical 3-D measurement techniques VI* (Vol. II, pp.16-22). Institute for Geodesy and Photogrammetry, ETH Zürich.

The Economist. (2008, April 12-18). Nomads at last: a special report on mobile telecoms. *The Economist.* Retrieved April 25, 2008, from http://www.economist.co.uk/specialreports/displaystory.cfm?story_id=10950394&CFID=3191279&CFTOKEN=31474922

The Economist. (2008, February 16). The electronic bureaucrat: A special report on technology and government. *The Economist.* Retrieved February 17, 2008, from http://www.economist.com/printedition/

The Pennsylvania State University. (1999). *Course policies for GEOG482.* Retrieved October 23, 2007, from https://courseware.e-education.psu.edu/courses/geog482/policies.shtml.

The World Bank. (2008). *WDRs: World development reports.* Washington, DC: World Bank. Retrieved March 23, 2008, from http://econ.worldbank.org/

Thorpe, M., Kubiak, C., & Thorpe, K. (2003) Designing for reuse and versioning. In A. Littlejohn (Ed.), *Reusing online resources: A sustainable approach to e-learning* (pp. 106-118). London, UK: Kogan Page.

Tooth, S. (2006). Virtual globes: A catalyst for the re-enchantment of geomorphology? *Earth Surface Processes and Landforms, 31,* 1192-1194.

Tufte, E. R. (2001). *The visual display of quantitative information* (2nd ed.). Cheshire, CT: Graphics Press.

U.S. Census Bureau. (2008) *American Factfinder.* Washington, DC: U.S. Census Bureau. Retrieved April 27, 2008, from http://factfinder.census.gov/home/saff/main.html?_lang=en

UK Data Archive. (2008). *Census.ac.uk: Moving you closer to the data.* Colchester: UK Data Archive, University of Essex. Retrieved April 27, 2008, from: http://www.census.ac.uk/

Underwood, J., & Szabo, A. (2003). Academic offences and e-learning: Individual propensities in cheating. *British Journal of Educational Technology, 34*(4), 467-477.

United Nations. (2008). *What's new.* New York: Population Division, Department of Economic and Social Affairs, United Nations. Retrieved March 23, 2008, from http://www.un.org/esa/population/unpop.htm

University of Leeds, University of Southampton, & Pennsylvania State University. (2008). *GIS online learning: MSc in geographic information systems.* Retrieved April 25, 2008, from http://www.geog.leeds.ac.uk/odl/

University of Leeds, University of Southampton, Penn State World Campus & Worldwide Universities. (2008). *GIS online learning.* Retrieved February 14, 2008, from http://www.gislearn.org/

University of Leicester. (2008). *Exploring online research methods incorporating TRI-ORM. ESRC Research Methods Programme 2004-6 and ESRC Researcher Development Initiative 2007-2009.* Retrieved February 17, 2008, from http://www.geog.le.ac.uk/orm/

University of Southampton. (2007). *Academic skills.* Retrieved March 20, 2008, from http://www.academic-skills.soton.ac.uk/

van Ossenbruggen, J., Hardman, L., & Rutledge, L. (2002). Hypermedia and the Semantic Web: A research agenda. *Journal of Digital Information, 3*(1). Retrieved October 23, 2007, from http://journals.tdl.org/jodi/article/view/jodi-61/77.

Vermote, E. F., Tanre, D., Deuzé, J. L., Herman, M., & Morcrette, J. J. (1997). Second simulation of the satellite signal in the solar spectrum, 6S: An overview. *IEEE Transactions on Geoscience and Remote Sensing, 35*(3), 675-686.

Vickers, D. (2006). *Multi-level integrated classifications based on the 2001 census.* Unpublished doctoral dissertation, University of Leeds. Retrieved February 20, 2008, from http://www.geog.leeds.ac.uk/people/old/d.vickers/thesis.html

Vickers, D., & Rees, P. (2006). *Introducing the National Classification of Census Output Areas.* Population Trends, 125, 15-29. Retrieved February 20, 2008, from http://www.statistics.gov.uk/downloads/theme_population/PT125_main_part2.pdf

Vickers, D., & Rees, P. (2007). Creating the National Statistics 2001 Output Area Classification. *Journal of the Royal Statistical Society, Series A, 170*(2), 379-409.

Vickers, D., Rees P., & Birkin, M. (2005). *Creating the national classification of census output areas: Data, methods and results.* Working Paper 05/1. School of Geography, University of Leeds, Leeds. Retrieved

February 20, 2008, from http://www.geog.leeds.ac.uk/wpapers/05-2.pdf

Wandersee, J. H. (1990). Concept mapping and the cartography of cognition. Journal of Research in Science Teaching, 27(10), 923-936

Weber, A. (2006). NOAA comes to second life. *Second Life Insider.* Retrieved March 27, 2008, from http://www.secondlifeinsider.com/2006/08/18/noaa-comes-to-second-life/

Weippert, H., & Fritsch, D. (2002). Development of a GIS supported interactive "Remote Sensing" learning module. *GIS, 14*(9), 38-42.

Weisgerber, D. W. (1997). Chemical Abstracts Service Chemical Registry System: history, scope, and impacts. *Journal of the American Society for Information Science, 148*(4), 349-360.

Weller, M. (2007). The VLE/LMS is dead. *The Ed Techie blog.* Retrieved February 18, 2008, from http://nogoodreason.typepad.co.uk/no_good_reason/2007/11/the-vlelms-is-d.html

Weller, M., Pegler, C., & Mason, R. (2003). *Working with learning objects—Some pedagogical suggestions.* Paper presented at the Association for Learning Technology conference. Retrieved March 4, 2005, from http://iet.open.ac.uk/pp/m.j.weller/pub/altc.doc

Wikipedia. (2008). *Wiki.* Wikipedia: The Free Encyclopedia. Retrieved April 28, 2008, from: http://en.wikipedia.org/wiki/Wiki

Wikle, T. A., & Lambert, D. P. (1996). The global positioning system and its integration into college geography curricula. Journal of Geography, 95(5), 186-193.

Wiley, D. A. (2000). Connecting learning objects to instructional design theory: A definition, a metaphor, and a taxonomy. In D. A. Wiley (Ed.), *The instructional use of learning objects.* Retrieved October 23, 2007, from http://reusability.org/read/chapters/wiley.doc

Wiley, D. A. (2002). Connecting learning objects to instructional design theory: A definition, a metaphor, and a taxonomy. In D.A. Wiley (Ed.), *The instructional*

use of learning objects. Bloomington, IN: Association for Educational Communications & Technology. Retrieved December 10, 2007, from http://reusability.org/read/chapters/wiley.doc

Wiley, D. A. (Ed.) (2000). *The instructional use of learning objects: Online version.* Retrieved March 27, 2008, from http://reusability.org/read/chapters/wiley.doc

Wilson, B.G. (Ed.). (1996). *Constructivist learning environments: Case studies in instructional design.* Englewood Cliffs, NJ: Educational Technology Publications, Inc.

Wong, D. W. S. (2002). Spatial measures of segregation and GIS. *Urban Geography, 23,* 85-92.

Wong, D. W. S. (2003). Implementing spatial segregation measures in GIS. *Computers, Environment and Urban Systems, 27,* 53-70.

Wong, D. W. S. (2004). Comparing traditional and spatial segregation measures: A spatial scale perspective. *Urban Geography, 25*(1), 66-82.

Wong, D. W. S. (2005). Formulating a general spatial segregation measure. *The Professional Geographer, 57*(2), 285-294.

Worldwide Universities Network (WUN). (2005). Worldwide Universities Network. Retrieved March 1, 2008, from http://www.wun.ac.uk/index.php

Worldwide Universities Network. (2008). *WUN: Worldwide Universities Network.* Retrieved April 25, 2008, from http://www.wun.ac.uk/

Wright, J. A., Treves, R. W., & Martin, D. (under review). Challenges in the re-use of learning materials: Technical lessons from the delivery of an online GIS MSc module. *Journal of Geography in Higher Education.*

Zakrzewski, S., & Steven, C. (2000). A model for computer-based assessment: The Catherine Wheel principle. *Assessment & Evaluation in Higher Education, 25*(2) 201-215.

Zemsky, R., & Massy, W. F. (2004). Why the e-learning boom went bust. *The Chronicle Review, 50*(4), B6.

About the Contributors

Phil Rees has been professor of population geography at the University of Leeds since 1990, having previously been reader (1980-90) and lecturer (1970-80). His research focuses on population analysis in a wide range of applications. Recently he has worked with John Parsons for the Joseph Rowntree Foundation estimating the socio-demographic makeup of the regions of the UK for 2010 and 2020 as part of a project on child poverty. This project included ethnic group projections for UK regions, which are being intensively developed in 2007-2009 with funding from ESRC. He has studied the social geography of the UK and the U.S. using population census data from 1960/1961 to 2000/2001. He assisted Daniel Vickers in producing the new 2001 Census Output Area Classification (OAC), which is being widely used by social scientists and practitioners as a convenient way of summarizing millions of data about small areas. From 1992 to 2002 Phil Rees coordinated the ESRC/JISC Census Programme, which delivered census data in electronic form, free at the point of use, to all UK HE and FE staff and students. In 2004 he was awarded a CBE in recognition of this work. Throughout his working career, Phil Rees has been an active teacher of undergraduates, master's students, and doctoral postgraduates. He uses computer-based practicals in his demographic methods course and employs e-learning materials on census analysis in both distance and campus-based masters' programmes. With David Martin and Paul Williamson he edited *The census data system*, Wiley (2002). Other edited books include *Population migration in the European Union*, Wiley (1996), *Elderly migration and population redistribution*, Belhaven (1992), *Migration processes and patterns: Volume 2,*" Belhaven (1992), "*Population structures and models*, Allen and Unwin (1986), *Regional demographic development*, Croom Helm (1979), and *Models of cities and regions*, Wiley (1976).

Louise Mackay is a research fellow in the School of Geography, University of Leeds. Since 2003 she has worked on the JISC Digital Libraries in Support of Innovative Approaches to Teaching and Learning in Geography project as a learning materials developer and tutor of online materials in earth observation to physical geography undergraduates. From 2003 to 2006 Mackay lived in Tokyo, Japan, whilst delivering online materials to students at the University of Leeds, a unique experience, which provided considerable insight to the nature of online teaching and learning. Her academic research focuses on land cover extraction and analysis from high spatial resolution earth observation data, supported by over 8 years experience in spatial analysis, geographical information systems and image processing. Mackay has an undergraduate degree in geography, a master's degree in applied remote sensing and a PhD in earth observation; throughout her career she has developed and delivered material for undergraduate, postgraduate, and industrial consultancy teaching in GIS and earth observation. For this book her experience of developing and delivering online materials, evaluation of online teaching, and the introduction of digital repository use as a means to collect materials for storage and appropriation provides the background material to several chapters.

Helen Durham has been a research officer in the School of Geography at the University of Leeds since 1993 and has worked on a variety of externally-funded projects. Her skill areas are in spatial analysis, geographical information systems and the UK Census. Durham has carried out research in internal migration and population dynamics in Europe with Professor Phil Rees and Dr. Marek Kupiszewski, co-authoring a series of working papers and journal papers from 1996 to 2001. From 2000 her focus has been in supporting and developing e-learning material, firstly for the JISC-funded Collection of Historical and Contemporary Censuses (CHCC) project and since 2003 on the JISC/NSF Digital Libraries in the Classroom Programme. Durham played a major role in the development of an online Census Atlas, a digital resource started under the CHCC project and completed as part of the DialogPLUS project. The Atlas allows the visualisation and exploration of Census data from the 1971, 1981, 1991 and 2001 UK Census, both for individual census years and also examining the change over time across the three-decade period. A paper describing the Census Atlas, which Helen lead authored, was published in *Area Journal* in 2006.

David Martin is professor of geography at the University of Southampton, and director of the Economic and Social Research Council's Census Programme, which provides access and support for UK census data use across higher and further education. Martin's research and publication over the last 20 years have continued to centre around the theme of his original text *Geographical information systems: Socioeconomic applications* (1996), with a particular focus on the use of GIS in census and health care. His research on automated zone design led to the creation of an entirely new output geography system for the publication of the 2001 Census in England and Wales. Martin has a longstanding interest in research-led geographical education, reflected in *Methods in human geography: A guide for students doing a research project* (2005), edited with Robin Flowerdew, and his role as a co-director of the ESRC National Centre for Research Methods. He also leads a number of e-learning projects within the School of Geography at Southampton.

****

Katherine Arrell is a lecturer in the School of Geography, University of Leeds, with research interests in environmental modelling. Arrell has carried out research on glacier melt modelling, landform morphology, and most recently environmental controls on prehistoric habitation in Laos. Arrell is a member of several organisations and is a steering group member of Geographical Information Science Research in the UK (GISRUK). Arrell became interested in online learning through her involvement with the online MSc in Geographical Information Systems delivered by the University of Leeds.

Christopher Bailey is a senior technical researcher within the Internet Development group of the Institute for Learning and Research Technology at the University of Bristol. Prior to that he worked as a research fellow within the Electronics and Computer Science Department at the University of Southampton until September 2006. He earned a PhD at Southampton in the fields of adaptive hypermedia systems, open hypermedia, and software agents. Since the completion of his PhD he has focused more towards Semantic Web technologies, Web-based educational tools, learning environments, accessibility, and usability issues. More recently Bailey has been involved in producing e-Portfolio tools, designing and constructing virtual learning environments in several different domains, as well as developing the pedagogic planner for the DialogPLUS project. Bailey has been involved in research for the last seven years and has presented at conferences, seminars, and workshops both nationally and internationally. He has published over 20 papers and journals.

Mike Clark, as director of the University of Southampton's GeoData Institute, has had a 25-year engagement with consultancy on practical environmental management. He also has a long-standing commitment to the processes of teaching and learning, and their associated technologies. Early work on educational film culminated in the presentation of the Gill Memorial Award of the Royal Geographical Society (for contributions to Geographical Educational Technology and Coastal Studies) in 1983. By 1985 he had been appointed as Chairman of the British Universities Film and Video Council (BUFVC), and continues to hold this position. The BUFVC (funded by HEFCE/JISC) is responsible for media strategy (including dissemination of moving image teaching and learning materials) in the UK Higher Education sector. Clark has received a number of significant research grants for educational projects (with sponsors including the UGC, Nuffield Foundation, UFC, TLTP, ESRC and JISC). He has played a core role in negotiating the geography programme within the Worldwide Universities Network (WUN), and was also a member of the team contracted (2003-2005) to produce online distance-delivered course modules in environmental and health GIS for an inter-institutional MSc through the UK eUniversities Worldwide. Other recent commitments have been the production of e-learning modules on fluvial geomorphology for Environment Agency professional training, and an ESRC grant focussed on public response to Web-delivered environmental risk awareness materials.

Gráinne Conole is professor of e-learning at the Open University, with research interests in the use, integration, and evaluation of information and communication technologies and e-learning and impact on organisational change. She was previously chair of educational innovation at Southampton University and before that director of the Institute for Learning and Research Technology, University of Bristol. She has extensive research, development, and project management experience across the educational and technical domains; funding sources have included HEFCE, ESRC, EU and commercial sponsors. Recently funded projects include the HEFCE-funded E-Learning Research Centre, the JISC/NSF funded DialogPLUS digital libraries project and the ESRC National Centre for Research Methods. She serves on and chairs a number of national and international advisory boards, steering groups, committees, and international conference programmes. She has published and presented over 200 conference proceedings, workshops and articles, including over 50 journal publications on a range of topics, such as the use and evaluation of learning technologies. She is co-editor of the recently published *Contemporary perspectives in e-learning* (RoutledgeFalmer). Conole's blog can be found at www.e4innovation.com.

Steve Darby is a senior lecturer in physical geography at the University of Southampton, with research interests in the development and use of numerical models in geomorphology, with a particular focus on river channel erosion and sedimentation. He previously worked as a research geologist at the U.S. Department of Agriculture National Sedimentation Laboratory in Oxford, Mississippi. He is associate editor of one of the leading international journals in geomorphology, *Earth Surface Processes and Landforms*. He has published over 30 refereed journal publications on a range of topics in fluvial geomorphology.

Hugh Davis is the head of the Learning Technology Research Group within the 5* research School of Electronics and Computer Science (ECS) at the University of Southampton. He is the university director of education with responsibility for e-learning strategy. He has been involved in hypertext research since the late 1980s and has interests in the applications of hypertext for learning, open hypertext systems, and architectures for adaptation and personalisation. He has extensive publications in these fields, and experience of starting a spin-off company with a hypertext product. His recent research interests revolve around Web and Grid service frameworks for e-learning, and he has a particular focus on the assessment domain. In his university role he is concerned with institutional change and how e-learning approaches can become embedded within the curriculum.

David DiBiase directs the John A. Dutton e-Education Institute within the College of Earth and Mineral Sciences. Institute personnel collaborate with faculty members in the college's five academic departments and three other institutes to design, develop and deliver online postbaccaluareate certificate and degree programs serving adult professionals across the country and around the world. As a faculty member in Penn State's Department of Geography, DiBiase manages its online master's of geographic information systems degree program. DiBiase's research publications include empirical studies of faculty workload and student satisfaction as well as e-learning strategy and policy. He has earned awards for educational innovation from Penn State, the Association of American Geographers, Environmental Systems Research Institute, and the University Consortium for Geographic Information Science.

Oliver Duke-Williams is a senior research fellow in the School of Geography, University of Leeds. He is the lead researcher in the ESRC funded Centre for Interaction Data Estimation and Research, and has research interests in population geography, census, and survey data quality issues and geocomputation.

Karen Fill has a joint honours degree in computer and management science and a master's degree in information systems. From Autumn 2001 to Spring 2007 she worked at the University of Southampton as an educational researcher with a specific interest in the innovative uses of technology to enhance teaching and learning in higher education. Her role on the DialogPLUS project was to support the academic staff as they designed online learning activities and to evaluate the impact of these on the student experience.

Mark Gahegan is professor in the School of Geography, Geology, and Information Science at the University of Auckland, New Zealand, where he also directs the Centre for e-Research. He was, until December 2007, associate director of the GeoVISTA Center at Penn State, USA and remains attached to the professional master's program offered there. His research interests span geographical information science, from semantics and knowledge management, through information and knowledge visualization and machine learning, to spatial data structures and algorithms. He is on the editorial board of seven related journals. Gahegan leads the GeoVISTA *Studio* (exploration and discovery) and *ConceptVista* (concept mapping and ontology) open-source software initiatives. He is also an active member of several ongoing cyber-infrastructure initiatives involving the geosciences, human-environment interaction, archeology, and plant pathology. He has specific interests in capturing and communicating meaning amongst collaborating scientists as they work to better understand Earth's complex systems.

Samuel Leung joined the School of Geography at the University of Southampton as a research assistant in 2004. He has been involved in the design, assemblage, and depositing of learning and assessment resources that are in compliance with the emerging IMS and ADL SCORM standards. Before his move to Southampton, Leung was previously a lecturer in cartography at Oxford Brookes University where he used ICT extensively in the teaching of mapping and topographic science. More recently, he has started working on the Geo-Refer project funded by ESRC to develop adaptive geospatial training resources for social science researchers

Stephen A. Matthews is an associate professor of sociology, anthropology, and demography (courtesy in geography) and a senior research associate in the Population Research Institute (PRI) at Penn State. In addition, Matthews is the director of the Geographic Information Analysis Core within PRI and the Social Science Research Institute (SSRI). Matthews' training and research experience in spatial demography have focused on the application of GIS technologies and statistical methods to health service utilization and accessibility, and epidemiology. His current research focuses on race/ethnic segregation in the U.S., the integration of GIS and ethnography in low-income families in diverse contexts, and a study of food environments, diet, and nutrition in low-income families. Matthews is the PI of GIS and Population Science Training Grants (http://csiss.ncgia.ucsb.edu/GISPopSci/). The primary mission of these training grants is to significantly promote the mastery and use of spatial methods in population research by the current cohort of young population scientists.

Ted Milton is a professor of geography at the University of Southampton, with research interests in quantitative remote sensing for earth observation. He has a particular interest in field spectroscopy and its use in the calibration of remotely sensed data, and was responsible for establishing a shared pool of spectroradiometric instruments in the UK. He was one of the co-founders of the GeoData Institute at the University of Southampton, a university-based research and consultancy group which specialises in environmental data management, analysis, and processing, and whose clients include public utilities, national, and European government organisations, charitable trusts, and commercial organisations. He is currently the co-ordinator of NCAVEO, a UK-based knowledge transfer network for the calibration and validation of earth observation data (www.ncaveo.ac.uk).

Rizwan Nawaz is currently director of an environmental consultancy firm, HydroRisk Ltd. Previously, he was a lecturer in physical geography at the University of Leeds. His teaching and research related interests are related primarily to river basin management. He has a keen interest in the role of information and communication technologies (ICT) in river basin management and has recently been awarded a grant by the European Commission to "twin" a river basin in a developing country to one in the UK to allow knowledge-transfer. ICT will play a pivotal role in the project, which seeks to enhance cooperation between river basin authorities in the developed and developing world. He has experience in preparing and delivering online material to both undergraduates and postgraduates. He is a member of several organisations including Web editor of the Royal Geographical Society's Climate Change Research Group.

Sally J. Priest is a research fellow at the Flood Hazard Research Centre at Middlesex University. She is currently working on the EU-funded FLOODsite project, which aims to bring together managers, researchers, and practitioners from over 30 partner organisations across Europe to achieve the goal of integrated flood risk management. Prior to that Sally was employed as a teaching fellow by the Dialog-PLUS project in the School of Geography at the University of Southampton. Her research interests are flood management and responses to flood risk (in particular, the economic benefits of flood warning and flood insurance), as well as ideas of risk and communication in the broader area of environmental management.

Terence R. Smith is professor of geography and professor of computer science at the University of California at Santa Barbara (UCSB). His research interests include the modelling of fluvial phenomena, concept-based computational modelling systems for scientific applications, digital libraries for scientific and geo-referenced information, and spatial database systems. He was chair of the Department of Computer Science at UCSB (1986-90) and associate director of the National Center for Geographic Information and Analysis (1988-90). He directed the Alexandria Digital Library Project at UCSB from 1994-2005. He is the author of over 130 published research papers.

Richard Treves is a learning technologist in the School of Geography, University of Southampton, United Kingdom. After an early career as a geologist he moved over to education at Sunderland University, part time at Leeds University and then for 3 years as a tutor manager in the technology faculty of the Open University. He has been responsible for writing and delivering a number of university level courses and numerous projects in staff development and student support. After completing a multimedia outreach project for the Hawaiian Volcanoes Observatory he moved to Southampton where he has designed Web systems to deliver MSc level GIS courses across multiple VLEs and has experimented with video podcasts as a teaching medium. Currently he is experimenting with the uses of Google Earth as a tool for teaching.

Jim Wright is a lecturer in geographical information systems (GIS) within the School of Geography, University of Southampton in the UK. Wright has an MSc in GIS and a PhD in GIS and human nutrition, both awarded by the University of Edinburgh. Wright has previously worked as a researcher in GIS for the Forestry Commission and as a research fellow for the Institute of Ecology and Resource Management at the University of Edinburgh. Wright's current research interests include environmental and health applications of GIS, particularly in quantifying the extent and impacts of global change. Beginning in 1996, Wright has taught modules in GIS for environmental management at both undergraduate and postgraduate level, initially at the University of Edinburgh and subsequently at University of Southampton.

Marcia Lei Zeng is professor of library and information science at Kent State University. Her research interests include knowledge organization systems (taxonomy, thesaurus, ontology, etc.), metadata and markup languages, database quality control, multilingual and multi-culture information processing, and digital libraries for cultural objects and learning objects. Her scholarly publications include over 60 published papers and four books. She was the P.I. of two National Science Foundation's (NSF) National Science Digital Library (NSDL) projects. Dr. Zeng has served on various committees of the International Federation of Library Associations and Institutions (IFLA), American Society for Information Science and Technology (ASIST), and the National Information Standards Organization (NISO).

324

Index

A

academic integrity (AI) xii, 22, 27, 64, 81, 82, 139,
140, 141, 142, 13, 143, 142, 144, 146, 147,
148, 149, 150, 151, 152, 153, 238
academic integrity, evaluating the benefits of 148
academic integrity, penalties and consequences 142
academic integrity, transgressions 141
academic integrity quiz 81, 144, 147, 148, 150, 151
Access Grid 41, 239
access management x, 13, 21, 32, 33, 34, 70
active learning 265
adaptive learning 206
ADEPT 186, 189, 190, 191, 192, 193, 194, 195, 197,
200, 201, 202, 203, 267
ADEPT DLE, current experiences in using in an
operational setting 201
ADEPT DLE, evaluating the efficacy of the 200
ADEPT DLE, formal evaluations of the efficacy of
the 200
Alexandria Digital Earth Project (ADEPT) 186, 189
alignment 241
ANGE 171
ANGEL 24, 28, 34, 144, 144, 145, 171, 178
animation 29, 79, 205, 206, 207, 208, 209, 210, 211,
212, 213, 214, 218, 219
assessment 247
assignment 10, 29, 81, 88, 90, 91, 106, 142, 144,
147, 150
Association of Teachers and Lecturers (ATL) 141
Athens , 32, 59, 60, 69
atmospheric correction 124, 125, 127, 128, 137
author 24, 29, 48, 130, 141, 175, 205, 207, 211, 212,
214, 217, 218, 238, 247, 323

authoring xiv, 24, 28, 29, 30, 92, 98, 135, 153, 163,
169, 204, 211, 239, 319
automation 153
autonomy and independence, theories of 10

B

benchmark statement 55
bidirectional reflectance distribution function
(BRDF) 127
bidirectional reflectance factor (BRF) 127
Blackboard 91
BlackboardTM 247
Blackboard VLE 82, 87, 88, 90, 125, 147
blended delivery 76, 113, 117, 119, 241
blended learning xii, 9, 35, 53, 55, 62, 68, 99, 115,
116, 119, 123, 133, 134, 135, 242, 247, 254,
256, 258, 259
blended learning module, in EO 133
blended teachers 241
blog 37, 137, 320
Bodington VLE 146
BRDF , 127, 128, 129
browser 129, 135, 167, 181, 190, 193, 211, 263

C

campus-based 4, 5, 93, 103, 119, 143, 144, 145, 256,
318
cartogram 59, 60, 61, 66
cascading style sheet (CSS) 27, 130
case study 1, 43, 44, 101, 113, 122, 133, 140, 143,
153, 202, 206, 284
CAS Registry 199
CASWEB 66

Copyright © 2009, IGI Global, distributing in print or electronic forms without written permission of IGI Global is prohibited.